Theory Change in Science

Monographs on the History and
Philosophy of Biology
RICHARD BURIAN, RICHARD BURKHARDT, JR.,
RICHARD LEWONTIN, JOHN MAYNARD SMITH
EDITORS

Theory Change in Science

Strategies from Mendelian Genetics

LINDLEY DARDEN

University of Maryland College Park

New York   Oxford
OXFORD UNIVERSITY PRESS
1991

Oxford University Press

Oxford   New York   Toronto
Delhi   Bombay   Calcutta   Madras   Karachi
Petaling Jaya   Singapore   Hong Kong   Tokyo
Nairobi   Dar es Salaam   Cape Town
Melbourne   Auckland

and associated companies in
Berlin   Ibadan

Copyright © 1991 by Oxford University Press, Inc.

Published by Oxford University Press, Inc.,
200 Madison Avenue, New York, New York 10016

Oxford is a registered trademark of Oxford University Press

Library of Congress Cataloging-in-Publication Data
Darden, Lindley.
Theory change in science : strategies from Mendelian genetics /
Lindley Darden.
p. cm. — (Monographs on the history and philosophy of biology)
Includes bibliographical references (p.   ) and index.
ISBN 0-19-506797-5 (cloth)
1. Science—Methodology. 2. Science—Methodology—Case studies.
3. Human genetics—Case studies. 4. Science—Philosophy.
5. Science—Philosophy—Case studies.
I. Title. II. Series.
Q175.D268 1991 502.8—dc20 90-22007

2 4 6 8 9 7 5 3 1

Printed in the United States of America
on acid-free paper

Preface

Mississippians, even former Mississippians who have come north toward home, write books. I was born in the same town as William Faulkner (New Albany) and I know the fun of trying to find the real life models for characters in a local's novel. Unfortunately, this book offers no such fun. It is a book about ideas, not personalities. Playing with ideas is fun too, but fun of a different sort. The "characters" in this book aren't the southern ladies in the garden club, but genetic characteristics and laws of their inheritance. Before plunging into the academic analysis, I'm going to be a bit self-indulgent and discuss characters in my life and autobiographical influences. Like many an author of a first book, I feel obliged to thank many people who have aided and influenced me over the years. Readers anxious to get on with the topic may prefer to skip to the beginning of the intellectual discussion in Chapter 1.

The book is dedicated to the memory of my father, Leslie Darden. He was a gracious southern gentleman and a lover of learning (as well as a "country lawyer," his own somewhat tongue-in-cheek appellation, issued from the podium of the state bar association). He awakened me to the excitement that comes from learning new ideas and provided a well-stocked library to explore. Moss Hill, the thirteen acres on which I grew up, provided the environment in which I came to appreciate the natural world. I followed my grandmother, "Mama" Nabors, around in her garden, and watered the grass with my mother, Inez Darden, and tasted the Lindley grapes grown by "Miss Josie," my other grandmother, who lived cross-town. The grapes, she claimed, were developed by some distant relative in England, who probably worked after our Lindley ancestors settled in Pennsylvania in the 1600s. My strong interests in biology are due to that early environment. Like all precocious but insecure students, the encouragement that I received from various teachers was important. One of my high school teachers, Hilda Hill, still serves as my role model for a woman in front of a class giving a lecture, because I had very few women professors in college or graduate school. Her poetic admonition still echoes: "you are a part of all that you have met."

My interest in philosophy of science developed in Larry Lacy's excellent course at Rhodes College (then named Southwestern at Memphis). We read Hanson's (1958) *Patterns of Discovery*. Hanson raised fascinating questions about reasoning in scientific discovery, but so obviously didn't solve them satisfactorily that I've been worrying about them ever since. The discovery of new ideas is exciting, yet methods for discovery have been too little explored. My graduate training in philosophy, biology, and conceptual foundations of science at the University of Chicago set me on my present course in the history and philosophy of science. Academic life in a congenial history and philosophy of science program at the University of Maryland College Park ended my search for a place where I, as a woman with scholarly interests, could feel at home.

At the University of Chicago, my dissertation analyzed the case study of the beginnings of genetics in the nineteenth century (and this book takes up where that case left off). Lasting influences were left by my teachers: Kenneth Schaffner, Stephen Toulmin, David Hull, Bill Wimsatt, the late Arnold Ravin, and especially Dudley Shapere. I'm sure that their work influences this book more than the actual citations indicate.

I've been working on Mendelian genetics as a case study for many years and have written a number of articles using it to address philosophical issues other than the ones taken up in this book. Earlier work on the case was supported by the History and Philosophy of Science Program of the National Science Foundation. This book received support from the American Council of Learned Societies, as well as the Graduate School and the College of Arts and Humanities at the University of Maryland College Park. I am grateful for that support. I have learned much from the historians who have written so cogently about genetics, especially Bob Olby, Onno Meijer, Garland Allen, E.A. Carlson, and Scott Gilbert. And, I am grateful to librarians and archivists for their help at the American Philosophical Society, the University of California at Berkeley, the California Institute of Technology, the Marine Biological Laboratory at Woods Hole, and the interlibrary loan office at McKeldin Library at the University of Maryland College Park.

In 1978 I attended a conference that Ken Schaffner organized on "Logic of Discovery and Diagnosis in Medicine." Hearing Bruce Buchanan's paper "Steps toward Mechanizing Discovery" (published in 1985) was another intellectual turning point for me. I realized that the field of artificial intelligence (AI) was addressing some of the same questions about scientific knowledge and reasoning as I was, from my more historical perspective. I am grateful to Bruce, his colleagues, and students at the Heuristic Programming Project at Stanford University (now the Knowledge Systems Laboratory) for their cordial welcome during my sabbatical visit in 1980. Following developments in AI and computational philosophy of science (as Paul Thagard has dubbed it) has been exciting. Many of the ideas in this book have been influenced by those developments. I am grateful to Roy Rada for his collaboration in constructing an AI LISP program to simulate the discovery of the chromosome theory of heredity. That work convinced me how difficult it is, with current AI techniques, to simulate scientific reasoning. Also, I have enjoyed learning about the work on abduction and diagnostic reasoning by B. Chandrasekaran, John Josephson, Dale Moberg, and their colleagues at Ohio State University. I look forward to further collaboration with them.

I'm sure that a number of my positions on various issues have influenced my analysis. It may be fruitful to mention some of my "biases." Just as that term is used in machine learning, my usage is not pejorative. Biases provide needed focus and direction. One can't do everything at once. Choices must be made, directions taken, within the space of possibilities. My basic interests are metaphysical and epistemological, but I am modest about what evidence I have for drawing conclusions in those grand philosophical subjects. I believe that science has been our most successful knowledge-gaining enterprise. An understanding of the nature of scientific knowledge and reasoning is an important approach to answering basic questions about the nature of reality and the nature of methods for knowing it. As a historian, I am an "internalist," concerned with ideas, not an "externalist," concerned with social and institutional aspects of science. If science doesn't produce reliable (though not infallible) knowledge, but is only a social construct that serves various human interests, then I am not interested in it. I might as well become a gardener, instead of a philosopher of science. If I take gardening more seriously, however, then I'll want to do a soil test to determine the scientific analysis of my soil. If I start breeding new species of grapes, then I will be using Mendelian principles and believing that they give me reliable knowledge. So . . . I simply can't escape being a realist about scientific knowledge.

In this book, I will not argue for internal as opposed to external approaches to science, nor will I argue for a realist as opposed to an antirealist interpretation. One book cannot do everything; others who see the need to debate those points will no doubt continue to do so. My task is to find strategies for theory change. I want to understand how to develop new scientific ideas and theories. I want to find the methods that scientists already seem to be using (although they usually are not explicit about their methods). And I want to suggest additional methods, such as more thorough and systematic methods that scientists can use, or methods that a computer can use more readily than humans. Abstract philosophical arguments for the justification of these strategies as methods for producing *true* scientific knowledge about the *real* world are beyond the scope of my discussion here.

I have discussed the ideas in this book with many people over the years. I have thoroughly enjoyed our lively intellectual discussions about these topics. In addition, a number of people read earlier drafts and provided me with insightful comments. I fear that I will forget to thank some of those whose ideas have left a mark here. At the risk of that, however, let me mention a few: Bill Bechtel, Myles Brand, John Clement, Stiv Fleishman, Jonathan Harwood, Jon Hodge, John Josephson, Josh Lederberg, Jerry Levinson, Jane Maienschein, Ed Manukian, Dale Moberg, Tom Nickles, Bob Olby, Phil Pauly, Moreland Perkins, Roy Rada, Bob Richardson, Pat Ross, Ken Schaffner, Dudley Shapere, Frederick Suppe, Paul Thagard, and Marga Vicedo. The members of our local Study Group in History and Philosophy of Biology have been very helpful, reading every chapter, sending me back to the drawing board at times, being encouraging at others. Perhaps to my peril, I have not always taken their advice. My heartfelt thanks to Pamela Henson, Joel Hagen, Joe A. Cain, and Bob Witkin. Thanks also to the students in my seminar on strategies for theory change for their comments on the manuscript, especially Fred Hickok, Ana Simoes, and Jim Antonisse.

I am grateful to Dick Burian for his extensive and knowledgeable editorial comments and for his support of my project with Oxford University Press. William Curtis of Oxford University Press was encouraging over the years; he expressed an early interest in the project and gently nudged me with queries about its progress when we met at conferences. He provided support during contract negotiations and manuscript preparation. My thanks to him. My thanks also to Stiv Fleishman, Pat Ross, and Liz Fannin for their help with the figures, bibliography, and manuscript preparation. Finally, various friends and relatives have been supportive during the years that I have been glued to my computer screen and tolerant of my plea that I would "do X after the book is finished": Shearer Rumsey; Natalie Schmitt; Virginia and George Moryadas; Tom, Betty, David, and Morgan Darden; and various folk dance friends at Buffalo Gap Camp.

I hope the reader will enjoy a quiet discussion with me while reading the book as much as I have enjoyed discussing ideas in it with others.

College Park, MD
July 1990

L.D.

Contents

Theory Change in Science

CHAPTER 1

Introduction

New ideas are exciting. Yet how they develop is mysterious. This book attempts to remove some of the mystery from the development of new scientific ideas.

Consider: If new ideas arise from old ones, then their origin isn't so puzzling. If new theories don't arise all at once, but instead develop piecemeal in incremental stages, then scientific theory formation is more comprehensible than it would appear at first glance. If theories often arise with vague new ideas, with implicit components and unclear implications, then the process of theory construction requires stages of refinement. And if theories undergo significant developments, driven by responses to anomalies, then theory change must be studied as an incremental process.

Theory change cannot always be viewed as, first, a brief moment of discovery when a theory springs full-blown from the mind of a single creative scientist, which, then, is followed by the logical task of justifying the fully formed new theory. Instead, the development of a new theory may occur gradually, over a significant period of time, either through the work of a single scientist or through a group process. Methods for effecting such development need to be studied.

My task is to explore the dynamics of theory change. The historical sections of this book focus on the development of a theory as evidenced in the public record, namely the published scientific literature, as the theory changes over time. The philosophical goal is to find *strategies of theory change*, reasoning strategies that *could have* produced the changes that did occur. At first glance, strategies based on the published record may not seem to offer much evidence about ways to develop new ideas. I hope to convince the reader otherwise.

One historical case of theory change will be examined in detail—the theory of the gene, in the period 1900-1926. In 1926, Thomas Hunt Morgan, a prominent American geneticist, succinctly stated the theory of the gene that had developed since the 1900 rediscovery of Mendel's 1865 work on heredity. By 1926, all the major theoretical components of the theory of the gene had been developed. This case is a well-circumscribed episode of the development of a successful theory, with numerous failed hypotheses along the way. The historical sections of the book trace the changes in the major theoretical claims within Mendelian genetics over approximately thirty years, from 1900 to about 1930. By 1930, they had reached the form that has now become textbook knowledge in genetics.

3

The historical changes provide the basis for an analysis of strategies of theory change that could have produced the historical changes that did occur. In suggesting strategies, I wear my philosopher's hat, not that of the historian. The claim is not a historical one, that any particular person consciously employed a specific strategy. Instead, the claim is that a change between a component of the theory at one time and that component at a later time "exemplifies" a strategy. The philosophical sections of the book suggest strategies that could have produced the historical changes that did occur. The strategies are named and are characterized sufficiently generally that they may be studied to see if they are applicable in other cases, either historical or contemporary. The names of strategies will be in **boldface** type to highlight them in the subsequent discussions.

What, the reader may now be asking (and may well have asked a while back), is a *strategy* for theory change? An example is helpful. A theory may change or be changed in order to resolve an anomaly. One example of a strategy for resolving an anomaly is to *complicate* an existing part of the theory. Suppose one component of a theory is the claim that there exist one-to-one relations between two types of things. The anomaly points (somehow, not specified in this example) to a problem with that simple assumption. One way of changing the theory is to complicate the one-to-one relation and to propose a many-to-one relation. Such a theory change exemplifies the strategy that I call **complicate an oversimplification**. Next, in order to assess the adequacy of such a change, the new theoretical component (many-to-one relations) would be assessed, using strategies for theory assessment, such as whether it adequately explains the anomaly. The strategy of **complication** is an example of a general strategy of taking a part of the theory and changing it slightly in order to account for anomalous data. Complicating existing ideas in small ways (stepwise refinement) is one of the easiest ways of getting (slightly) new ideas; however, other strategies are needed when the change in theory must be more drastic. Assessment of **explanatory adequacy** is one strategy often used in evaluating theories; that is, a theory is usually expected to explain a domain of phenomena. As Chapter 2 will argue, how and when to apply criteria of theory assessment are strategic decisions; thus, such criteria are labeled "strategies." Chapter 2 also will discuss my use of the term "strategy" more extensively.

The historical analysis here aims at an accurate characterization of historical changes in the components of the theory of the gene over time. The philosophical analysis aims at finding some of the general strategies that are exemplified in those conceptual changes. Such strategies can be divided into three types: (1) strategies for producing new ideas, such as the use of analogies or the use of specific interrelations to entities or to processes studied by another field; (2) strategies for theory assessment, such as the assessment of a theory's predictive adequacy or its consistency with other accepted theories; and (3) strategies for anomaly resolution and expansion of scope, such as generalizing, specializing, or complicating a hypothesis.

These types of strategies for theory change cut across, and so are not easily mapped into, the categories of discovery and justification, categories which are often discussed in philosophy of science. It is indeed the case that some strategies for producing new ideas may be characterized as discovery strategies, and most of the strategies for assessing theories play a role in justification. However, strategies for assessment may also function in producing new ideas, as we will see. Moreover, use of some strategies for producing new ideas obviates some ways of assessing the theory. Similarly, using one strategy at an early stage may guide the use of a particular anomaly resolution strategy at a later stage. Furthermore, strategies for anomaly resolution and expansion of scope are not easily categorized as either discovery or evaluative methods.

In explaining what I take my current task to be, namely finding strategies of theory change, it is also useful to say what I am not doing. My task as a philosopher of science is not to do psychology, not to delve into the private mental life of the creative scientist. Furthermore, I am not trying to test a general model of scientific change, such as that of Kuhn (1970), Lakatos (1970), Toulmin (1972), or Laudan (1977), or provide an alternative one. Much of the work in the interdisciplinary area of history and philosophy of science has focused on applying these general models to specific cases and showing the ways they apply, or, more often, do not apply. I am not optimistic about any single model of scientific change being adequate. A piecemeal consideration of numerous types of claims made by the models is a more worthwhile task, as Laudan and others (Laudan et al., 1986) have recently argued. Some of those claims, such as the relations of theory and data, will be mentioned in the context of discussing the gene case. The purpose of this work, however, is not to test those claims. I am engaged in a different enterprise.

Another thing I am not trying to do is to place scientific ideas in their social context, as is the fashion in recent externalist history of science (e.g., Sapp, 1987) and sociology of knowledge (e.g., Latour and Woolgar, 1986). I have chosen to approach the material in an internalist way. I recognize that social and cultural factors influence actual scientific developments, but, as a *philosopher*, my interests focus on scientific reasoning and the development of scientific ideas. In so far as my search for reasoning strategies is successful, those strategies will transcend the social context in which they are found in this one historical case. They can be used by scientists in other social contexts and, possibly, by computer programs that do scientific reasoning relatively independent of a social context.

In addition, my primary purpose is not to write a new historical narrative or provide new historical interpretations of this case, although some new historical interpretations are presented here. My choice of episodes from the history of genetics is selective, chosen to serve my purpose of tracing the most important empirical and conceptual developments leading to the theory of the gene. In Chapter 4, more will be said about my philosophical approach to the history.

The aim of the historical reconstruction is to be faithful to the historical record. I want to portray what the actual data and theoretical components of Mendelism were during specific periods, not what they should have been, given some model of scientific change. My account lays out actual historical changes. The aim of the philosophical analysis is then to find general strategies, which I claim are "exemplified" in such historical changes. The strategies are my own proposals of methods that could have produced those changes. In brief, the historical changes provide my data and the strategies are my working hypotheses about methods for producing such changes. In order to emphasize these differences, the historical sections of the chapters are separate from the philosophical analyses of strategies.

Thus, this book approaches the study of the development of new ideas, not by searching for a single logic of discovery nor by studying the psychology of creative people nor by looking for the social influences on scientists' behavior, but by searching for reasoning strategies for producing new scientific theories.

A guide through the topics covered in the chapters may be helpful:

Chapter 2 discusses philosophical issues, including prior work on the three types of strategies: strategies for producing new ideas, strategies for theory assessment, and strategies for anomaly resolution. It also contains an introduction to some of the metascientific vocabulary to be used in this analysis. The final section introduces an oversimplified diagram of stages and strategies of theory change, a diagram that will be refined in subsequent chapters.

Chapter 3, "The Problem of Heredity," introduces the scientific problem confronted when trying to understand the hereditary process. It was this problem that the theory of the gene solved. The chapter aims at enlivening the reader's historical imagination. It provides a hypothetical path of reasoning to several of the important early discoveries. It challenges the reader to rethink the problems that might have faced the scientists at the time and to engage in a creative problem-solving process so as to reach solutions. It also serves the function of introducing the reader to early Mendelian genetics.

Chapter 4, "Historical Introduction," switches from the hypothetical mode of discussion found in Chapter 3 to actual history. After a brief introduction to the historiographical method that I will be using, the chapter presents recent historical interpretations of Mendel's work and of the beginnings of genetics in 1900. It is based on both primary sources and recent historical interpretations of Mendel's work.

Chapter 5, "Mendelism, 1900-1903," presents an analysis of the theoretical components of Mendelism in the period 1900-1903 and justifies that analysis by extensive evidence from the literature of the period. It thus sets the stage for tracing the changes that followed. Tables 5-1 and 5-2 summarize the domain to be explained and the theoretical components of Mendelism that explain that domain, as of 1903.

Chapters 6 to 10 trace the important changes in Mendelism in the period from 1900 to 1926, including the discovery of multiple factors and multiple alleles, the development of the chromosome theory, the tests of the generality of segregation, the discovery of Mendel's second law of independent assortment as a separate law, the discovery of linkage and crossing-over, and the development of the new problems about mutation and the physical nature, location, and functioning of genes. Failed hypotheses along the way include the presence and absence theory, the hypothesis that genes contaminate each other, three-dimensional (as opposed to linear) linkage relations, the reduplication hypothesis, and the idea that genes were autocatalytic colloidal particles. Separate sections of each chapter analyze strategies that are exemplified in the changes documented in the historical sections.

Many of the developments discussed in Chapters 6 to 10 were driven by efforts to resolve empirical anomalies. In contrast, Chapter 11, "Unit-characters to Factors to Genes," focuses on conceptual problems, problems about the nature of a new theoretical entity—the gene. Conceptual developments included issues about appropriate genetic symbolism and appropriate terminology. Also at issue was the question of what properties were to be ascribed to the newly postulated theoretical entity. The field began with a vague claim about unit-characters and gradually developed explicit claims about the existence of a new theoretical entity. No single moment of discovery produced the new idea of the gene. Changes in the lists of properties ascribed to the gene are depicted. The chapter concludes with a discussion of strategies for finding and solving conceptual problems.

Chapter 12, "Exemplars, Diagrams, and Diagnosis," introduces a more diagrammatic representation of the theory of the gene, in contrast to the analysis in the previous chapters in which the theory is stated in sentences. A contrast is made between diagrammatic and sentential methods for representing theories. The diagrammatic representation shows how exemplary patterns can function in explanation and how "model" anomalies function to provide new patterns. Furthermore, the chapter discusses how this alternative, diagrammatic representation is useful for localizing anomalies that the theory faced, as well as giving indications as to how to resolve them. Analogies between diagnostic reasoning and strategies for anomaly resolution are also explored.

Chapter 13, "Genetics and Other Fields," summarizes the problems that the theory of the gene had solved by 1926 and the problems remaining, both old ones left unresolved and new ones that the theory itself raised. These problems are analyzed in terms of the successful and unsuccessful interrelations between genetics and other fields, including cytology, evolutionary studies, embryology, and biochemistry. The final section of the chapter summarizes and extends the strategy of using interrelations between different bodies of knowledge.

Chapter 14, "Summary of Strategies from the Historical Case," summarizes the strategies exemplified in the development of the components of the theory of the gene between 1900 and 1926. This chapter can provide a "quick read" for the major changes in theoretical components and the strategies they exemplify, which are discussed in Chapters 6 to 13.

Chapter 15, "General Strategies for Theory Change," takes a more "theory-driven" approach to discussing strategies. The previous chapters use a "data-driven" approach: they discuss strategies that emerged from an analysis of historical data supplied by the gene case. Chapter 15 places those strategies into more systematic lists of types of strategies. The more systematic lists were generated by relating the strategies from the gene case to previous work in philosophy of science, cognitive science, and artificial intelligence (AI). The oversimplified diagram of stages and strategies introduced in Chapter 2 is "complicated" in the light of the preceding analysis.

Chapter 16, the conclusion, discusses implications of the results of this case study for analyses of theory structure, theory development, and strategies for theory change. It also discusses possible uses of strategies by scientists, science teachers, and researchers in AI.

The readers who will be most familiar with the problems I am trying to solve are philosophers of science, especially those of us who are called "friends of discovery" (Gutting, 1980; Nickles, 1980a). Still, the book is written for a wider audience. I hope that scientists will find the strategies suggest methods for dealing with current problems. Readers not familiar with Mendelian genetics but interested in the broader issues about strategies for theory change are introduced to Mendelian genetics in Chapter 3.

Historians of genetics will be familiar with this case study; the historical sections of the book serve as an updated version of older histories of genetics, incorporating recent historical interpretations of Mendelian genetics and extending those interpretations in some places.

I have found the work of cognitive scientists and AI researchers helpful. For example, I used some results and terminology from their study of general reasoning strategies as I puzzled over what strategy was exemplified in a historical change and then tried to devise terminology for naming that strategy. Chapter 6 discusses the strategy of generalizing and specializing so often used in studies of machine learning (that is, AI work to devise computer programs that can learn and improve their own performance). Chapter 11 attempts to develop a frame-structured representation for the properties attributed to the gene. Frames are one of several structures that AI has developed for representing knowledge so that computer programs can use it to perform tasks to simulate the reasoning of human experts. Chapter 12 draws on the AI literature in diagnostic reasoning and functional representations, in order to devise strategies to diagnose and fix faults (anomalies) in the theory of the gene. Exciting new methods are emerging in the new area called "computational philosophy of science" (Thagard, 1988), which combines cognitive science, AI, and philosophy of science. Colleagues and I have done some preliminary AI experiments to explore ways of implementing the strategy of **using interrelations**. Although discussion of computer implementations of strategies is, for the most part, outside the scope of this more historical book, I hope my less precise characterizations of strategies here will be suggestive for that more formal work. AI implementations hold promise for transforming philosophy of science into an experimental field, as will be discussed in Chapter 16.

Furthermore, I have developed the idea of "interfield theories" (Darden and Maull, 1977; Darden, 1980b) more extensively in this book. Those who found that work suggestive for analyzing interdisciplinary interactions may wish to concentrate on the strategy of **using interrelations** in hypothesis formation, which is discussed in Chapters 7, 9, 10, 13, 14, and 15.

I believe this work has some implications for science education and the teaching of philosophy of science, both by human teachers and via instructional software. I will mention those implications in the conclusion.

Much of my discussion of general strategies in Chapter 15 is preliminary and suggestive. My hope is that others in diverse disciplines will find this effort of interest and will provide constructive criticisms, additional cases of the use of these strategies, and refinements and additions, as we explore methods for developing new ideas.

CHAPTER 2

Philosophical Preliminaries

2.1 Introduction

This book will discuss three types of strategies for theory change: strategies for producing new ideas, strategies for theory assessment, and strategies for anomaly resolution. This chapter briefly discusses some of the previous work relevant to each type of strategy, makes clearer the goals of this analysis, and introduces some of the terminology to be used. The final section of the chapter introduces a diagrammatic representation of stages in theory construction that will be refined as the analysis proceeds.

2.2 Strategies for Producing New Ideas

Some literature in the philosophy of science exists that discusses strategies for producing new ideas, but most work on this subject has been done outside twentieth-century philosophy of science. Such lack of attention is a consequence of philosophers' skepticism about the possibility of producing a logic of discovery. Philosophers of science in the twentieth century have distinguished sharply between the logic of discovery and the logic of justification (or falsification). Most have concluded that there is no (or, more strongly, there cannot be) a logic of discovery. Hence, the task of philosophy of science, on this view, is to analyze methods for assessing the warrant of knowledge claims, in other words, to provide criteria for the justification of theories (e.g., Popper, 1965; Laudan, 1977; Losee, 1987).

The hypothetico-deductive method has dominated many characterizations of science: a hypothesis is guessed (somehow) and a consequence is deduced. If the prediction holds, then the hypothesis is confirmed (or not falsified); otherwise, it is to be discarded and a new one (somehow) proposed (e.g., Hempel, 1965). This characterization, however, leaves as a complete puzzle the reasoning to hypotheses, assuming that they arise in a fully developed form ready for testing. Furthermore, the view that the testing of theories is an all-or-none act neglects the important information that anomalous data may provide for the next stage of theory construction. On a Popperian view, a falsifying instance (an anomaly) requires that the theory be discarded and that a new trial and error process be started to find a replacement (Popper, 1965). This procedure

9

of "conjectures and refutations" neglects the information that the prior refutation can play in guiding the construction of the next conjecture.

Philosophers of science concerned with understanding scientific change have advanced the discussion by postulating relations among successive theories. Lakatos (1970), for example, discussed progressive "research programmes" that consist of a succession of theories, each an improvement on its predecessor. Kuhn (1970) argued for a view of science as successive "paradigm" changes, provoked by an anomaly that (somehow) causes a crisis. Laudan (1977) analyzed changing "research traditions" and suggested adding a "context of pursuit" between discovery and justification. However, none of these philosophers concerned with conceptual change has proposed methods for the development of new ideas for the next stage. Nor have they indicated how anomalies at one stage can guide refinements. Laudan (1980) even argued that a search for such methods should not be the concern of philosophers.

N. R. Hanson was one of the few philosophers who tried to analyze reasoning in discovery. He argued that scientists do not start from hypotheses, as the hypothetico-deductive method suggests. Instead, discovery begins with "problematic phenomena requiring explanation" (Hanson, 1961, p. 34). Reasoning in discovery, Hanson claimed, required a logic different from either deduction or induction. He called this logic "abduction" (using Peirce's term) or "retroduction":

> Schematically, [retroductive reasoning] can be set out thus:
> (1) Some surprising, astonishing phenomena p-1, p-2, p-3 . . . are encountered.
> (2) But p-1, p-2, p-3 . . . would not be surprising were a hypothesis of H's type to obtain. They would follow as a matter of course from something like H and would be explained by it.
> (3) Therefore there is good reason for elaborating a hypothesis of the type of H; for proposing it as a possible hypothesis from whose assumption p-1, p-2, p-3 . . . might be explained. (Hanson, 1961, p. 33)

Hanson said little about the reasoning that occurred between Steps 1 and 2, that is, how the puzzling data suggested a type of hypothesis. He did indicate that analogies might play a role (Hanson, 1961, p. 25). Furthermore, he continued: "We can have good reasons, or bad, for suggesting one kind of hypothesis initially, rather than some other kind" (Hanson, 1961, p. 21). "A logic of discovery, then, might consider the structure of arguments in favor of one *type* of possible explanation in a given context as opposed to other types" (Hanson, 1961, p.30).

Hanson thus gives tantalizing hints about reasoning in discovery, especially his suggestions that puzzling facts provide a beginning point for theory construction and that analogies point to *types* of hypotheses. His work is exciting in that it poses the problem of discovery as a philosophical one. Nevertheless, Hanson's view is not without problems. First, he placed too much emphasis on the role of data and too little on ideas from other sources in hypothesis formation. Second, Hanson omitted discussion of constraints on the generation of possible hypotheses, other than the requirement that the hypothesis explain the puzzling data. Achinstein (1987) criticized retroduction as allowing "crazy" hypotheses that are wildly implausible, based on accepted background knowledge. For example, he proposed, suppose you see that I am happy about news I just received. Suppose you have the background information that receiving news of winning the Nobel prize makes people happy. According to the retroductive schema, since my happiness would not be surprising if I had received news of winning the Nobel prize (Step 2), then you have good reason for elaborating the hypothesis that I received news of winning the

Nobel prize (Step 3). Yet additional background knowledge that I am a philosopher and Nobel prizes are not given for work in philosophy makes the hypothesis wildly implausible. Achinstein's critique shows that additional constraints must be placed on hypothesis formation, other than that the hypothesis explain the puzzling data. Consistency with other accepted claims is one such constraint. More will be said about recent work on abduction in Chapter 15, in the context of developing the strategy of **assess a theory in relation to its rivals.**

The most serious failing of the retroductive schema as a method for reasoning in discovery is that a hypothesis, or at least a type of hypothesis, must already be available (Salmon, 1967, p. 113). The step of constructing such a hypothesis occurs between Hanson's Steps 1 and 2, but actual theory construction was omitted in Hanson's scheme. In his discussion, Hanson did hint that analogies might play a role, but he did not develop methods for using analogies in hypothesis (or hypothesis-type) construction. Schaffner (1980) correctly criticized Hanson for failing to distinguish between (1) a logic of generation and (2) a logic of preliminary evaluation.

Another problem with Hanson's view is that he dichotomized science too sharply into reasoning in discovery and reasoning in justification. Types of inference that play a role in the construction of a theory may also serve to make it plausible (Schaffner, 1974a). Criteria for assessing a theory may be introduced early to provide constraints on generation (Buchanan, 1985). Furthermore, the term "logic of discovery" is misleading. The use of "logic" implies that only one type of inference is found in discovery; also, a comparison with deductive logic suggests that a method guaranteeing the certainty of its conclusions might exist. Strategies for generating possible or plausible hypotheses are more numerous and weaker than the logic Hanson sought. Further, the term "discovery" implies a single event rather than a gradual process of developing a theory over time.

When logic of discovery is recast as the search for strategies for theory development, the work of other philosophers becomes relevant. Shapere's (1974b; 1974a) analysis of the way domain items can provide reasons for proposing a theory of a certain type can be viewed as providing strategies for theory construction; he called such strategies "principles." Philosophers who have studied the role of analogies and metaphors in hypothesis formation, such as Hesse (1966) and Boyd (1979), also can be viewed as discussing strategies for producing new ideas, new hypotheses, and new theoretical terms. Reasoning by analogy will be discussed briefly in Chapters 9 and 15.

The idea that theories may begin as vague ideas that are developed in stages has been suggested by several recent philosophers, such as Shapere (1974b), Monk (1977), Boyd (1979), Gutting (1980), and Nickles (1987b). Moving away from the idea that theories always spring full-blown from the (unconscious?) mind of a lucky guesser to the idea that they may develop in steps over a period of time makes the process of theory development more amenable to analysis. More than the suggestions by philosophers, work by historians and cognitive scientists analyzing the notebooks of scientists provides evidence for the view of an incremental process of theory development in the work of a single scientist. Gruber (1974), Schweber (1977), and Kohn (1980) studying Darwin's notebooks, and Holmes (1985) studying Lavoisier's, have shown the sometimes tortuous twists and turns that were taken before an adequate theory emerged. Similar conclusions have been reached by those studying Faraday's notebooks (Tweney, 1985; Gooding, 1990). Pieces of theory from earlier stages are retained, but new ideas are added along the way.

Furthermore, from the work of computer scientists with interests in artificial intelligence (AI), a new method is emerging to study scientific discovery. The work of Langley, Simon, Bradshaw, and Zytkow showed that methods for discovery could be found that were computationally adequate for rediscovering empirical laws. They said, "Large discoveries take place by the cumulation of little steps, and it is the understanding of these steps and of the processes by which they are accomplished that strips the larger discovery of its aura of mystery" (Langley et al., 1987, p. 58).

Of course, it is an open question to what extent similarities exist between (1) more routine problem solving, (2) an individual scientist struggling to construct a new theory, (3) the incremental development of a theory by a group of scientists over a period of years, and (4) computational methods for reproducing past discoveries. This study of the gene case, an example of (3), will provide further evidence for this new incremental view of discovery.

No one person discovered the theory of the gene. It did not spring full-blown from the mind of a genius or lucky guesser. It began with vague and implicit components that were changed and improved over time by many biologists. Changes in the components of the theory were constrained and directed by a number of factors that show a very intimate relation between the discovery process (or the process of "constructing a new hypothesis," as I prefer to call it) and the process of justification or assessment. Anomalous results often provided strong directives as to where and how to modify the theory. Anomalous data did not supply all that was needed for changing the theory, however. Other factors played roles in suggesting new ideas, and sometimes brought with them a certain amount of justification.

Subsequent chapters will suggest strategies for producing new ideas, all of which are examples of the reasoning pattern: take an old idea and change it to satisfy constraints. Chapter 15 provides a systematic discussion of those strategies.

2.3 Strategies for Theory Assessment

Both discovery and justification will be analyzed as employing strategies. The traditional view, with which I disagree, holds that discovery has only heuristics, while justification has strict rules for deciding the adequacy of a theory. The analysis here will try to make plausible an alternate view: both discovery and justification can be characterized by strategies that may be useful and yet be weaker than a universal method or criterion.

Losee (1987) has called for a descriptive analysis of the strategies for theory evaluation that scientists actually employ. As will be seen, during some of the disputes about modifications to the theory of the gene, different scientists appealed to different criteria of theory assessment. Thus, this case study begins the kind of descriptive analysis that Losee advocates. Determining what criteria a scientist actually employed is easy if explicit appeal is made to a criterion in the published work; at other times, it is necessary to infer what criteria the preferred hypothesis satisfies.

Many criteria of theory assessment have been proposed by philosophers of science; these criteria have been of use in my descriptive task of forming (metascientific) hypotheses about what criteria are exemplified in a given episode. However, the criteria have usually been proposed by philosophers as prescriptions for how to evaluate good scientific theories. The criterion most often proposed has been that a theory be testable or falsifiable (e.g., Popper, 1965). A list of criteria was given by Newton-Smith, which he called the "good-making features of theories." A good theory has: (1) observational nesting—preserving the observational successes of its predecessors; (2) fertility—having scope for further development; (3) a track record—

having a record of past observational success; (4) intertheory support—from being consist with other accepted theories to being able to reduce one to the other; (5) smoothness—making adjustments smoothly in the face of failure; (6) internal consistency; (7) compatibility with well-grounded metaphysical beliefs; and (8) simplicity. Simplicity is a problematic criterion for several reasons, including lack of a criterion for relative simplicity and lack of evidence that simplicity has been a good sign of long-term success in past scientific cases (Newton-Smith, 1981, pp. 226-232). Newton-Smith stresses the role that judgment must play in applying these criteria: "Reasonable men [& women] may be expected to have reasonable disagreements...." That does not mean that anything goes, but at certain times there may be no "knock-down proof of superiority" of one theory over another at those times when a choice must be made (Newton-Smith, 1981, p. 234).

Newton-Smith's discussion is representative, in contemporary philosophy of science, of prescriptions for criteria of theory assessment, although it is a bit more comprehensive than most (see, e.g., McMullin, 1982; Thagard, 1988, ch. 5; Thagard, 1989). Philosophers usually assume that such criteria are to be used to evaluate a well-developed theory at a given time. Like most philosophers, he assumes that theory assessment begins when a theory has been fully developed; like some others, he fails to consider that assessment may play a role in the construction of vague, not yet fully formed, theories. Although Newton-Smith mentions the possibility that compatibility with well-grounded metaphysical assumptions may act as a constraint in theory construction, for the most part he has little to say about the construction of theories.

This book takes a different view. The emphasis here is not the usual one of a philosopher attempting to find prescriptions for how to justify a theory. One of my purposes is to analyze how criteria of theory evaluation can function in strategies for the generation of new ideas. On the one hand, such criteria can be imposed as "constraints in the generator" (a phrase from AI) to eliminate the generation of hypotheses not consistent with them. On the other hand, the evaluation criteria can be invoked subsequently, after the generation stage, thereby producing a less constrained generation process. The former method produces a few plausible hypotheses. The latter method produces a greater number of hypotheses, but they are hypotheses that are less plausible than those produced by the former method. Such relations between generation and assessment will be one focus of this discussion of strategies for theory assessment. Another role played by strategies for theory evaluation is to generate anomalies for the theory, such as empirical anomalies, conceptual problems, or problems of relations to other accepted theories. How strategies for evaluation function in producing different kinds of anomalies also will be addressed. In Chapter 15, I will develop my own list of strategies for theory assessment, a list that shares much with the prior work in philosophy of science.

2.4 Strategies for Anomaly Resolution and Change in Scope

Resolving an anomaly for a theory entails generating a new hypothesis, but that generation occurs in a more constrained context than de novo hypothesis construction. Moreover, the nature of a failure can provide guidance as to where and how to modify a theory. Like reasoning in discovery, methods for anomaly resolution have received comparatively little attention in philosophy of science. Popper (1965), e.g., concentrated on falsifying instances as indicators of the inadequacy of a theory, but gave no hints how to use an anomaly to localize and correct the problem to produce an improved version of the theory. Kuhn (1962) discussed the accumulation of anomalies as sometimes provoking crises which, in some mysterious way, led to the proposal of a new theory or "paradigm." Humphreys (1968, p. 248) argued that historical cases support the view that science often changes through the "exploration, definition, and explanation of anomalies," but he had little to say about strategies for resolving them. Laudan (1977) discussed an interesting set of categories for classifying ways anomalies become solved problems, but he provided no strategies for generating such solutions. Instead of searching for a method of localizing a problem within a theory (as a first step to generating a solution), Laudan argued for spreading the blame for an anomaly evenly among the parts of the theory (Laudan, 1977, p. 43). Laudan's concern was only for how to weight the anomaly in theory assessment, not for how to generate new hypotheses to solve it and produce an improved version of the theory. My discussion of the role that an anomaly may play in *generating hypotheses* contrasts with the usual discussion of the role of anomalies in theory assessment, such as Laudan (1977) discussed.

The first step in anomaly resolution is localization; potential sites of failure within the theoretical components need to be found. Some philosophers have been pessimistic about the ability to localize anomalies, since isolating one faulty component in a complex theory appears too difficult (e.g., see Laudan, 1977, and Quinn's, 1974, discussion of Duhem). My discussion will show that for the gene case such pessimism is not warranted, whatever may be the case for knowledge considered from a more global epistemological perspective. The idea of localization of problems within a theory or a theoretical model, in the light of an anomaly, has received some recent attention in philosophy of science (e.g., Glymour, 1980; Nickles, 1981; Darden, 1982b; and Wimsatt, 1987). Furthermore, researchers in AI are developing methods for determining which parts of a complex, explanatory system are involved in the explanation of particular data points. Their techniques for doing "credit assignment" are relevant to the problem of localizing plausible sites (or even all possible sites, given an explicit representation of all knowledge in the system) for modification, in the face of a particular anomalous data point (Charniak and McDermott, 1985, p. 634). Finding one or more locations in such ways is only the first step in resolving an anomaly. Localization must be followed by the redesign of the failing components.

After localization, the next stage in anomaly resolution is the generation of the new hypotheses to replace the failing components. Shapere (1974b) suggested that when theoretical inadequacies arise, simplifications made in the early stages of theory development are likely areas for hypothesis formation. Wimsatt (1987) suggested how mechanical and causal models might aid in forming hypotheses for resolving anomalies that arise for such models; his analysis extends that of Hesse (1966). She suggested that unexplored areas of an analogy used in the original construction of a theory might function in forming hypotheses to resolve an anomaly at a later stage. AI work on diagnostic reasoning (e.g., Sembugamoorthy and Chandrasekaran, 1986) and on redesign in the light of failures (e.g., Karp, 1989) provides fruitful analogies for reasoning in anomaly resolution (see Chapter 12).

In his delightful imaginary dialogues in *Proofs and Refutations*, Lakatos (1976) discussed "monster-barring," a strategy to improve mathematical conjectures in the light of counterinstances. Lakatos proposed monster-barring as a way of preserving a generalization in the face of a purported exception: if the exception could be barred as a monster, that is, shown not to be a threat to the generalization after all, then it could be barred from causing a change in the generalization. Lakatos was concerned with distinguishing between legitimate exception-barring instances and any illegitimate barring of instances when such instances really did require a change in the conjecture. I will discuss Lakatos's monster-barring, in subsequent chapters, when counterparts to his work in mathematics emerge from the gene case.

When Lakatos (1970) switched from discussions of mathematics to natural science, he did not attempt to discuss such strategies. Instead, he spoke vaguely of the "positive heuristic" associated with each "research programme" and thus implied that the heuristics were domain-specific. (For critiques of Lakatos's positive heuristic, see Newton-Smith, 1981; Nickles, 1987b.) In contrast to Lakatos, I am trying to devise general, nondomain- specific, strategies for scientific theory change. I wonder why Lakatos abandoned that task when he moved from mathematics to science. Certainly domain-specific strategies can be more powerful than vague general ones (Langley et al., 1987, p. 46), but I believe general ones also can be developed.

The focus in this analysis will be on finding general strategies for anomaly resolution. The strategies provide methods for localizing the theoretical component(s) at fault and for providing hints about how to modify the faulty component(s). The gene case, as we will see, provides several episodes in which, first, an anomaly was localized in one or more theoretical components, then alternative hypotheses were generated and tested, and then the theory was changed. It is a rich case for developing strategies for anomaly resolution. A series of steps to follow in resolving an anomaly will be developed in subsequent chapters and summarized in Chapter 15. Although the strategies are extracted from the analysis of a single case, they are not specific to genetics. Some instances of their use in other cases will be discussed in Chapter 15; furthermore, Chapter 16 will suggest that they may be amenable to being programmed into AI discovery systems.

2.5 Descriptive, Hypothetical or Normative Strategies?

A question to ask about the strategies I will be proposing is whether they are *descriptive* of reasoning strategies actually used by scientists or whether they are *prescriptive* of the strategies that scientists should have used (or should use in the future). Neither descriptive nor prescriptive is quite the right characterization; I prefer to call the strategies to be discussed in subsequent chapters "hypothetical." They are my (metascientific) hypotheses about strategies that could have produced the historical changes that did occur. They are not descriptive of strategies that I claim a given scientist *consciously* followed (and I would have no idea how to describe what someone does *unconsciously*). Hence, my goal is not to find descriptively adequate strategies (although those are good to have when possible). The extant historical evidence in the gene case does not provide enough evidence to argue that the strategies describe the actual reasoning processes of scientists. Nonetheless, the strategies may serve as hypotheses about what reasoning strategies scientists might have used. Alternatively, they may be considered hypotheses about what strategies could be followed in such circumstances, if scientists explicitly attended to their methods of reasoning, rather than implicitly using some method or other.

Furthermore, I will not argue that a given strategy was *necessary*; some other strategy might have produced the result. I am exploring some of what may be characterized as the "space of possible strategies." I would like to be able to argue that a given strategy is *sufficient* for producing the change. A good proof of sufficiency could be made if computer simulations could be done, showing that the strategy was sufficient for producing the change (more on this topic in Chapter 16). Discussion of such a rigorous proof of sufficiency, however, is beyond the more historical analysis in this book. My more limited goal here is to argue that a given strategy is a plausible hypothesis for a reasoning method that could have contributed to the change that did occur.

Nickles (1987c; 1987d) outlines a view of methodology that shares some features with mine. (Some ways in which we differ will be mentioned in the next section.) He suggests that methodology should be "descriptive, critical, and advisory." Methods should be descriptive of what has actually worked in the past; they should be critically assessed for their efficiency; they should provide guidance to inquiry (Nickles, 1987c, pp.126-127). I am less optimistic about finding descriptively adequate strategies than Nickles is; he underestimates the difficulties of using extant historical evidence to find the actual strategies used by scientists. Nonetheless, I share with him the belief that the critical examination of past scientific work can yield valuable insights about reasoning strategies (Darden, 1987). (What I call "strategies," he calls "heuristics"; the next section will discuss the relation between my usage of strategy and his use of heuristic.)

Moreover, I agree that finding inviolable prescriptive rules should not be the goal of philosophy of science. Probably no infallible logic for producing new theories exists; certainly none has yet been found. The search is not for strategies that invariably produce correct hypotheses, but for strategies that produce a range of plausible ones, given the constraints at a particular time. The search is also for strategies that aid in narrowing the range of plausible candidate hypotheses, that is, for strategies for hypothesis assessment, as well as hypothesis generation.

An evaluation of heuristics, as proposed by Nickles, is also an important part of the "normative theory of discovery" outlined by the AI researchers, Langley, Simon, Bradshaw, and Zytkow, in their book *Scientific Discovery: Computational Explorations of the Creative Processes*. They said:

> We think normative statements can be made about the discovery process. Such statements will take the form of descriptions of good heuristics, or of evaluative statements about the relative merits of different heuristics or other methods. The evaluative statements can be generated, in turn, either by examining historical evidence of discovery or failure of discovery or by constructing computer programs that incorporate the heuristics and then testing the efficacy of the computer programs as machines for making discoveries. (Langley et al., 1987, p. 54)

Some evaluation of strategies will be done in subsequent chapters when I believe I have adequate evidence to make an evaluation. (More will be said in Chapter 4 about my use of history in doing this.) The normative task of critically evaluating strategies, as well as the advisory task of arguing for the future use of good ones, are not the primary tasks of this analysis, however. These important tasks must await further work, as Chapter 16 will suggest. I have found it sufficiently challenging for this project just to construct hypothetical strategies that I claim are "exemplified" in the historical changes in the gene case.

Another question to ask about the strategies to be discussed here is the sense in which they are "general" strategies. A contrast can be made between domain-specific strategies and domain-independent ones. Nickles said:

> . . . a useful methodology will go beyond the vapid generalizations characterizing all scientific activity (and indeed almost all intellectual activity) which have dominated recent philosophy of science. This means that methodology no longer will be a single, unitary subject but will, at the more interesting levels, break down into domain- and context-specific rules, practices, and advice. (Nickles, 1987c, p. 127)

Certainly, domain-specific strategies can be powerful ones. Yet I believe more general strategies also can be formulated. They may, nonetheless, be context-specific—for example, in this kind of context, employ this kind of strategy (or perhaps one of several kinds). The strategies to be discussed here are not domain-specific, in the sense of being of use only in Mendelian genetics.

Instead, they are formulated in a general way so that they can serve as candidates for use in other, relevantly similar problem contexts. The strategy of **complicate an oversimplification**, for example, is exemplified in many historical cases; however, it will be of use only in a context in which an oversimplification is (or may be) responsible for anomalous results. Thus, the strategies to be extracted from this case are general in the sense that they are hypotheses about plausible reasoning methods that may be used in analyzing other historical cases or that may be used in solving problems in relevantly similar contexts.

This search for general strategies is not a general model for scientific change of the kind proposed by, for example, Kuhn (1970) and Lakatos (1970). In so far as any general model of science is being developed here, it is the view of science as a problem-solving activity. Scientific changes are viewed as responses to problems; context-specific, explicit, problem-solving strategies are being sought. To say that is not to do much more than make another of what Nickles called "a vapid generalization" about all of science. What is interesting is the search for problem-solving strategies, an open-ended search this book only begins.

2.6 Metascientific Vocabulary

A few words about words are in order because so many (but not all, I think) philosophical disputes hinge on the meaning of terms. Current philosophy of science has a zoo of terms for discussing science. The old dichotomy between "theory" and "observation" has proved too impoverished and problematic to label all the parts of science that we wish to discuss (Suppe, 1977). Out of the possibilities, I am going to be using "domain," "hypothesis," "theory," "anomaly," "exemplar," "field," and "interfield theory." I have also found it (perhaps unfortunately) necessary to appropriate a few more ordinary language terms and give them my own technical meanings: "theoretical component," "interrelation," and (most importantly) "strategy."

Now, some clarifications and worries about the words I will be using:

Domain: A useful term for denoting the data or generalizations that a theory is expected to explain is Shapere's term "domain" (Shapere, 1974b). The domain is a set of items, which, on the one hand, may be data produced by observations or experiments. On the other hand, domain items may be general claims that have previously been labeled "empirical laws" or "explanatory theories"; they become part of the domain when they are judged to be well confirmed and are themselves in need of explanation. All the items of a domain are currently accepted as "facts," are somehow grouped together, and are judged to be in need of a single theory to explain them. The scope of the domain (and thus the explanatory scope of the theory) refers to the number of (kinds of) items in the domain. A more general theory has a domain with a larger scope; a more specific theory a smaller one. As we will see, it may be a matter of dispute which items should be inside and which items should be outside the domain of a given theory at a given time.

Hypothesis, Theory and Theoretical Component: The term "theory" is used in two very different ways in common parlance. First is the sense of being hypothetical, not yet proven, as in "It's only a theory." Second, it can refer to a well-supported, general, explanatory claim, as in "The theory of the gene was widely accepted by 1930." I will use "theory" in the second sense.

Philosophers have taken a very strict attitude about what counts as a change from one theory to another. For some, those with a view that theories can be explicitly stated, any change in any claim is usually characterized as a change from T1 to T2. This notation obscures what has remained the same from one version to the next, and so relations between stages of theory formation are not apparent. Some higher level unit, such as a research program or research tradition, then needs to be introduced, and the different theories are placed in it to show that they are related. I will not take such a strict attitude; changes can occur in components of a single theory without its becoming a different theory.

Another problem arises about what counts as a single theory in analyses of science. Neither philosophers nor scientists are very careful about the level of generality a claim must have to be called a theory; what some might call a theory, others might see as a component of a more general theory. Universal criteria for individuating theories are difficult to formulate; in each case, judgments have to be made about the components of a single theory and how much they can change before a new theory has arisen.

In order to deal with some of these problems in characterizing scientific theories, I will introduce my own usage of several terms. I want to discuss the development of a very general theory over time, so I will be using the term "theoretical component" to discuss parts of the theory that change over time. Some of these theoretical components were historically called theories themselves. For instance, with respect to the presence and absence theory (discussed in Chapter 6), I am doing a little violence to scientists' usage by calling this a "hypothetical theoretical component of Mendelism." I will use the term "hypothesis" for a proposed alternative to a theoretical component. The principal theory that I will be discussing is the theory of the gene; the early ideas that developed into it will be termed "theoretical components of Mendelism." No fully articulated alternative to the theory of the gene was ever formulated, although it was challenged by those who wanted a cytoplasmic theory of heredity and development, as we will discuss. I shall characterize most of the debates discussed here as debates about "hypotheses" that were candidates for being "theoretical components" of the "theory" of the gene.

Thus, I will discuss changes that occurred during the development of Mendelism as changes in theoretical components. All the components were subjected to criticisms and tests during the development of the theory; some components changed more than others because of the debates. The way specific components function in relation to specific items in the domain will be discussed. Some of the components were modular, that is, relatively independent of other components. Other components were not independent but were systematically connected in complex ways, as we will see. The *modular* nature of the theoretical components will be very important in allowing the localization and resolution of anomalies. The *systematicity* among the components limits their modularity, but serves to give unity and coherence to the theory. The trade-off between modularity and systematicity in the relations among theoretical components will be a topic of discussion.

Anomaly: An "anomaly" is a problem that is a difficulty for an existing theory. Different ways of assessing a theory can produce different kinds of anomalies for it. An anomaly is often generated by data, when a prediction made by the theory fails, that is, when the strategy of assessing a theory as to its predictive adequacy is applied and the theory fails. Other strategies for theory assessment, however, can produce anomalies that do not result from a failed prediction. Laudan (1977) recognized that problems often arise from factors other than anomalous data. Laudan called these "empirical" and "conceptual" problems. The way I am using "conceptual problem," in contrast to Laudan's usage, will be discussed in Chapter 11. In addition to anomalies that suggest an incorrectness in a theory, a theory also can be incomplete (Shapere, 1974b); several instances of incompleteness versus incorrectness will be discussed.

Exemplars, Exemplary Patterns, and Mechanistic Explanation: Kuhn (1970; 1974) discussed the importance of "exemplars," which he characterized as concrete problem solutions in which a formalism (such as a mathematical equation) is applied and given empirical grounding. Chapter

12 will discuss the exemplars of Mendelian genetics and expand Kuhn's analysis. It will be argued that exemplars may serve in the construction of abstract explanatory patterns or schemas (Kitcher, 1981; Schank, 1986; Thagard, 1988). The patterns abstractly characterize mechanisms, which, when they are operating, produce observable data-points as output. Thus, fitting an observation into a pattern is a way of explaining it. A set of exemplary patterns constitutes the explanatory repertoire of Mendelian genetics; diagrams of such patterns provide one way of representing the theory of the gene, as Chapter 12 will discuss.

Field, Interrelation, and Interfield Theory: Philosophers have recognized the need for discussing metascientific units larger than a single theory. Among the many terms, all embedded within a general model of scientific change, are the following: Toulmin's (1972) "discipline"; Kuhn's (1962) "paradigm" and his later (1970) refinement, "disciplinary matrix"; Lakatos's (1970) "research programme"; Laudan's (1977) "research tradition." Although I did not embed it in a general model of scientific change, I contributed to this menagerie the term "field" (Darden, 1974; Darden and Maull, 1977). These terms are not coextensive and deciding which is appropriate in a given historical case has proved difficult. In fact, I think the research program of testing the adequacy of the models and their terms to find an adequate general account of scientific change is a degenerating exercise in history and philosophy of science, even though articles continue to be published doing so. Furthermore, historians and sociologists have coined their own terms for the social units associated with these conceptual units, including such terms as "research school," "speciality," "invisible college," and "discipline" (see, e.g., Geison, 1981; Crane, 1972; Kohler, 1982). Sociologists have tried to devise precise criteria to delineate instances of these groups, such as citation analysis (e.g., Price, 1965). In addition, hierarchical relations between the units have been introduced; one example of such sociological relations in science is that specialities make up disciplines (Zuckerman, 1988). How the sociologists' categories relate to the philosophers' conceptual units is often problematic (Kupferberg, 1989).

I believe that actual areas of science are too various, too fluid, at too many different levels of organization to expect that a single term (or even a hierarchy of terms) can be defined with necessary and sufficient conditions that will be adequate for discussing all of science. I am not trying to give a general model of the organization of all of science, but I need a term to apply to genetics, to cytology, and to more loosely organized areas, such as evolutionary studies and colloidal chemistry studies. I am going to continue to call the conceptual aspects of these areas "fields." In the sense developed in Darden and Maull (1977), a field has: a central problem; a domain to be explained; techniques and methods (unique to it or shared with other fields); concepts, laws, and theories; special vocabulary; and more general assumptions and goals more or less shared by those scientists using the techniques in trying to solve the central problem.

One of the most important ideas in Darden and Maull (1977) (I now believe with the benefit of hindsight), was the list of kinds of relations between two bodies of knowledge, such as identity, part-whole, structure-function, and causal. I will use the term "interrelation" for such relations. Analogical relations will not be designated interrelations; when analogical borrowing of models or techniques from another field occurs, that is a weaker relation between the fields than the more "ontological" relations that I designate interrelations. Analogical relations provide a much weaker form of relation between scientific fields than do specific interrelations: an analogy merely indicates that there is some similarity between them. More will be said, in subsequent chapters, about the contrast between analogy and specific interrelations.

When the two bodies of knowledge can be adequately classed as fields, then the interrelation between them may be called an "interfield" relation or "interfield theory" (Darden and Maull,

1977; Staats, 1983; Bechtel, 1984, 1986, uses "interdisciplinary"). The chromosome theory of heredity is an example of an interfield theory that relates genetics and cytology by claiming that genes are parts of chromosomes. If the two bodies of knowledge can be analyzed as being at different levels of organization in some sort of hierarchy, then the interrelation is an "interlevel" relation (Darden, 1986a; 1986b). Hypothesizing an interrelation between two bodies of knowledge and using one as a guide to form hypotheses in the other is a very powerful strategy for producing new hypotheses, as we will see.

Strategy: A strategy is a method, a procedure, a practice, a principle. Strategies can be codified and taught. In military parlance, most of the strategies I will discuss are a bit more like tactics; that is, I discuss the more localized maneuvers to achieve local ends, rather than general strategies for winning the war or forming a complete theory. Nonetheless, I do not wish to use the infelicitous phrase "tactics for theory change." Strategy has the nonmilitary usage of "method" that is more appropriate here. Pushing the military metaphor a little further, consider that, for example, General Lee's maneuvers can be analyzed in retrospect to find successful and unsuccessful military tactics or strategies. What actually occurred is documented in the historical record. The history "exemplifies" and can be used to extract strategies or tactics, whether Lee consciously planned the maneuvers using such strategies or tactics. (My thanks to Moreland Perkins for this analogy.) Similarly, the theoretical changes, recorded in published scientific literature, can be analyzed to devise and assess strategies for theory change, independently of what occurred in scientists' minds. I will use the term "exemplified" to designate the relationship between a strategy and the historical change that could have been produced by it.

The relation between "strategy" and "heuristic" is problematic. The *Oxford English Dictionary* attributes the first usage of heuristic in English to Whewell in 1860: "If you will not let me treat the Art of Discovery as a kind of Logic, I must take a new name for it, Heuristic, for example" (*OED*, 1971, p. 259). Polya too characterized heuristic as a "branch of study," and said: "The aim of heuristic is to study the methods and rules of discovery and invention." He added: "Heuristic reasoning is reasoning not regarded as final and strict but as provisional and plausible only, whose purpose is to discover the solution to the present problem" (Polya, 1957, pp. 112-113).

The term "heuristic" has come to refer, not to a field of study, but to a reasoning method that produces plausible results. Buchanan discussed heuristic computer programs for forming plausible scientific hypotheses in his "Steps Toward Mechanizing Discovery":

> The traditional problem of finding an effective method for formulating true hypotheses that best explain phenomena has been transformed into finding heuristic methods that generate plausible explanations. The problem of giving rules for producing true scientific statements has been replaced by the problem of finding efficient heuristic rules for culling the reasonable candidates for an explanation from an appropriate set of possible candidates. (Buchanan, 1985, p. 110-111)

Sometimes heuristics are contrasted with algorithms; algorithms are guaranteed to produce correct results, heuristics are not. On the other hand, heuristics that are implemented in computer programs may be considered algorithms (in another sense) because they are procedures that terminate, although they do so without guaranteeing a correct result. (see Barr and Feigenbaum [eds.], 1981, pp. 28-30 for more on "heuristic" in AI).

My use of "strategy" shares much with the use of "heuristic" by AI researchers. I have been reluctant, however, to use the term "heuristic" and have chosen the more neutral "strategy." My strategies are often more general than the heuristics that are built into computer programs. One could imagine different heuristics to implement a given strategy and different algorithms to implement the heuristic. For example, one strategy for theory assessment is predictive adequacy. A more specific heuristic is to test the theory by looking at extreme cases within its purported domain in order to find its limits (Lenat, 1976). Different algorithms might be used in different computer programs to implement the heuristic of looking at extremes.

Another reason for my use of "strategy" rather than "heuristic" is that Nickles made a sharp distinction between heuristics and epistemics. According to Nickles, heuristics are rules of thumb for guiding research that may be more or less efficient. Epistemics are methods for evaluating the correctness of a theory. The relations between these two are complex (see Nickles, 1985; 1987a; 1987b; 1987d). I am not sure that I want to accept Nickles's distinction. The strategies for theory assessment that I will discuss can function as constraints in the process of producing new ideas and can thereby make the search for new ideas more efficient. Assessment strategies also can function in evaluating the adequacy of a well-developed theory (that is, determining whether it is worthy of becoming part of the body of currently accepted scientific knowledge), but assessment strategies do not provide strong prescriptions for infallibly assessing epistemic warrant. If heuristics are claimed to play a role in discovery or in choosing a line of research, but not in evaluating a theory, then they play more narrow roles than some of the strategies I will discuss. I am interested in the interplay among the methods of (1) searching for new ideas, (2) assessing them, and (3) improving them. "Strategies" is a term for discussing methods for all three processes.

I do not believe that inquiry starts from a set of terms defined by necessary and sufficient conditions and with methods that are completely specifiable via explicit rules in advance. I begin with rough and ready concepts and develop them in the light of historical data and the problems that turn up as the scope is expanded; for example, I have done that with the term "scientific field" in my previous work. I am also pursuing that method here with "strategy." As we will see, progressive refinement is a strategy for developing scientific concepts, as well as metascientific ones. The meaning of "strategy" thus will be refined as the analysis proceeds.

2.7 Stages and Strategies

Figure 2-1 depicts an oversimplified set of stages in theory development. The first stage is a problem or a domain to be explained. The problem arises in some already established context, based on past scientific work and expectations. Strategies for producing new ideas provide one or more seminal, perhaps vague, ideas, which serve as beginning points for hypothesis formation. These ideas are subjected to various constraints provided by both the problem situation and the strategies for theory assessment. Imagine two extremes: (1) a very freewheeling, brainstorming stage in which few constraints are applied, thereby generating many seminal ideas and wild hypotheses, versus (2) a much more constrained generation process, employing many of the criteria that must be satisfied in the assessment process, thereby generating only a few plausible hypotheses. Hence, in Figure 2-1, dotted lines go from strategies for theory assessment to the stage of producing seminal ideas. The lines show that, in AI terminology, few or many "constraints can be put into the generator" (Buchanan, 1985).

Next comes the process of applying strategies of assessment to the hypothesis to determine its plausibility. This stage will include assessments that can be made prior to testing, such as internal consistency. Assessment also involves making predictions and testing them empirically, if the theory is part of a field with experimental techniques. For more historical and observational sciences, predictive adequacy will give way to other means of assessment, such as postdictions (predicting what will be found in the historical record) or assessing the number of different (types of) items in the domain explained by the theory. Applying strategies for assessment may result in success or failure: either (1) one or more hypotheses with evidence in its favor may have been constructed or (2) a set of failed hypotheses may have been examined and there is a need to return to the first stage to begin hypothesis generation again (or even to rethink the formulation of the problem). Deciding which criteria of theory assessment to apply and how to weight their relative importance is, itself, a reasoning strategy. Different scientists apply different assessment criteria to the same hypothesis, as we will see (especially in Chapters 8 and 9).

New plausible hypotheses are often vague and in immediate need of refinement; exploration of their scope is needed; some anomalies may have resulted from the assessment process, even for hypotheses judged plausible and worthy of pursuit. Thus, strategies for anomaly resolution and change of scope may next play a role. If resolving an anomaly requires producing new ideas to replace faulty parts, then a return to the beginning stage occurs, with the additional constraints of preserving the successful theoretical components that are not currently seen as at fault.

As the arrows pointing back to earlier stages indicate, a failure at any one stage may require returning to an earlier stage: all the way back to the beginning to reformulate the problem, or to the generation of new seminal ideas, or to the development of another idea into a plausible hypothesis, or to the testing of another plausible hypothesis, or to minor modifications of a hypothesis with evidence in its favor. During the construction of a successful theory, many "runs" of the stages of theory development will occur, producing many hypotheses as candidates to

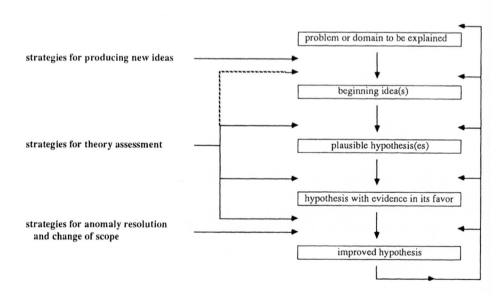

Fig. 2-1. Stages and strategies of theory development.

become theoretical components. The ongoing processes of beginning with vague ideas, evaluating them, refining them, connecting them with other components, testing them, improving them to resolve anomalies—these continue during the development of a theory.

We will see to what extent this oversimplified depiction of stages in theory development is exemplified during hypothesis formation in the gene case. The major task in this analysis is to find strategies of all three types: for producing new ideas, for theory assessment, and for anomaly resolution and change of scope. The diagram (Figure 2-1) will be refined as analysis of strategies proceeds.

CHAPTER 3

The Problem of Heredity

A discussion of the origin of Mendelism cannot start by following the reasoning that led Mendel and others to the important early discoveries in Mendelism; few historical records exist to provide evidence about such discoveries. This chapter is not a historical account of how any person actually discovered what came to be called "Mendel's first law." It is not an exercise in trying to rethink the thoughts of Mendel or some other historical person. Instead, it proposes a possible reasoning path to important Mendelian discoveries. It is a speculative, problem-solving exercise designed to enliven the reader's historical imagination. The chapter asks the reader to pretend to be in a historical period, to pretend not to know what happened later, to formulate problems about heredity, and to consider the use of certain problem-solving strategies.

This chapter also serves the purpose of introducing Mendelian genetics to readers with sketchy or no knowledge of Mendelian genetics. Even those familiar with textbook accounts may find the conjectural steps of problem-solving of interest. This chapter is written in a "Socratic method, school teacher" voice. That voice allows me to pose queries and make suggestions to the reader. This is the only chapter written in this voice and in a speculative style. Subsequent chapters will be more constrained by the historical record and will not employ this literary device.

Why are new theories constructed in science? Surely, for a variety of reasons. One reason might be to solve a problem or a set of problems posed by data. How can data pose problems? Again, in a variety of ways. Sophisticated science is done in a context, a context in which observations have been made, prior theories proposed, problems developed, guiding assumptions adopted. Some problems seem quite esoteric; to trace their origins would require much historical background. The problem of heredity, however, readily lends itself to speculations as to why it was considered to be a problem. People easily notice that children tend to resemble their parents. For instance, red hair runs in families, a trend that is easily perceived. Yet, exceptions occur. Seeds that are saved from good-tasting cantaloupes tend to produce good-tasting cantaloupes the next year, but not always. Sometimes a child resembles a grandparent rather than its parents. Regularities and exceptions to regularities naturally give rise to puzzles. A puzzle demands an explanation: what accounts for regularities and what happened to produce an exception? Scientific theories provide explanations. An explanation often takes the form of a postulated mechanism, such that, if the mechanism is operating properly, the regularities result. Exceptions to the regularities indicate that the mechanism is somehow failing.

Instead of beginning with puzzling data, an investigation might begin with problems needing solutions and then the guessing of some hypothesis as a way to solve them. What problems might lead to an investigation of hereditary phenomena? The problem might be—what happens during the process of fertilization in sexual reproduction? What is contributed by the female and by the male? Alternatively, the general problem context might be questions about the origin of species: What happens when two varieties of organisms are crossbred? Can crossing itself produce new forms? Will those new forms breed true? Can a new species arise by such a cross? When parents with different characters are crossed, do the offspring show a blend form or one or the other of the parental characters? Are there "units" of heredity that can be investigated by doing crosses? Numerous other problems might lead one to form a hypothesis related to hereditary phenomena.

Suppose you decide to investigate the problem of how characters are inherited in plants. Suppose you know something about sex in plants; for instance, that pollen must fertilize the egg in order to get seeds. Then one way to experimentally manipulate hereditary phenomena is to do artificial breeding, crossing one variety with another and noting the results in the offspring. Suppose you choose to investigate differing varieties of peas. Normally pea flowers are self-pollinating, but they can be cross-pollinated if a breeder puts the pollen from one variety on the flower of another. Hybrids are produced in this way.

Some decisions must be made prior to collecting data. Data may be collected in different ways. One way of collecting data is to note whether the offspring resemble, *on the whole*, one parent rather than the other, or whether they look like a grandparent or some more distant ancestor. On the other hand, instead of looking at the organism as a whole, you might focus on *individual characters*, such as the color of the flower or the seed. Then you could follow that single character through the generations. You might develop such a concept of discrete characters because you have some prior hypothesis or just because you note that flowers differ in flower color. Deciding the categories to be used for collecting data, for example, "resembles the mother" versus "shows the red hair color of the mother," is an important preliminary step in the investigation.

Let's say you choose differently colored peas, yellow ones and green ones. Pea plants that differ in the color of peas they produce are to be crossbred. You want to see what happens in the subsequent generations of peas produced from that one original cross. A count is to be made of the number of peas of each color that are produced by the hybrid plants in the next generation. Then, the hybrid peas will be planted and the flowers on the plants that grow from those peas will be allowed to self-fertilize. The peas in the next generation will again be planted, the plants allowed to self-fertilize, and the numbers of differently colored peas counted, and so on for several generations. You want to see what effects the original hybrid cross produced on the characters of pea plants in subsequent generations.

Before a single piece of data has been collected, a lot of preliminary work has been necessary. A general problem or a preliminary hypothesis has pointed to a choice of a more particular area of investigation. It is hypothesized that plants can be viewed as consisting of separable characters, which can be studied separately. With that in hand, a technique, artificial breeding, has been chosen. Furthermore, you have designed steps of experiments and chosen methods of collecting numerical data about discrete characters.

Before doing the experiments, you can speculate about what results you may find. You might hypothesize that the hybrid will show a blended character qualitatively "located" between those of the two parents, for instance, yellows crossed with greens would produce chartreuse peas. You might also form the hypothesis that all the offspring from the hybrids will be chartreuse, so too their offspring, and so on. You might think that mongrels (offspring produced by hybrid crosses) will always show some effect of the cross, in contrast to their purebred parents.

Still other hypotheses might occur to you. You might speculate that the hybrid will show either yellow or green (any idea which?). Yet another possibility is that the offspring of the hybrids will show a totally new color, say pink. Does that sound plausible? No? Then think about what constrains your hypothesizing. You expect some sort of continuity between the characters of the parent and the offspring because you have previously observed regularities. You come to the experiments with expectations based on prior observations and hypotheses. It is useful to be as explicit about biases and expectations as possible; they can have large effects on hypothesis construction. However, uncovering these implicit assumptions may be difficult.

That said—on to the experiments! For your artificial breeding experiment, you choose a pea plant that produces only yellow peas and that comes from plants that produced only yellow peas for several previous generations. In more technical language, the plant is a "pure breeding yellow." You put the pollen from that plant on the flower of a plant that produces only green peas. You also take the pollen from a pure breeding green to put on flowers of the yellow variety. You do this for a large number of flowers on a large number of plants. All the hybrid peas produced from these cross-pollinated flowers are yellow; no trace of green appears in any pea. You plant these hybrid peas and allow them to self-fertilize. In the next generation, some green peas appear. Counting the number of yellow and the number of green peas in that generation gives the following: 258 plants yield 6,022 yellow peas and 2,001 green peas. No chartreuse. The plants resulting from these seeds are allowed to self-fertilize. The greens produce only greens, but the yellows again produce both yellow and green: 166 plants (raised from the yellow peas) give only yellow peas, while 353 behave as hybrids, subsequently producing yellow and green peas.

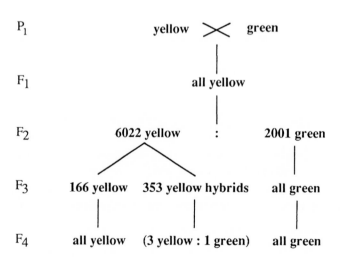

Fig. 3-1. Results from crosses.

Some new terminology will help to keep the various generations straight. The pure-breeding forms used in the first cross are designated the P (for parents) generation. The initial yellow hybrids are symbolized by F_1 (for first filial generation); the offspring produced by self-fertilizing the hybrids are the F_2 and so on to F_3 (or more, with continued self-fertilizations). Figure 3-1 shows the observed results.

There are puzzling phenomena here, but what you regard as puzzling will depend on your prior expectations and hypotheses. Why are all the hybrids in the F_1 generation yellow, and why were they entirely yellow, even though one parent was green? If you expected chartreuse peas to appear, then entirely yellow peas in the F_1 are a surprise. Even more puzzling is the disappearance of green in the F_1 generation and its reappearance in the F_2. How did the hybrid peas in the F_1, all of which were yellow, nevertheless give rise to peas that were green in the F_2? How can the green grandparental character disappear in the hybrids but reappear in their offspring? Disappearance and reappearance is a puzzling phenomenon calling for explanation. There are others: Can the green peas in the F_2 produced by the hybrid yellow parents really breed true forever more? Will the green really never show the influence of having been produced by that hybrid yellow, either by making a chartreuse pea or by producing an occasional yellow?

Before seeking an explanation for these findings, you need to see how much your data shows you. The numerical data can be analyzed to show ratios between the different characters. (Why might it occur to you to analyze data into ratios? Maybe you teach high school mathematics in addition to doing experiments with peas.) The ratio of yellow to green in the F_2 generation was about 3:1 (6,022 to 2,001). Results in the F_3 generation (166 pure yellow to 353 hybrids) show that the "3" of the F_2 consisted of two classes: 1 pure yellow to 2 hybrid yellow. Thus, the F_2 ratio was actually 1 pure breeding yellow: 2 hybrids: 1 pure breeding green, even though the yellows looked just alike. You thus may hypothesize that you have discovered an "empirical regularity" or "empirical law" of 3:1 ratios in the F_2 generation. This can be tested by performing another cross between tall and short varieties of pea plants. Results provide additional evidence for the hypothesized law: the results are 3 tall to 1 short. Other crosses with other characters can be done. How many do you need to give you confidence that 3:1 ratios will always appear? Let's say you have data showing 3:1 ratios for seven different characters. Are you convinced that you now have discovered an empirical law?

But empirical laws do not provide mechanistic explanations. Suppose you ask: why do 3:1 ratios occur? And someone answers: they always (at least in the seven cases examined so far) occur. Would you accept that as an explanation? If not, then the next task is to go beyond the data to find an explanation of the regularity. Any ideas? Even vague preliminary ideas of how to start? Here is one possible path.

It may occur to you, especially if you remember your high school mathematics, that the formula: $(A + a)(A + a) = 1AA + 2Aa + 1aa$ produces 1:2:1 ratios. Once you have introduced symbols you can ask: what might the A and a stand for in the experimental situation? Well, back at the beginning you took pollen grains and put them on the female part of the flower. You know from cell theory that the pollen will fertilize one egg cell in the female plant (or, perhaps, that was one of the hypotheses you were investigating). Connecting your newly introduced symbols with knowledge obtained from the field of cytology (the microscopic study of cells), you might conjecture that A and a denote the different types of pollen and egg cells in the hybrid that combine in a one-to-one way during the fertilization process. A stands for pollen or egg cells that produce yellow in the offspring, and a for ones that may produce green. Because all the pollen cells look just alike (the egg cells are difficult to see, even under a microscope), you have taken a bold step—

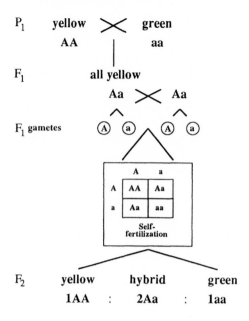

Fig. 3-2. Symbolic representation of crosses.

postulating an unseen difference between observationally identical pollen grains and egg cells.

This move from the level of observable characters, such as yellow in peas, to the level of unobservable differences among pollen and egg cells is an important step in theory construction. It involves bringing in an idea not explicitly in the data about pea color—the idea of different types of pollen and eggs—in order to explain that data. Because pollen was manipulated at the outset, introducing hypotheses about the nature of the pollen is a plausible step. An analysis of the data, a use of symbols, and inquiring about the referents of the symbols—these steps could lead to ideas about unobservable differences.

Reasoning just a bit more, you can say that, in the formation of the pollen and egg cells in the hybrids, each is of either the A type or the a type. Thus, in the hybrid, symbolized by Aa, there is a splitting or a segregation so that each germ cell is one or the other type. This separating of Aa into A and a is necessary if the formula $(A + a)(A + a)$ is to apply. You have arrived at a formulation of what came to be called "Mendel's first law" or the "law of segregation." You do not yet have the idea of discrete material particles called "genes" that segregate in the formation of germ cells. All you have is the idea of different types of germ cells. But asking what makes the germ cells different is a plausible next step. Coming to believe that they contain different material is one plausible answer, but certainly not a necessary one. Various possibilities may occur to you as to how different types of germ cells can later produce different characters. Figure 3-2 refines the diagram in Figure 3-1 by depicting the hypothetical step of segregation. The segregation step results in two types of germ cells in the hybrids that combine randomly during self-fertilization.

How confident are you about your plausible hypothesis of two different types of germ cells? How could you test the claim that the A and a are segregating in the formation of the germ cells in the hybrid? Figure 3-3 shows a cross to test the hypothesis. One of the hybrids, Aa, is "back crossed" with the pure green, aa. The cross may be symbolized: $(A + a)(a + a) = 1Aa + 1aa$. Such a symbolic representation yields the prediction of one-to-one ratios of yellow to green. In fact,

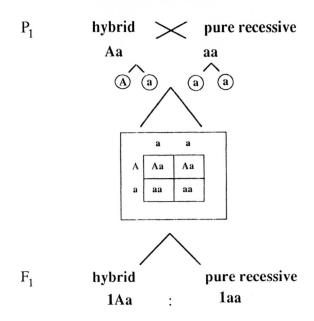

Fig. 3-3. Test cross.

such a cross gives approximately such ratios. This result provides evidence in favor of the symbolism and the hypothesis of segregation.

New discoveries beget new terms. Let's call the character that appears in the F_1 generation, for example, yellow, the "dominant" character, and the one that reappears in the F_2, for example, green, the "recessive." Germ cells (pollen, sperm, and eggs) are also called "gametes." A fertilized egg is called a "zygote." Zygotes formed from the same type of gametes (AA or aa) are called "homozygotes" and those formed from different types of gametes (Aa) are "heterozygotes." For now, you will refer to the A and a vaguely as differing "factors," because nothing is known about their physical nature, nor why one dominates over the other, nor how they cause characters. You have formed the hypothesis that "segregation" occurs during the formation of germ cells of hybrids, thereby producing germ cells that have one or the other of the two factors; you have discovered "purity of the gametes" in hybrids (as Bateson, 1902, called it.)

As a consequence of such thinking, such experiments, and such results, many more questions, problems, and experiments may now occur to you. Do characters other than yellow and green and tall and short in pea plants also produce 3:1 ratios? How general is the empirical regularity? Does it occur sufficiently frequently that it should be called an empirical "law"? How else can the segregation explanation be tested besides via the one test cross done above? How general is the explanatory law of segregation? What happens when two different traits such as pea color (yellow or green) and height of pea plant (tall or short) are observed together? This list of questions serves to show how a promising line of investigation can begin from the discovery of an empirical regularity and the proposal of a hypothetical explanation. The only one of these inquiries we will pause to consider here is the last one, and we do so by following what happens in a cross with two different traits, called a "dihybrid" cross.

If you cross tall pea plants that produce yellow peas with short pea plants that produce green peas, then, in the F_1, all the peas are yellow and all the plants are tall. Self-fertilization produces 9 yellow tall: 3 yellow short: 3 green tall: 1 green short. Figure 3-4 shows the explanation in terms of combinations of the different types of pure germ cells. Thus, segregation also occurs in a dihybrid cross, just as it does in the monohybrid case.

Consider the steps in problem-solving that your reasoning to the discovery of segregation entailed. A general problem or a prior hypothesis led to a choice of a way of analyzing the phenomena into units that could be counted. Doing experiments produced numerical results that were analyzed into ratios. Regularities in the data led to the introduction of a mathematical symbolism. Questions about the referents of the symbols led to the introduction of knowledge from another field, in this case, cell theory, and that knowledge was used to postulate unobserved differences at another level of organization. A mechanism operating at that unobservable level was constructed in order to account for the observed phenomena. These methods for the introduction of ideas obtained from some other field—methods such as using a mathematical formalism, making analogies to combining particles, or utilizing detailed empirical information at another level of organization—are strategies for producing new ideas illustrated in this line of reasoning. They provide ways of going beyond the data to find an explanation for that data. Although some small jumps were made in the midst of stepwise problem-solving, jumps such as introducing the binomial formula and postulating the idea of different types of germ cells, nothing mysterious occurred.

Once an explanatory hypothesis had been formed about differing types of pure germ cells, then the new formalism could be used to make predictions about the results of new kinds of experiments. The hypothesis was tested by carrying out the new experiments to see whether the predictions held. Testing predictions is one of the most common strategies for evaluating a hypothesis. In the very neat set of experiments you have considered, the predictions were confirmed. No anomalies have yet been encountered for the hypothesis of segregation. This very

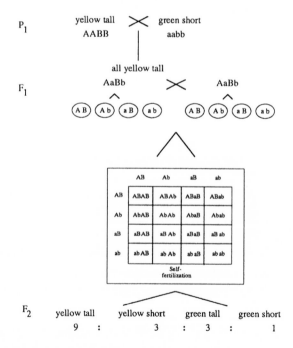

Fig. 3-4. Representation of cross with two different characters.

clean hypothetical discovery path to segregation can be constructed, given the clarity of hindsight; no wrong turns were made. Actual reasoning to hypotheses is usually not so straightforward. Very likely, such a straightforward path was not followed in the actual historical discovery of segregation. In new discoveries, wrong paths may be pursued; backtracking may be necessary.

Yet, even in such a clean path to the discovery of segregation, consider all the steps: finding the problem, choosing the kind of data to be collected, setting up a sequence of experiments, analyzing the results, and constructing an explanatory mechanism. In a bit more detail, the steps were:

1. thinking of characters as units, not as parts of the "essence" of the species;
2. doing the specific cross in which parents differed only in one character, then allowing the hybrids to self-fertilize;
3. counting the characters in each generation;
4. analyzing the numbers to find ratios and rounding them to whole number ratios;
5. investigating the generality of the ratios;
6. asking for a theoretical explanation for the 3:1 ratios;
7. analyzing the 3:1 ratios into 1:2:1 ratios by making inferences based on the F_3 data;
8. the biggest jump—introducing the formalism of $(A + a) (A + a) = 1AA + 2Aa + 1aa$;
9. assuming that $(A + a) (A + a)$ could be used to symbolize self-fertilization of hybrids to produce $1AA + 2Aa + 1aa$ in the F_2 generation;
10. assuming that the account for one pair of characters could be extended to account for two pairs when 9:3:3:1 ratios were found; and
11. making a very general, guiding assumption that biological phenomena can be analyzed mathematically and will be likely to exhibit simple, repeating, quantitative patterns.

In Figure 3-2, a number of assumptions about the process of germ cell formation are implicit, and we have not found it necessary to discuss them explicitly in reaching a problem solution. Diagrams often contain much implicit information. In not quite a thousand words, the implicit ideas present in the picture in Figure 3-2 include assumptions that (1) the hybrid can be symbolized by Aa; (2) no new forms are produced by the hybrid, for example, no new blended forms; (3) A and a do not influence each other while they are together in the hybrid; (4) A and a separate or segregate completely (into pure parental types) when the germ cells of the hybrid are produced; (5) the types of germ cells symbolized by A and a are produced in equal numbers; (6) when self-fertilization occurs in the hybrid, the different types combine randomly; and (7) all the different zygotes produced are equally viable. All of these assumptions play roles in an explanation of the 3:1 ratios that postulates the segregation of parental forms in the formation of germ cells in the hybrid.

As we shall see, no one person in the period from 1900 to 1903 explicitly laid out all of these steps and implicit assumptions. Biologists struggling at the frontier, who have no idea of the end to be reached, must thread their way among numerous alternative paths to find a single one analogous to the simple one discussed here. The next chapter will turn to the actual historical case of the discovery of segregation. Although few extant records exist about the discovery pathway, it is safe to assume that it was more messy than the rather clean pathway traced in this chapter.

A refinement of Figure 2-1 in the light of the problem-solving processes discussed in this chapter produces Figure 3-5. The problem that marks the beginning stage in the diagram might be considered as general as the problem of explaining how heredity occurs. Alternatively, the

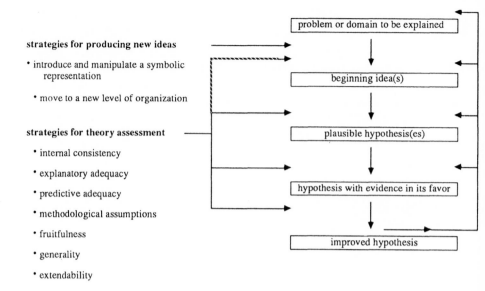

strategies for producing new ideas

* introduce and manipulate a symbolic
  representation

  * move to a new level of organization

strategies for theory assessment

  * internal consistency

  * explanatory adequacy

  * predictive adequacy

  * methodological assumptions

  * fruitfulness

  * generality

  * extendability

Fig. 3-5. Refined stages and strategies, II.

path to segregation might begin with more specific problems, such as the following: What happens during fertilization? Can hybridization give rise to new and different forms? How are characters inherited in hybrid crosses? Once the crosses have been done, the data collected, and the 3:1 ratios found, then the problem becomes to explain the 3:1 ratios.

Two strategies for producing new ideas emerge from this speculative problem-solving account. The first is the **introduction of a symbolic representation**, that is, the $(A + a)(A + a)$ symbolism, and its manipulation to show how 3:1 ratios result. Then the second strategy plays its role: **introduce ideas from another level of organization**, perhaps supplied by another field of science. The A and a symbols are claimed to represent different kinds of germ cells. Thus, the level of organization changes from organisms and their visible characters to unobserved, postulated differences between germ cells. The new idea of different kinds of germ cells is then used to construct a plausible hypothesis about the types of germ cells in the hybrid, along with hypotheses about their numbers and combinations during fertilization. When such a mechanism is operating (represented in Figure 3-2), then the 3:1 ratios are produced. Thus, the strategies of introducing a symbolic representation and a new level of organization play a role in constructing the plausible, explanatory hypothesis of segregation.

In the next stage, the hypothesis is assessed using several strategies for theory assessment. First, the hypothesis satisfies the minimal requirement of being **internally consistent**, that is, it contains no contradiction. Second, it satisfies the criterion of **explanatory adequacy**, because it explains the 3:1 data. Third, it satisfies the demand for **predictive adequacy** because it produces correct predictions about the results in test crosses. Moreover, it satisfies the more **general methodological demand** of being a quantitative hypothesis, namely that it provide numerical predictions that can be tested by experiment. Finally, the hypothesis also can be assessed via the strategy of determining whether it is **fruitful** in suggesting additional experiments. Furthermore, new experimental crosses between varieties with two differing pairs of characters showed the hypothesis to be extendable to those types of crosses also. Thus, the hypothesis also satisfied the criterion of **extendability**. These various strategies for assessing a hypothesis are summarized in Figure 3-5.

Figure 3-5 is a refinement of Figure 2-1. In addition to the strategies for producing new ideas and strategies for theory assessment that are depicted in Figure 3-5, Figure 2-1 contained the category of strategies for anomaly resolution. Anomalies that the hypothesis of segregation faced have not yet been discussed, so strategies for anomaly resolution are omitted from Figure 3-5. Strategies for anomaly resolution will be discussed in following chapters, as actual episodes from the history of genetics illustrate the resolution of anomalies.

CHAPTER 4

Historical Introduction

4.1 Introduction

The task in Chapters 4 to 14 is to discuss those strategies of theory change exemplified in an actual historical case, rather than discuss a hypothetical problem solution, such as in Chapter 3. This chapter begins with a discussion of why the theory of the gene is a good historical case for tracing the development of a new theory. In addition, this introductory section discusses the philosophical approach to history that I will use in analyzing this case. Section 4.2 is a brief note on recent reinterpretations of Mendel's work, showing how little of the theory of the gene was present in Mendel's 1865 paper. Section 4.3 discusses the rediscovery of Mendel's work in 1900, again a topic of recent historical reinterpretations. Finally, Section 4.4 discusses the emergence of the new field of Mendelian genetics and Bateson's role in that emergence.

The theory here examined is the theory of the gene, with emphasis on its development between 1900 and 1926. In 1926, Thomas Hunt Morgan gave a succinct statement of this theory:

> The theory [of the gene] states that the characters of the individual are referable to paired elements (genes) in the germinal material that are held together in a definite number of linkage groups; it states that the members of each pair of genes separate when their germ-cells mature in accordance with Mendel's first law, and in consequence, each germ-cell comes to contain one set only; it states that the members belonging to different linkage groups assort independently in accordance with Mendel's second law; it states that an orderly interchange—crossing over—also takes place, at times between the elements in corresponding linkage groups; and it states that the frequency of crossing-over furnishes evidence of the linear order of the elements in each linkage group and of the relative position of the elements with respect to each other. (Morgan, 1926; p. 25)

Various parts of this statement may not be clear to the reader unfamiliar with genetics; this may be remedied somewhat in the following chapters, where each will be the subject of much discussion. Some components of the 1926 theory were present in 1900, but most were either not present or present only in a rudimentary form. How these components were introduced or developed is the historical case to be studied.

Various challenges were to beset the theory of the gene after 1926, but at the time it was a preeminently successful theory. Much of Morgan's theory of the gene is found today in textbooks

34

of Mendelian genetics (e.g., Strickberger, 1985). Given Morgan's statement, one goal of my analysis is clear: to find strategies for producing each component of the theory, as stated by Morgan in 1926, starting from the theoretical components of Mendelism extant in the period 1900-1903. The theoretical components of 1900-1903 will be discussed in Chapter 5.

The development of the theory of the gene has several advantages as a historical case in which to examine theory development. While some theories have been developed by one scientist working alone, others were developed over a long period of time through the efforts of many investigators. The development of the theory of the gene is intermediate between these two extremes. The theory's development was far from complete in 1900; it developed gradually over an approximately thirty-year period. No single biologist developed all the theoretical components, but the number of important geneticists were few. Another advantage is that the period from 1900 to 1926 is well-circumscribed. The rediscovery of Mendel's work forms a beginning point and Morgan's book is a reasonable choice for an end point.

Furthermore, the published record for Mendelian genetics is easily accessible and contains the basis for an account of the debates about alternative modifications to Mendelism. Recent historiography traces many of these developments and provides accounts of the lives of several principal geneticists. The philosophical task of finding the strategies for theory change exemplified in this case is aided by the good primary and secondary sources. The goal here is not to study the private mental life of geneticists, but to devise strategies that characterize the conceptual relations between publicly debated hypotheses at different stages; therefore, the lack of unpublished notebook evidence relevant to the development of the theory of the gene is not a serious drawback for this analysis. Moreover, the theory was an important, new, scientific development, solving old problems and suggesting new ones. Its development includes the postulation of a major new theoretical concept in twentieth-century science, namely the concept of the gene. For these reasons, the history of Mendelian genetics serves my purposes well.

Historical accounts in the introductory chapters of scientific textbooks create myths about the heros of the field. The accounts look back with the clarity of hindsight, interpreting the work of predecessors from the current perspective of what is accepted. Professional historians of science, in contrast, have moved away from writing this sort of "presentist" or "whiggish" history that merely traces the path to the present. They do not merely look for precursors of currently accepted scientific ideas; instead they examine ideas in the context of the historical period in which the ideas arose. Instead of praising the ones who "got it right" and blaming those whose ideas are not currently accepted, contemporary historians of science try to understand scientists' thinking by viewing their problems and knowledge within the appropriate historical context (Kragh, 1987).

The method that I will use in the discussion of the gene case will not be completely whiggish but neither will it neglect completely the eventual fate of plausible hypotheses posed at a given time. The rest of this section tries to make clear the nature of this method.

It is not always easy to put aside one's current scientific knowledge and come to understand the problems that historical figures faced, what they knew at the time, what was plausible within their context. One must develop one's historical imagination in order to try to understand both the problems faced and the solutions attempted by the early geneticists. Further, we need to try to understand why, when many of these solutions were plausible, only some were eventually accepted.

As with any historical account, it is necessary to be selective, to start somewhere and focus on episodes relevant to the goal of the analysis. It is difficult to decide where to begin to trace the development of the theory of the gene. All Western thought, it seems, can be traced to Greek roots, and so pervasive was Aristotle's influence that discussions of biological theories through the middle of the nineteenth century generally require one to discuss Aristotle's theories. Some histories of genetics begin with Greek thought (e.g., Stubbe, 1972), but for our purposes, a starting point in the late nineteenth century is sufficient. By then, there were several traditions important to a study of the origins of the field of genetics, including empirical hybridist traditions, cell theory and the study of fertilization, and general theories of heredity. These traditions formed the background to the emergence of the field of genetics in 1900 and have been discussed in other work (e.g., Dunn, 1965a; Darden, 1974; Robinson, 1979; Farley, 1982; Olby, 1985). The starting point for discussion of the development of genetics is 1900 (as did Carlson, 1966). Mendel's work was rediscovered that year. Thus, my focus will be primarily on the twentieth-century developments. After the brief note on Mendel in Section 6.2, the nineteenth-century developments will be discussed only when they are relevant to views of biologists after 1900.

The historiography of genetics has blossomed in the last twenty-five years. The early whiggish accounts of Sturtevant (1965) and Dunn (1965a) were improved on by Carlson (1966), who chronicled the debates about ideas in genetics, both those later judged to be incorrect as well as the successful ones. More recent synoptic accounts of the history of genetics are those of Provine (1971), Allen (1975, chs. 3,5), and Mayr (1982, pt. 3). Recent biographies of Morgan (Allen, 1978; reviewed by Darden, 1980a), Muller (Carlson, 1981; reviewed by Darden, 1983) and Wright (Provine, 1986), as well as many articles, contribute to the wealth of material on the history of genetics. Most of this work has concentrated on developments in Britain and the United States, with occasional mention of biologists in the Netherlands or other European countries. The history of genetics in Germany and France is just now being written (e.g., Saha, 1984; Sapp, 1987; Harwood, 1984, 1985; Burian et al., 1988).

In tracing the development of the theory of the gene, I will use both published primary sources and this wealth of secondary sources. Along the way, I will occasionally provide new interpretations or comment on current historiographic disputes, in some of which I have participated. Some discussions, moreover, fill gaps in other historical accounts, such as the discussion of colloid chemistry in Chapter 13.

Little of the unpublished material relevant to this case reveals the reasoning individual biologists used to obtain new hypotheses. Genetics lacks the wealth of notebook evidence available, for example, to Darwin scholars to study the development of Darwin's theory of natural selection (e.g., Herbert, 1980; Kohn, 1980; Ospovat, 1981). It has been a matter of dispute whether some of Bateson's fragmented notes represent an attempt to form a nonparticulate theory of heredity or something else entirely (Coleman, 1970; Darden, 1977; Cock, 1983). Morgan cleaned out his files every five years (Allen, 1978, p. 413). Although a few extant laboratory notebooks of Morgan's student, Sturtevant, record data concerning *Drosophila* crosses, they do not contain evidence about steps in hypothesis development (Sturtevant, unpublished).

My task here is not to trace the private work of geneticists as might be revealed in notebooks, however. My historical analysis is at a less fine-grained level. The development under study is that of a theory during a thirty-year period. The theory of the gene was developed as a collective enterprise by numerous scientists over many years. Alternative hypotheses were suggested as candidates for the components of the theory, and debates about those hypotheses occurred in the published record. The hypotheses of greatest interest for our study of this development are those that were put forward for consideration in published form, subsequently subjected to tests, and then either accepted, abandoned, or revised. The historical analysis thus focuses on published hypotheses and debates.

Ziman, in discussing what he called "public knowledge," argued that, in the development of a scientific idea, the form in which such knowledge is presented in the first published accounts, along with the "subsequent criticisms and citations from other authors and the eventual place that it occupies in the minds of the subsequent generation—these are all quite as much part of its life as the germ of the idea from which it originated or the carefully designed apparatus in which the hypothesis was tested and found to be good" (Ziman, 1968, p.103). For the gene case, I will try to locate such first published accounts of new ideas, as well as subsequent debates in the public forum and their eventual resolution.

Thus, I will be working at the level of public disputes about alternative hypotheses. I will chiefly use published, primary sources, including articles, book-length monographs, and textbooks. Monographs written by the principal disputants and textbook accounts of those disputes are significant resources for tracking developments. I am especially interested in looking at disputes about alternative hypotheses and in following the development of ideas— from the working hypotheses proposed in articles to the views accepted at the time a book was written. Morgan, Castle, Bateson, and Punnett were the principal geneticists contributing to the development of the theory of the gene. Each wrote books in which current advances were discussed. They often distinguished the well-established from the uncertain or unknown. Their books often went through several editions; Morgan is reputed to have said that the only book worth writing was one that was soon out-of-date. Following the claims about alternative hypotheses through these editions is a way of tracing the hypotheses' fates and thereby a way of following the steps in the development of genetics. Among the sources used in the present study are the works of Bateson (1902; 1909; 1913); Punnett (1911; 1927); Castle (1911; 1916a; 1920; 1924); Morgan, Sturtevant, Muller, and Bridges (1915; 1922); and Morgan (1919; 1926; 1928).

It would be interesting to have information about what books were used as textbooks in courses given during the early twentieth century, both genetics courses and other biology courses with some genetic content. Such books might be either the original monographs written by leaders in the field or more synoptic accounts written by textbook writers who were not themselves active researchers. Occasionally, biographical information is preserved about what textbooks a scientist used as a student or assigned as a teacher. Muller, for example, studied Lock's *Recent Progress in the Study of Variation, Heredity, and Evolution (1906)* in Wilson's course at Columbia University (Carlson, 1981, p. 27). In general, knowledge about the usage of textbooks is difficult to obtain. The attempts I have made to find historical evidence about what books were used in college genetics (or biology) courses from 1900 to 1930 (or how many copies were sold and to whom) have met with little success. Such information is ephemeral; neither university archives nor book publishers tend to keep such records. A more thorough search for such material must remain a future task for me or for other students of the history of genetics to undertake. (See Selden, 1989, for a discussion of eugenics in high school textbooks.)

Surveying the state of knowledge in a given field at a given time is not an easy task. Ziman made a penetrating comment about an inadequacy along these lines in contemporary published scientific writing:

In my view the gravest weakness in the organization of modern science [1968] is the lack of systematic exposition of the consensus at the stage between the review article and the undergraduate textbook. Experimental practitioners of any field of research will tell you what ideas are well understood and accepted by everyone, and what is still speculative and uncertain, but there is reluctance to set this down on paper. In principle, this is the role of the treatise or monograph; but all too often such volumes are mere compendia of review articles, which do not give [a] judicious general account of the subject as it now exists.

The importance of such a general account is that it turns *information* into *knowledge*. The separate pieces of information in the separate primary papers need to be joined to one another... welded into a coherent intellectual machine, which may be used as a whole, whether for material benefit or for further scientific exploration. (Ziman, 1968, p. 123)

The books written by Bateson, Punnett, Castle, and Morgan during the development of genetics played something of the role that Ziman wanted from contemporary scientific monographs. A question we will not pause to consider is why that form of writing by scientists has become more rare. Ziman continued by saying that about every five years there occur "quite new insights, quite new modes of comprehension and logical connection, which do not 'revolutionize' the subject, but which are as important as major new discoveries in making the pattern of things clear" (Ziman, 1968, p. 124).

The historical account to be given in this study of the development of the theory of the gene will focus on episodes every few years that had such an impact on the development of the field. My choice of episodes is guided by the statement of the theory of the gene in Morgan's comprehensive monograph in 1926. Thus, I am using this end point, this public knowledge of the theory, as a way of selecting episodes. This analysis involves some hindsight and thus departs somewhat from the method used by those historians of science who want to avoid all whiffs of whiggishness. Still, my selections avoid the whiggish error of trying to show only a single path of "correct advances" leading to the currently accepted theory. My account of the development of the theory of the gene does not neglect the ideas that were discarded along the way.

It is important to look at the discarded as well as the accepted hypotheses in order to understand the range of plausible alternatives that were explored at the time. Furthermore, I will even occasionally go beyond the historical record to ask whether other hypotheses that were not proposed at a given time could have been, given the constraints at that time or given the particular strategies that I propose. In AI terminology, I am interested in exploring the "search space" of plausible hypotheses at a given time (see, e.g., Buchanan, 1982), that is, the range of alternatives that could have been proposed or could have been generated using the strategies to be discussed.

After recounting the historical changes and debates as evidenced in the public record, I will then ask: what strategies could have produced the changes that did occur? Strategies for producing plausible, but subsequently rejected hypotheses, are as important as strategies that produced the subsequently accepted hypothesis. A question that I will sometimes raise is whether some strategies for forming and improving hypotheses are better than others, in the light of hindsight.

Morgan's 1926 theory was a successful theory, one that is now part of textbook accounts of Mendelian genetics (e.g., Strickberger, 1985) and so constitutes current public knowledge. I will be exploring whether strategies exemplified by the development of subsequently confirmed hypotheses were different from the strategies exemplified by the failures. Thus, there will sometimes be an evaluative component to this analysis that might provide a basis for normative recommendations; for instance, analysis might produce the recommendation that, when faced with a problem of a certain type, a particular strategy would be worth trying. The present analysis of the development of the theory of the gene thus occasionally aids in answering in the following question—can we learn about good strategies for doing good science from past instances of successful cases?

Sometimes historians have looked back at an episode and made assessments (e.g., Carlson, 1966, pp. 56-57; Allen, 1978, p. 212). But their tasks were not carried out in order to formulate strategies based on those assessments, as mine will be. Taylor (1983, pp. 210-212), for example, in trying to assess why so many of Castle's hypotheses proved to be wrong, made the following suggestion: Castle searched for hypotheses to explain anomalies that had appeared in his own data, but he was not sufficiently concerned about the general applicability of those hypotheses to other items in the domain of genetics. The details of Castle's hypotheses and their fate will be discussed in subsequent chapters. My point here is to show how, from Taylor's strictly historical assessment, a general strategy can be formulated—a hypothesis for resolving an anomaly should be assessed for its general applicability, as well as its adequacy for explaining the anomaly. (This strategy, of course, will sound familiar to philosophers, because it is a usual prescription for avoiding ad hoc hypotheses.) This example shows my method: I wish to use historical assessments, based on a hindsight understanding of the eventual fate of the postulated hypotheses, to extract strategies for doing good science. Thus, as a philosopher, I am looking back with a bit more hindsight than some of my resolutely non-whiggish, historian colleagues. On the other hand, I have fewer presentist concerns than the writers of textbooks who wish only to trace the earlier steps to the present, or than the philosophers who only use successful theories as examples and omit any consideration of failed competitors.

4.2 A Note on Mendel

We begin the historical case study with 1900. This indicates that understanding the ways that biologists working after 1900 read and interpreted Mendel's 1865 paper is more important than understanding Mendel's own historical context. Nevertheless, a note on both Mendel and recent analyses of his work will be useful in setting the stage for understanding how little of the theory of the gene was actually present in Mendel's own work or in that of his "rediscoverers."

The historical account of the origins of genetics, an account familiar to the typical biology student, is the following: Gregor Mendel was a monk and school teacher isolated in Brno (now in Czechoslovakia). Working alone, he experimented with peas in his monastery garden in order to solve the problem of heredity. He discovered the property of dominance in pairs of related, phenotypic, unit characters, and he discovered the 3:1 ratios of dominant-to-recessive characters in the F_2 generation for seven traits in peas. One example is the 3:1 ratio of dominant yellow to recessive green in seed color in peas. He explained those results by postulating the existence of paired material elements or factors, later called "genes." One dominates over the other and is symbolized with a capital letter, A; the other is the recessive, symbolized a. The genotype of the hybrid is designated as Aa. A and a segregate in the formation of germ cells, which then combine randomly at fertilization to give offspring according to the formula: $1AA + 2Aa + 1aa$. Since those with AA or Aa appear yellow (A in this case is dominant yellow seed color), this mechanism explains the 3:1 ratios. By crossing plants with two different traits (two different character pairs), such as yellow, tall peas with green, short ones, Mendel discovered 9:3:3:1 ratios of characters among the offspring. He explained the ratios by postulating that the material factors (later called "genes") assort independently. He thus discovered two fundamental laws of heredity: Mendel's first law, the law of segregation, and Mendel's second law, the law of independent assortment. Mendel presented these results in 1865 to the local natural history society, where they went unappreciated. They were published in the obscure proceedings of the society in 1866 and did not receive wide circulation. When he tried to extend his results to another species, *Hieracium*, he failed to find the same results (Mendel, 1869).

The story continues: Mendel's data and explanatory laws remained virtually unknown until they were independently rediscovered by three biologists, who published their results in 1900: Hugo de Vries, Carl Correns, and E. Tschermak. The rediscoverers recognized the importance of Mendel's laws for solving the general problem of heredity. Work began to confirm these laws. In 1905, William Bateson named the new field "genetics." In 1909, W. Johannsen introduced the term "gene" as a replacement for the vague terms "element" and "factor"; also, he clarified the difference between "genotype" (all the genes in an organism) and "phenotype" (all the characters caused by the genes). Beginning in 1910, T. H. Morgan and his students, working with the fruit fly *Drosophila*, extended Mendel's work; they discovered linkage among groups of genes on the same chromosome, having made use of the discoveries in cytology since Mendel's time. By the 1920s, with the development of the Mendelian/chromosome theory, Mendel's second law had been refined. That is, the second law, the law of independent assortment, had been found to apply only to genes in different linkage groups; Morgan's "beads on a string" model postulated that linked genes were situated ("strung") along the chromosome.

Almost all the claims in this account about Mendel and the rediscovery have been criticized by historians of science. Debates continue about the historical interpretations that should replace the above account. As yet, a complete consensus has not emerged among Mendel scholars. A means of choosing among competing historical interpretations is often difficult, because the extant historical evidence is insufficient and underdetermines any one account. This problem is, of course, not unique to Mendelian scholarship. Discussions of the development of Darwin's theory of natural selection, for example, are replete with competing accounts by the Darwin scholars working with the same notebook evidence (e.g., Herbert, 1971; Ruse, 1973a; Ruse, 1973b; Gruber, 1974; Kohn, 1980; Ospovat, 1981).

Several aspects of Mendel's work are at issue in the critical debates. What were the intellectual and social contexts in which he began his work? What problem was Mendel trying to solve in his research? What data did he collect? How accurately does his published account represent his actual experiments? Were his results "too good?" What was his explanation of his results in peas? To what extent were the results for peas found in other plants? What was Mendel's essential discovery, that is, was there one, most important discovery made by Mendel? I will not attempt to give complete answers to all these questions nor will I assess all the evidence for the answers I do discuss. Much of the work on Mendel is still in progress and a thorough discussion of this material would delay us too long in the mid-nineteenth century. A few brief answers and indications of some of the relevant literature will serve as pointers for the reader who wishes to pursue recent historiography on Mendel in more detail.

Mendel scholars now agree that Mendel was not a lone monk working independently of the scientific problems and context of his time. Brno was in a region with active work in agricultural breeding, and Mendel was sent to study with eminent scientists in Vienna. He was, therefore, aware of the latest developments and controversies in the biology of his day (Orel and Matalová, 1983; Orel, 1984). Mendel scholars have disagreed about what general problem Mendel was trying to solve, but they have usually agreed that he was not trying to find general laws of the transmission of hereditary characters, the problem to which his work is today considered relevant. Wunderlich (1983) and Orel (1984) argued that Mendel was primarily concerned with the problem of fertilization, i.e., the problem of whether one pollen cell fertilizes one "germinal cell" (containing the egg). Olby (1985) and the Sandlers (1985), on the other hand, placed Mendel within the nineteenth-century context of those who studied hybrids and who asked the old question, going back at least to Linnaeus, of whether new species could arise by hybridization and thereafter breed true.

As to the question of the extent to which Mendel's work was within the broader context of the 1860s debates about evolution and Darwin's theory of natural selection, there is little historical evidence upon which to base an answer. Mendel had read Darwin's *On the Origin of Species* sometime between 1861 and 1865 (Orel, 1984, p. 69). He probably accepted the fact of evolution; however, his specific views on Darwin's theory of natural selection are unknown. No good evidence exists to explain what Mendel had in mind when he said in the introduction to his 1865 paper that his hybridization experiments were relevant to "the one correct way of finally reaching the solution to a question whose significance for the evolutionary history (*Entwicklungs-Geschichte*) of organic forms must not be underestimated" (Mendel, 1865, p. 2).

In his 1865 paper, Mendel reported data from many experiments with garden peas. He analyzed the numerical data for the "first generation from hybrids" (Mendel, 1865, p. 10) into ratios that approximated 3:1. In 1936, Fisher used statistical methods to analyze Mendel's data and found that some results were closer to 3:1 than would be expected to occur in such experiments (Fisher, 1936). This analysis has given rise to the claim that Mendel fudged his data. Numerous responses have been made in Mendel's defense (see Olby, 1985, pp. 209-211, for an evaluation of some of them). One of the more interesting is that of Meijer, who argued that evidence exists that Mendel was taught in Vienna to repeat experiments to reduce error and to choose from all the results the one with the least error. In addition to this methodology, which, of course, differs from modern methods, Meijer claimed that Mendel believed in a deterministic universe and so would have expected to get closer to the truth as more observations were made (Meijer, 1983, p. 128).

The most important issue for our purpose of understanding Mendelism in 1900 is the analysis of Mendel's explanation for his 3:1 ratios. Most Mendel scholars now agree that Mendel did not explicitly claim that material particles existed in the germ cells. Olby (1985) argued that Mendel worked primarily at the level of the observable characters; consequently, Mendel did not provide an explanation that postulated paired units for all characters, units that were mutually exclusive in the formation of germ cells. Olby supports this view, first, by pointing out that Mendel used single letters in his formula: $A + 2Aa + a$. Mendel did not use the modern notation of double letters for homozygous forms, that is, AA and aa. Thus Olby argued that this notation provided evidence that Mendel did not have the concept of "one element for one character" for all characters.

Furthermore, Mendel discussed bean crosses at the end of the 1865 paper in which the hybrid did not show dominance, but a whole range of colors. Mendel's attempted explanation is puzzling and has been interpreted by some historians as showing that Mendel proposed multiple genic interaction, a later conception. Olby, however, interpreted the symbols for bean color, $A1$ and $A2$ and a (as well as some additional evidence from unpublished notes), as indicating that Mendel could conceive of *three* contrasted characters that could exist together in the hybrid and were mutually exclusive in the formation of the germ cells (Olby, 1985, p. 247; see also Mayr, 1982, p. 717). Olby questioned whether Mendel was a Mendelian:

If we arbitrarily define a Mendelian as one who subscribes explicitly to the existence of a finite number of hereditary elements which in the simplest case is two per hereditary trait, only one of which may enter one germ cell, then Mendel was clearly no Mendelian. On the other hand, if by Mendelian we mean one who treats hereditary transmission in terms of independent character-pairs, and the statistical relations of hybrid progeny as approximations to the combinatorial series, then Mendel surely was a Mendelian. (Olby, 1985, p. 254)

Other historians have disagreed with Olby's interpretation. Meijer, for example, agreed that Mendel proposed no concept of morphological material particles, but, Meijer speculated, Mendel did have a concept of the nature of underlying elements in the different types of germ cells. Meijer suggested that Mendel thought in terms of pairs of fluid substances that united at fertilization. For characters that segregated, the fluids did not mix, but separated out in the formation of germ cells (Meijer, 1983, p. 147). These differing interpretations show how difficult it is to determine exactly what Mendel thought were the differences among the types of germ cells that he symbolized with A and a.

Another claim of the textbook histories of Mendel has been called into question. Mendel did not explicitly formulate a second law called "independent assortment." While he did do crosses that involved two or three pairs of characters and applied his scheme to them, he did not explicitly state the 9:3:3:1 ratio (although he reported numbers that showed that ratio). Nor did he claim that assortment according to this ratio was a separate, second law. The law that held for one pair of characters also held when two or more pairs were followed through crosses. Independent assortment was not formulated as a second law by Mendel nor by the rediscoverers nor by others in the period from 1900 to 1903 who discussed the Mendelian conceptions (Monaghan and Corcos, 1984). As will be seen in Chapter 9, the second law was stated as such only when exceptions to it were found.

Mendel crossed varieties of other species in order to test the generality of the law that he had found for seven character-pairs in peas. Yet he had difficulty in finding 3:1 ratios in most of the other cases. His second paper (Mendel, 1869) did not confirm the generality of the results from peas. It is only with hindsight that Mendel's 1865 work is chosen for acclaim, while *Hieracium* work is neglected. Later work explained away the *Hieracium* results as an unusual case in which some plants were produced by hybridization and others developed parthenogenetically (without the influence of a second parent).

Meijer (1983) argued that there was no "essence" of Mendel's discovery. Mendel presented a cluster of ideas about segregating and constant hybrid forms. Meijer is correct that many of Mendel's own ideas have been difficult to understand from our current perspective because it is so difficult to refrain from using the clarity of hindsight when reading his papers. There is much less present in Mendel's 1865 paper than later scientists and historians have tended to read into it. And Mendel's 1869 paper discussed constant, not segregating, hybrids.

Not only do contemporary scientists and historians read with a presentist perspective, even Mendel's rediscoverers in 1900 read his 1865 paper with their own hindsight, sharpened by the developments in the late nineteenth century. They too read more there than Mendel had probably stated himself. Our task now is to imagine the setting in 1900, not that in 1865, in order to understand the beginnings of Mendelism, whether or not Mendel himself should, in any way, be considered a "Mendelian."

4.3 Rediscovery of Mendel's Work

Between 1865 and 1900, a number of developments had occurred that were important to the emergence of Mendelism as a field of research. For instance, studies of cell structures and the fertilization process were lively areas of research in cytology. (These developments will be discussed in Chapter 7.) Debates about the nature of variability, an important concept in evolutionary theory, raised the issue of whether small "continuous" variations or larger "discontinuous" variations were more important. (For a discussion of those distinctions, see Darden, 1977, pp. 104-106.) Artificial breeding was a common technique, used both for practical purposes and to investigate more theoretical issues about variability and the origin of species (Meijer, 1983; Olby, 1985).

Numerous speculative theories of heredity were proposed in the nineteenth century. Most were theories of embryological development as well as theories of heredity. None received support when tested and none was widely accepted by biologists at the time. I will briefly discuss Darwin's and de Vries's theories here and Weismann's in Chapter 7. They were three among many. (For further historical discussion of these speculative theories of heredity see, e.g., Morgan, 1926, pp. 26-31, who labeled them as "speculations"; Dunn, 1965a, ch. 3; Darden, 1976; Robinson, 1979; Olby, 1985, ch. 3.)

Darwin (1868) subscribed to a view common at the time that, in hybrids, the characters of the parents usually blend, rather than showing "prepotency" of one over another. ("Prepotency" probably refers to what Mendel called "dominance.") Blending was one of many phenomena that Darwin attempted to explain with his "provisional hypothesis of pangenesis." Darwin proposed the existence of material units, called "gemmules," which were formed in cells throughout the body (hence, *pangenesis*). In hybrids, Darwin argued, hybrid gemmules formed and were passed to the offspring along with latent parental gemmules. The gemmules were transmitted to offspring during fertilization and grew into cells in the developing embryo (Darwin, 1868, v. 2, ch. 27). Darwin thus had a particulate, blending theory of heredity and development. The particles literally came from, and grew into, parts of the body. No means existed for inferring or determining the exact numbers of units in hybrid offspring, either numbers of pure parental units or of new hybrid units. There were as many types of gemmules as there were individual cells. Whether the hybrid gemmule or one of the grandparental (or earlier ancestral) gemmules developed in the offspring from hybrids was dependent on the number and strength of the gemmules for that type of cell (see Darden, 1976, for further discussion).

In 1889, de Vries explicitly modified Darwin's theory of pangenesis in the light of recent work to eliminate the claim that units formed and circulated throughout the body. His primary problem was to find a general theory of heredity. To this end, he proposed that "pangens" were located in the nucleus of cells. Each nucleus contained all the pangens. He considered the "chromatic thread" in the nucleus as the likely location for the pangens (de Vries, 1889, p. 201). During embryological development, pangens multiplied and some passed into the cytoplasm to become active. Thus, he called his theory "*intracellular* pangenesis." (For further discussion of the ways de Vries modified Darwin's theory, see Darden, 1976.) No such movement of particles from the nucleus to the cytoplasm had been observed in microscopic preparations. Nonetheless, de Vries argued, recent evidence of the importance of the nucleus in cell function was sufficient to make it likely that such movement took place, albeit below the level of current microscopic observations (de Vries, 1889, pp. 201-207). De Vries was especially impressed with the supporting evidence provided by the experiments with protozoa. When protozoa were cut, the only section that could regenerate a new organism completely was the one with the nucleus (de Vries, 1889, p. 188).

Types of pangens influenced types of cell parts and processes. De Vries explicitly argued for the concept of unitary hereditary characters and for a view of a species as composed of many such characters. The pangens were the material bearers of the individual hereditary characters. De

Vries's work during the 1890s involved experimental manipulation of these unit-characters. In a retrospective account of this work, de Vries said that he was investigating the main principle of pangenesis, namely, "the conception of unit characters. This led on the one side to the theory of the origin of species by means of mutations, and on the other to the description of the phenomena of hybridization as recombinations of these units" (de Vries, 1924a, p. 133). The latter experiments, de Vries claimed, led him to the discovery of segregation. De Vries claimed he had "completed the majority of his experiments and had deduced from them" the explanations that he published in 1900, prior to reading Mendel's paper (de Vries, 1900, p. 110).

In a paper in 1976, I took de Vries at his word and argued that the research program that de Vries outlined at the end of his *Intracellular Pangenesis* in 1889 could plausibly have led him to discover the 75 to 25 percent proportions (as he reported 3:1 ratios). Further, I boldly continued, his theory of material pangens in the nucleus guided his discovery of their segregation in the formation of the germ cells of hybrids. Although I still believe such a pathway was possible, extant historical evidence has not been found to confirm that view.

Much recent work has tried to assess de Vries's claim that he first learned of the existence of Mendel's paper only after he had discovered 25 percent latent forms (Mendel's "recessives") in the offspring of numerous hybrid crosses and after he had formulated an explanation in terms of segregating pangens. De Vries scholars now disagree about the extent of de Vries's originality and independence from Mendel. Recent work has discussed these issues in detail (e.g., see Heimans, 1962 and 1978; Kottler, 1979; Meijer, 1985; Darden, 1985). My current view is that these issues will never be resolved and we will never know how much of de Vries's claim to independence is correct. I still believe that his 1889 theory and his subsequent research program, plausibly, could have led him to an independent discovery of both 3:1 ratios and the segregation explanation.

Besides the questions raised about de Vries's claim to be an independent rediscoverer, the claim to independent discovery of Mendel's results by Tschermak (see letter in Roberts, 1929, ch. 11) has also been questioned. Stern and Sherwood (1966) and Olby (1985) dismissed Tschermak as a rediscoverer. A reading of Tschermak's 1900 paper shows that he mentioned the 3:1 ratio in the context of many other investigations and gave no explanation of it, neither in terms of segregation nor in any other way. Correns (1900), as did de Vries, published not only data with 3:1 ratios, but also an explanation in terms of segregating units, which he called "*anlage*," a term developed by embryologists for the rudimentary basis of a part or organ. Work on the materials at the Correns archives is just beginning (Saha, 1984; Meijer, 1987). I do not know whether any historical evidence exists to substantiate his retrospective claim (letters in Roberts, 1929, pp. 335-338; reprinted in Stern and Sherwood, 1966, pp. 134-138) that he discovered both the ratios and their explanation before reading Mendel's paper.

Historians have also questioned that part of the usual story that states that Mendel's work was unknown, uncited, and unappreciated until the three 1900 "rediscovery" papers. A treasure hunt found numerous copies of, and citations to, Mendel's 1865 paper. More and more continue to be discovered (e.g., Weinstein, 1977; Olby, 1985, pp. 216-220). Thus, it appears that Mendel's work was neither as unknown nor as uncited as the usual story suggests. All three rediscoverers managed to locate it prior to their 1900 publications. It still seems correct to say, however, that no one, before 1900, had published results showing that they had read Mendel's 1865 paper and appreciated both the significance of his numerical ratios and his explanation in terms of different types of germ cells. (For further discussion of possible reasons for neglect of Mendel and his work, see Sandler and Sandler, 1985.)

Yet, if Mendel did not discover that material units segregated in the formation of germ cells of hybrids, and, further, if none of the purported rediscoverers discovered it prior to reading it into Mendel's paper, then who discovered segregation? Actually, no one person did. Recall that at the end of Chapter 3, segregation was analyzed as consisting of a series of related claims. No one person singly discovered and explicitly stated all of them. Hybridization results are many and varied and can produce a bewildering array of ratios (or percentages) of characters in the offspring from hybrids, depending on the varieties crossed and the characters chosen for study. Researchers doing hybridization experiments did crosses in many ways and obtained varied results. A simple, clean set of experiments, such as Mendel's, resulting only in 3:1 and 9:3:3:1 ratios, was not easy to produce. Finding the explanation of those ratios in terms of segregation is no easier.

The ideas constituting the segregation explanation were numerous and complex: that one material unit causes one character, that hybrids have paired units, and that the pairs separate (segregate) in a pure, uncontaminated way in the formation of germ cells of hybrids. Darwin had postulated that hybrid units were found in the hybrid organism, along with parental units; that hybrids contain only pure parental units was unexpected, from this Darwinian perspective. As we will see, the idea of the purity of the germ cells was questioned and resisted by some biologists in the period from 1900 to 1910. It is indeed surprising that some characters of hybrid offspring would never again show the effects of having been produced by a hybrid cross. The idea that mongrels (hybrids) showed influence from both parents was an old one; the idea that, *with regard to one character*, some of the second generation hybrid offspring would never show their mongrel past was a new idea. Working out all the conditions necessary during the formation of germ cells in the hybrid to produce this result was a process that took some time in the early days of genetics.

Given the complexity of both finding the surprising data and formulating all the components of the segregation explanation, it is not so surprising that no one person discovered it all. Misreading Mendel may have aided the rediscoverers in making sense of the array of results from their numerous different experiments (e.g., Olby, 1985, p. 241 on Correns's interpretation of Mendel; Olby, 1985, pp.120-124 for Tschermak's confusion even after reading Mendel; and Meijer, 1985, pp. 220-221, for one interpretation of what de Vries got from reading Mendel).

Despite the details of their discovery moments, about which we will likely never have adequate historical evidence, none of the rediscoverers saw the law of segregation as a promising line of research for developing a general theory of heredity. According to Saha, Correns spent the period from 1900 to 1905 investigating the scope of the Mendelian results. He eventually decided that they were too limited and had too many exceptions to supply the foundation for a general theory encompassing heredity, "cytology, evolution, sex determination and development" (Saha, 1984, p. 162).

In contrast to Correns and other German biologists (Sapp, 1987), de Vries showed throughout his career little concern for finding a theory for embryological development. Instead, his concern was to find a general theory of heredity and to establish its role in the origin of species. In his proposal of a general theory for the origin of new species, de Vries argued that segregation did not occur for the important, species-forming, "progressive" mutations (de Vries, 1901-3). He had discovered a mutation in the evening primrose, *Oenothera*, and he used it as the model organism on which he based his mutation theory (Darden, 1976). In the years following 1903, de Vries pursued his own theory of mutation, not Mendelism, as a general theory for the origin of new species. The genetics of *Oenothera* proved to be a quirky anomaly and a disastrous choice as a model organism on which to base a general theory. Mendelian geneticists spent many years explaining the anomalies posed by the evening primrose, as we shall see. Such a bizarre and unique genus would not have received such attention had de Vries not founded an important theory on it. (In the language to be introduced in Chapter 12, it was a "monster anomaly.")

4.4 Bateson and the Emergence of Genetics

In contrast to Correns and de Vries, Bateson did see the promise of Mendelism for producing a general theory of heredity. In order to understand Bateson's judgment of its promise, it is necessary to understand his problem-context. What problem was he trying to solve? What methods did he advocate for its solution? What did he believe constituted an adequate theory of heredity? Answers to such questions require inferences, based on the historical evidence. I have discussed Bateson's assessment of promise elsewhere (Darden, 1977) and will only summarize my conclusions here.

Bateson's primary interest was in evolution. He began his career within a framework of studying embryological stages in order to understand phylogenetic pathways in evolution. Early in his career, however, he changed his approach and began explicitly studying heredity and variation. In 1894, Bateson argued that what he called "discontinuous" variants were especially important in evolutionary change. During the late 1890s, Bateson shifted from field work to experiments with artificial crosses in order to try to follow the appearance and inheritance of discontinuous variations, that is, ones that arose in a single step between parent and offspring (Coleman, 1970). In an 1899 address, Bateson advocated a program of careful breeding between varieties, following the offspring of hybrids through several generations, with the results to be examined statistically. He thus was well-prepared to understand de Vries's and Mendel's papers when they came to his attention (see Olby, 1987, for a discussion of which he read first).

The "new conceptions" of Mendelism, as he called them, were seen by Bateson as promising for the following six reasons (Darden, 1977, pp. 97-98):

1. The Mendelian conceptions allowed Bateson to reinterpret old data about hybrid crosses (if the relevant crosses had been done and the numbers of characters had been counted), as well as put the previously confusing results into the Mendelian framework.
2. These conceptions dealt with phenomena, namely discontinuous variation, that Bateson saw as relevant to solving a larger problem, namely the type of variation important in producing new species.
3. Mendel's conceptions solved certain problems about the phenomena themselves, namely how newly arisen variants could be maintained. If new variations blended rather than segregated, then new, discontinuous variants would be swamped out in a population. The problem of swamping was one that Darwin faced in accounting for the maintenance of new, adaptive variations. Mendelism helped to solve that problem, since pure parental characters segregated and thus could be maintained in a population without being swamped out.
4. The new conceptions were determined by the precise techniques of artificial breeding and numerical analysis that Bateson had already advocated. These techniques provided a method for a new line of research to determine the generality of the law of segregation.
5. Bateson, in contrast to de Vries, judged that the Mendelian conceptions held the promise of being very generally applicable to the problem of heredity.

Bateson believed that an adequate theory of heredity would not only explain discontinuous variation, but also explain embryological development as well. Sandler and Sandler (1985) have argued that Mendel's work was neglected in the period between 1865 and 1900 because biologists during that period expected any theory of heredity also to be a theory of development. They are correct that nineteenth-century speculative theories of heredity were usually also theories of development. This was true of Darwin's hypothesis of pangenesis and Weismann's theory of the germplasm, but less so of de Vries's theory of intracellular pangenesis. In 1900, however, Bateson was still holding such a view. In fact, throughout his career as a major proponent of Mendelism, he continued to search for a general theory encompassing both (Coleman, 1970; Darden, 1977; Cock, 1983). The desire for a general theory to explain both heredity and development did not preclude Bateson from seeing Mendelism as promising. Thus, the Sandlers' (1985) argument that the desire for such a theory in the late nineteenth century was the primary reason for Mendel's neglect should be reassessed. Because Mendelism never fulfilled its promise of being both a theory of heredity and a theory of development, it was rejected by later critics, as we shall see.

Bateson did not see Mendelism as necessarily implying a commitment to material particles in the germ cells. He believed such material particles could not adequately explain the myriad changes in the developing embryo. Coleman (1970) and I (Darden, 1977; Darden, 1980b) have argued that Bateson tried, unsuccessfully, to develop an alternative theory in terms of vibrations, vortices, and forces. Bateson's attempt at developing this alternative theory drew on analogies to entities and agencies postulated by late nineteenth-century physical theories, theories proposed as alternatives to particulate, atomic theories. In contrast to our interpretation, Cock (1983) has argued that Bateson was not anti-materialist and that his vibratory theory was not an attempt to formulate a general theory of heredity. Evidence for Bateson's attempts to develop such a theory is found primarily in his unpublished papers, although he made some comments about it in print. This dispute about the nature of Bateson's theory building will be discussed in Chapters 7, 9, and 10, as we discuss Bateson's opposition to the chromosome theory.

I believe that another reason for Bateson's assessment, in 1900, of the promise of Mendelism was (6) that it did not entail a commitment to the hereditary particles, to which he was opposed.

Between 1900 and 1903, Bateson became the advocate for promoting knowledge about and research on the new conceptions of Mendelism. He is rightly called the founder of Mendelian genetics. He coined new terminology, such as "homozygote" and "heterozygote" (Bateson and Saunders, 1902), and introduced the term "genetics" (Bateson, 1905) to name the new field. He argued against an opposing theory, that of the biometricians, who advocated the importance of smaller scale variations and did not accept the claim of segregation of pure characters (Provine, 1971). Bateson's advocacy of Mendelism was a key factor in converting Mendel's work from an interesting curiosity about the results of hybrid crosses in a few plant species into a line of research on which to found the new field of genetics. Other scientists joined Bateson in redirecting their own research to use the methods of Mendelism to tackle various biological problems of the early twentieth century.

The next step in this analysis of the development of Mendelism is to examine the published historical record for the period from 1900 to 1903. From it will be extracted the data and empirical generalizations in the domain to be explained, as well as the new theoretical components of Mendelism that were proposed to explain them. The historical evidence for such a reconstruction is the published work of Bateson, Correns, and de Vries, including material showing how they read and interpreted Mendel's paper (the translating and reprinting of which, Bateson supported).

Further evidence for claims in this analysis is provided by the work of others who read and published about Mendelism after 1900. The most important of these are Davenport and Castle. As we shall see, only a few of the components of Morgan's 1926 theory of the gene were present by this time.

Mayr, commenting on the beginnings of genetics, said:

> Mendel did not, by a single stroke, create the whole modern theory of genetics. He did not have a theory of the gene, but neither did his rediscoverers.... However, Mendel's various discoveries (segregation, constant ratios, independent assortment of characters), combined with the new insights acquired between 1865 and 1900, led, one is tempted to say automatically, to the theory quite legitimately called Mendelian. (Mayr, 1982, pp. 717-718)

We will now see how much of the theory of the gene was present in the published work in the period 1900-1903 and then follow the changes that produced that theory by 1926. The degree to which those changes were (or were not) "automatic" will become apparent.

CHAPTER 5

Mendelism, 1900-1903

5.1 Introduction

Regardless of the details of the historical rediscovery, the period from 1900 to 1903 marked the years of publication of the 3:1 ratios and their explanation in terms of segregation. The work of several biologists publicized the new discoveries to the scientific community. This work included publications by de Vries (1900), Correns (1900), Bateson (1900, 1902), Davenport (1901-2), Cuénot (1902), and Castle (1903). Their work, as well as the English translation of Mendel's 1865 paper (made available by Bateson in 1901), created the public knowledge of Mendelism in the period 1900-1903.

No single statement of Mendelian theory as succinct as Morgan's 1926 statement (quoted in Chapter 4) appears in the literature in this earlier period. Prior assumptions, experimental results, and theoretical components were interwoven in the early papers. Some reconstruction is required to tease apart these different elements and to explicitly separate the data from the theoretical components. Tables 5-1 and 5-2 give my reconstruction of the domain to be explained and of the theoretical components for Mendelism at its beginnings. I have made an effort to stay very close to the language in the original sources.

(Readers who want to see quickly where we are heading may wish to turn to Table 14-1 in Chapter 14 to see how the theoretical components of Mendelism in Table 5-2 had changed by 1926. Their changes are what we will be tracing in Chapters 6 to 13, along with devising strategies for producing those changes. Chapter 14 summarizes the changes and strategies from this case.)

Table 5-1 gives the empirical generalizations to be explained, along with one example of data for each generalization. Mendel's pea examples are familiar ones. A cross between, for instance, a variety of pea that produces yellow peas and another that produces green peas yields hybrid peas, all of which are yellow. Yellow is said to be the dominant, green the recessive (See Item 1). Bateson labeled the hybrid generation the "F_1" (Bateson and Saunders, 1902). Item 2 states the results in the next, F_2, generation: when plants grown from such hybrid peas self-fertilize, they produce peas that approximately show the ratio of 3 dominant to 1 recessive. Thus, on the average, peas in the F_2 generation are 75 percent yellow and 25 percent green. Growing additional generations from the F_2 peas shows that the 3:1 ratio consists of 1 part pure breeding dominant, 2 parts hybrids, and 1 part pure breeding recessive, which decomposes the 3:1 ratio into 1:2:1.

Item 1 Dominance-recessiveness

In artificial breeding experiments involving crosses between varieties of animals or plants differing in the characters of one trait, one dominates over the other in the hybrid (F_1) generation (e.g., yellow x green peas yield hybrids all of which are yellow).

Item 2 3:1 Ratios

When hybrids from crosses such as those described in Item 1 are allowed to self-fertilize or are crossed with each other, on an average they produce a ratio of 3 dominants to 1 recessive in the next generation (F_2) (e.g., yellow hybrid x yellow hybrid yields 3 yellow to 1 green). When the recessives from the F_2 cross are self-fertilized, all offspring in the F_3 generation are recessive (e.g., green x green yields all green). When the dominants from the F_2 cross are self-fertilized and followed for successive generations, then one-third yield pure dominants while two-thirds again behave as hybrids (e.g., yellow x yellow yields 1 pure yellow: 2 hybrid yellow). The 3:1 ratio in the F_2 thus resolves into 1 pure dominant: 2 hybrids: 1 pure recessive.

Item 3 9:3:3:1 Ratio

In other experiments involving crosses between varieties differing in characters of two traits, the characters behave independently, giving on an average an F_2 ratio of 9:3:3:1 (e.g., yellow tall x short green peas gives hybrids that are yellow and tall; when these self-fertilize the F_2 gives, on an average 9 yellow tall: 3 yellow short: 3 green tall: 1 green short).

Table 5-1. Domain of Mendelism, 1900-1903.

When two different pairs of characters are involved in the cross, such as yellow-green and tall-short, the ratio is 9:3:3:1, as stated in Item 3.

Testing the generality of these results for other species was an important line of research for the early Mendelians. De Vries's 1900 paper reported data for fifteen crosses among varieties of different species of plants; the offspring in the F_2 generation all yielded, on average, 25 percent recessives. By 1902, Bateson and Saunders extended the list to include characters in poultry, the first animals demonstrated to produce 3:1 ratios. Thus, by 1903, the empirical generalizations stated in Table 5-1 had been formulated and had been shown to have a good bit of empirical support.

Examples of the data and empirical generalizations (see Table 5-1) were usually stated in a straightforward way in the early Mendelian papers (listed in the first paragraph of this chapter). As will be discussed later, all the papers discussed Item 2, the 3:1 ratio. On the other hand, they differed in their claims about the generality of the phenomenon of dominance (Item 1). They also differed as to whether they mentioned the 9:3:3:1 ratio. Thus, the 3:1 ratio was the primary domain item to be explained.

In contrast to the empirical generalizations, the theoretical components of Mendelism (see Table 5-2) were less clearly stated in the published record. Although widespread agreement existed about the central importance of segregation and purity of the germ cells, different biologists stated in somewhat different ways the important "conceptions" or "principles" or "law" of Mendelism. One person stated a theoretical component that was absent in another person's paper, while another stated explicitly an assumption that was only implicit in the writings of another. Specific differences will be discussed later in the sections that analyze each component. Thus, the set of theoretical components in Table 5-2 is not merely a set of quotations; it is my reconstruction of the theoretical components, based on the published accounts. All the

C1. Unit-characters

C1.1 An organism is to be viewed as composed of separable unit-characters.

C2. Differentiating pairs of characters

C2.1 In varieties of organisms, the traits by which they differ are antagonistic or differentiating pairs of characters.

C3. Interfield connection to cytology

C3.1 The connections between generations are the germ cells, also called "gametes" (i.e., pollen and egg cells in plants; sperm and eggs in animals).

C4. Dominance-recessiveness

C4.1 In a hybrid formed by crossing parents that differ in a single pair of characters, there is some difference such that one character dominates over the other; thus, the character in the hybrid resembles one but not the other of the parents. (Let A symbolize the dominant character which appears; a the recessive which is not visible.)

C5. Segregation

C5.1 In the formation of germ cells in a hybrid produced by crossing parents that differed in a single pair of characters, the parental characters segregate or separate, so that the germ cells are of one or the other of the pure parental types. (Each germ cell has either A or a but not both; called "purity of the gametes.")

C5.2 The two different types of germ cells form in approximately equal numbers. (Equal numbers of germ cells with A and a.)

C5.3 When two hybrids are fertilized (or self-fertilization occurs), the differing types of germ cells combine randomly.
$[(A + a)(A + a) = AA + 2Aa + aa$; appearance $3A:1a]$

C6. Explanation of dihybrid crosses [later called "independent assortment"]

C6.1 In crosses involving parents that differ in two pairs of differentiating characters, the pairs behave independently. Symbolically, $(AB + Ab + aB + ab)(AB + Ab + aB + ab) =$ complicated array that in appearance reduces to $9AB:3Ab:3aB:1ab$.

Table 5-2. Theoretical components of Mendelism, 1900-1903.

terminology in the theoretical components comes from the period; the set is just a bit more explicit and complete than discussions found in any one of the early papers alone, as we will see.

In producing such a reconstruction, it is difficult not to let the clarity of hindsight affect the account. (I myself was guilty of too much hindsight in an earlier attempt to state the 1900 version of Mendelian theory: Darden, 1982b, Table 2.) A careful reading of the early literature shows that a marked lack of clarity existed. The biologists did not make clear exactly what they were postulating to explain the numerical data. Much of the early theoretical discussion was in terms of "characters," rather than in terms of the unobservable causes of those characters, sometimes called "factors," even later called "genes." Geneticists gradually came to recognize the need to be explicit about the distinction between, on the one hand, the observable character, and, on the other hand, its unobservable cause carried by the germ cells. Subsequent clarifications were made, as we will see. I have tried to state the theoretical components in this early stage without explicitly postulating factors that cause characters. Thus, the reconstruction in Table 5-2 of the 1900-1903 theoretical components avoids explicit claims about unobservable factors that cause characters. Instead, segregation is stated in terms of "characters." The quotations that follow will show that, at the beginning of the field, biologists themselves used the term in just such a confused way.

Sections 5.2 through 5.7 discuss each of the theoretical components in Table 5-2. Quotations from the original papers provide evidence to justify the inclusion of each component as part of the 1900-1903 theory. Components 1 to 3 state preliminary, general assumptions that are used in Components 4 to 6. Components 4 to 6 are directly connectable with the domain items in Table 5-1. As we will see, the modularity of Components 4 to 6 proved important in later developments. An important point to note is how much in early Mendelian theory was vague, unclear, and implicit. Section 5.8 discusses some additional claims made in one or a few of the early papers but not included in Table 5-2. Section 5.9 begins the analysis of the relations between domain items and theoretical components; the section provides a diagram (Figure 5-1) to represent the way the theoretical items "cover" the domain items. Finally, the conclusion points the way to the following chapters, which trace the changes in these theoretical components between 1900 and 1926.

(The rest of Chapter 5 serves the purpose of providing the historical evidence for my claim that the statements in Table 5-2 are adequate characterizations of the theoretical components in 1900-1903. The reader who is not interested in textual justification for the theoretical claims can skip to the beginning of the discussion of strategies in Chapter 6. In passing, it would be useful to glance at Figure 5-1 because changes in the theoretical components and domain items will be depicted similarly.)

5.2 Component 1. Unit-characters

C1.1 An organism is to be viewed as composed of separable unit-characters.

Before the collection of data, a decision was made to regard organisms as composed of separable unit-characters. This view of a plant is implicit in Mendel's work, in which he investigated sevsen separable characters of peas, such as color of pea and height of plant. The unit-character view of an organism is an alternative to the view of an organism as exhibiting the whole essence of the species. Lewontin (1974) argued that Darwin and Mendel shared an important difference from their predecessors: they both concentrated on individual variations in characters as important objects of study. They, thus, rejected the older "holistic" view, in which a "whole" organism was viewed as representative of a species type. This "essentialist" view was held, for instance, by Naudin, a hybridist in the 1850s.

Olby (1985, pp. 47-51) discussed how Naudin's view differed from Mendel's. Naudin believed that the species had a "specific essence" and that after a hybrid cross the maternal and paternal "essences" separated. Although Naudin observed results of hybrid crosses similar to Mendel's, he "believed that the species segregates as a whole" (Olby, 1985, p. 51). Thus, Naudin's alternative view of segregation illustrates the importance of the unit-character concept in (1) the kind of data that Mendelians collected, (2) the empirical generalizations that they made, based on the data, and (3) the theoretical explanations that they constructed. All three depended on viewing the organism as composed of separable unit-characters, namely characters that could be counted separately and that could be claimed to segregate independently. It may be a safe conjecture that Naudin could not have discovered the independent assortment of two or more pairs of characters (in which the maternal and paternal pairs of unit-characters were mixed), given his holistic view of maternal and paternal essences (see Olby, 1985, p. 51, for further discussion).

In his 1859 discussion of the origin of species, Darwin, as did Mendel, focused on the individual variations of characters. Mayr argued that Darwin "could not have arrived at a theory of natural selection if he had not adopted populational thinking," that is, abandoned essentialism (Mayr, 1982, p. 47). Indeed, some nineteenth-century biologists made this an important conceptual shift: no longer was the species viewed as having an essence that behaved in a holistic way. Instead they viewed organisms (or groups of organisms) as having individual *characters*, whose *variations* within a population were important objects of study.

In contrast to Darwin's discussions in the *Origin of Species* in 1859, in his 1868 hypothesis of pangenesis he did not focus specifically on individual characters. According to Darwin in 1868, each hereditary unit (a "gemmule," as he called them) was produced by one cell, was passed on at fertilization, and grew into that type of cell in the offspring. His hereditary units, thus, did not cause *types* of characters, such as red color of flowers. Instead, a gemmule grew into, for instance, a single *cell* in the flower petal. As many gemmules were needed as there were uniquely located cells in the organism. Data about characters provided no evidence about *types* of gemmules. Darwin's hereditary theory, thus, was based on the concept of *one unit-one cell*, not the *one unit-one character* concept of Mendelism.

In 1889 de Vries proposed the hypothesis of intracellular pangenesis, which he argued was a modification of Darwin's hypothesis of pangenesis. De Vries made an important conceptual shift that brought him closer to Mendel's viewpoint than to Darwin's 1868 view. According to de Vries, "pangens" (as he renamed the hereditary units) produced cell parts, such as pigment granules; for example, copies of the pangen for red pigment were found in the nuclei of all the cells of the petals of a red flower. During the growth of the flower, the pangens moved from the nucleus into the cytoplasm of the cell; in the cytoplasm the pangens produced the pigment granules, the granules responsible for the red flower color. Thus, visible characters, such as red color in poppies, could be counted and could be used to make inferences about *types* of pangens. The unit-character concept, implicit in Mendel's work and stated explicitly by de Vries, was an important conceptual shift. It affected the kind of data that was collected; it also affected the kinds of inferences about unobservable hereditary units that were made, based on that data.

De Vries's 1900 paper, in which he presented his rediscovery of Mendelian segregation, began with the viewpoint that he had advocated in his 1889 *Intracellular Pangenesis*:

> According to pangenesis the total character of a plant is built up of distinct units. These so-called elements of the species, or its elementary characters, are conceived of as tied to bearers of matter, a special form of material bearer corresponding to each individual character. (de Vries, 1900, p. 107)

De Vries's statement made explicit not only the view that the observable organism was to be considered as composed of separable characters, but also his assumption of the existence of underlying material particles, the "material bearers" "corresponding to" (causing?) those characters. Others, however, were not committed to an underlying material cause of the unit-characters. Bateson, in particular, stressed that the nature of the cause of the characters, whether material or not, was completely unknown (Bateson, 1902, p. 5). Correns (1900) explicitly postulated an "anlage" for each character, although he was silent about the possible nature of such an underlying basis of the character.

Should the assumption of the unit-character be considered one of the theoretical components of Mendelism? This question might reasonably be answered in different ways. On the one hand, one might argue that it was a concept that was formed prior to, and needed for the collection of, the data, and thus should not be considered a component of the explanation of that data. For example, if the phenomenon to be explained is that a percentage of offspring inherited red color, then saying that one should regard the flower as having a unit-character of color seems like a prior assumption. Furthermore, such an assumption seems to provide no help in constructing an explanation for the data about red color.

On the other hand, the unit-character concept developed into the claim that genes cause characters. This change was a key part of the development of the theory of the gene. Consequently, the unit-character concept can reasonably be considered one of the general theoretical components of early Mendelism. Given the lack of statements by biologists in 1900-1903 about underlying units, Component 1 cannot be accurately stated, for example, as "factors in the germ cells cause characters during subsequent development" or something similar. (This conclusion represents a criticism of my earlier attempt at describing the state of Mendelism in 1900, in Darden, 1982b)

This intimate relation between, on the one hand, a concept important for collecting data, and, on the other hand, its later development into a claim about theoretical units illustrates the close relation between data and theory. In other words, the unit-character concept shows the close relation between categories for data collection and theoretical concepts used in the explanation of the data. (For a discussion of relations between conceptual changes and categories for data collection, see Scheffler, 1967, pp. 36-44.) The role played by the unit-character concept illustrates how the nature of the data categories serves to constrain (or perhaps to provide guidance for determining) the range of possible explanations for that data.

Thus, I have included the unit-character concept as Component 1 in Table 5-2, as the first of the theoretical components of Mendelism in the period 1900-1903.

5.3 Component 2. Differentiating Pairs of Characters

C2.1 In varieties of organisms, the traits by which they differ may be grouped in antagonistic or differentiating pairs of characters.

This component would be clearer if it were stated in terms of "traits" and "characters." Pea color, for example, would be a trait; green or yellow would be characters or "character states." In the historical literature, however, this (type/token) clarification was not made.

Although one might question the early Mendelians' ability to clearly differentiate among unit-characters, they developed a workable criterion: unit-characters were individuated by noting whether the characters could vary independently. The early geneticists were dependent on the occurrence of natural variation to supply their experimental material. Bateson, in discussing the "new conceptions" of Mendelism, stated explicitly:

> Each such character, which is capable of being dissociated or replaced by its contrary, must hence forth be conceived of as a distinct unit-character; and as we know that the several unit-characters are of such a nature that any one of them is capable of independently displacing or being displaced by one or more alternative characters taken singly, we may recognize this fact by naming such unit-characters *allelomorphs*. (Bateson, 1902, p. 22)

Thus, Bateson's term "allelomorph" (which later was shortened to "allele" and used to refer to alternative states of a gene) was originally introduced to refer to one of a pair of alternative, observable unit-characters.

Correns, too, introduced the concept of the unit-character by discussing the pairs that were found: "The *traits* which differentiate the varieties of peas, can, as in all other cases, be grouped into *pairs*, each member having an effect on the same trait, one in one and the other in the other one of the varieties...." (Correns, 1900, p. 121). De Vries called the pairs of characters in German "*antagonistischen Eigenschaften*." Davenport (1900-1), reporting de Vries's results, called them "antagonistic peculiarities."

In sum, two important conceptual developments in early Mendelism were Component 1, seeing the organism as composed of unit-characters, and Component 2, seeing varieties as having different, yet paired, characters.

5.4 Component 3. Interfield Connection to Cytology

C3.1 The connections between generations are the germ cells, also called "gametes," i.e., pollen and egg cells in plants and sperm and eggs in animals.

In the period from 1900 to about 1905, the segregation explanation (Components 5.1 to 5.3 in Table 5-2, to be discussed later) was formulated in terms of types of germ cells, not in terms of units within the germ cells. By 1900, cytologists had shown via microscopic analysis that fertilization consists of the combination of a male and female germ cell (Wilson, 1896, p. 130). In this way, the work in cytology made available to Mendelians information about the microscopically observable germ cells. The Mendelians, in turn, formulated hypotheses, based on breeding data, that made claims about yet smaller, unobservable differences among types of germ cells.

The claim in Component 3.1 (the germ cells are the material passed from parents to offspring) might be considered background knowledge of the early Mendelians. Thus, one might argue that it should not be an explicit theoretical component of Mendelism; however, the subsequent development of Component 3.1 resulted in the chromosome theory of Mendelian heredity (see Chapter 7). (By the late nineteenth century, cytologists had identified chromosomes as threadlike bodies within the nuclei of cells and had investigated their behavior during cell division and fertilization; see, e.g., Wilson, 1896; Voeller, 1968.) Furthermore, the early Mendelians made claims about germ cells in the formulation of the segregation explanation; the level of the germ cell was the level of organization used in the hypothetical claims of early Mendelism, as we shall see. Including Component 3.1 as a theoretical component will serve my purpose in tracing the development of the chromosome theory of heredity.

Although the 1900-1903 papers used germ cells integrally in formulating the segregation explanation, none mentioned chromosomes. The discussion was at the level of the germ cells, as Mendel's had been. De Vries (1900) used "pollen grains" and "ovules"; Castle (1903) "germ cells." Correns (1900) pushed the discussion to the level of the nuclei within germ cells, which he called "reproductive nuclei." Thus, they all freely drew on knowledge developed by cytology in the late nineteenth century about the nature of germ cells and the combining of germ cells (and their nuclei) at fertilization.

It is a plausible historical conjecture that Mendelism would have been impossible before the discovery of the cellular level of biological organization. The segregation explanation was the key Mendelian discovery: it was formulated in terms of types of germ cells in hybrids. The pollen cells and egg cells were the explanatory factors whose postulated behavior explained the 3:1

ratio. The idea that certain scientific problems get solved at certain levels of organization will occupy us in subsequent chapters (especially in Chapter 15, Section 15.1.4).

Should the claims in Components 1 to 3 be considered *theoretical* components of Mendelism in 1900? I do not believe a definitive answer can be given to that question because no canonical statement of 1900 Mendelian theory exists. I believe, however, that these claims may plausibly be considered theoretical components, based on the evidence presented earlier. To some extent, I am including Components 1 to 3 because, with hindsight, we can see how they developed during the later stages of theory formation. It is interesting to note that assumptions, which later developed into theoretical components, were often implicit in the 1900-1903 papers. Uncovering implicit assumptions, clarifying them, sometimes modifying them—these are aspects of theory development that changes to Components 1 to 3 will illustrate in subsequent chapters.

5.5 Component 4. Dominance-recessiveness

C4.1 In a hybrid formed by crossing parents that differ in a single pair of characters, there is some difference such that one character dominates over the other; thus, the character in the hybrid resembles one but not the other of the parents. (Let *A* symbolize the dominant character which appears; *a* the recessive which is not visible.)

In all the seven character pairs of pea varieties that Mendel (1865) examined, one character, but not the other of a pair, was visible in the F_1 hybrid. Mendel introduced new terminology that was adopted by his successors: the visible character was dominant and the other recessive. Thus, in the pea color cross, yellow was dominant over green because all the F_1 hybrid peas were yellow. Evidence from the subsequent cross, which produced about 25 percent green peas in the F_2 generation, provided evidence for the green's being called recessive, or hidden, in the hybrid.

The claim in Component 4.1 is not an explanatory, theoretical component because it provides no explanation of why one character dominates over the other. It is merely a promissory note: some difference will be found between the characters to account for this relationship. Nor can it predict the dominant character: to find out which of two new characters will dominate over the other, the cross must be done and the results observed. Furthermore, the early statements of dominance were in terms of "characters," not in terms of postulated explanatory "factors," one of which somehow causes a dominant character.

The early geneticists differed as to how much stress they placed on dominance-recessiveness in their statements of Mendelism. De Vries stated it as one of his two primary conclusions: "Of the two antagonistic characteristics, the hybrid exhibits [*tragt*] only one, and that in complete development. Thus, in this respect the hybrid is indistinguishable from one of the two parents. There are no transitional forms" (de Vries, 1900, p. 110). Davenport took his account primarily from de Vries's, so he too stressed dominance and recessiveness as a central claim of Mendelism, although he recognized the promissory nature of the claim: "What determines which character shall be dominating is still unknown, and the determination of this point offers an enticing field of inquiry" (Davenport, 1900-1, p. 308).

In contrast, Correns, Bateson, and Castle stressed that dominance was not a universal generalization. Correns was puzzled that de Vries did not report any exceptions to dominance. Even in varieties of peas, Correns found that not all characters showed complete dominance

(Correns, 1900, p. 121). Bateson listed the different kinds of forms found in the hybrid: "(a) dominant and recessive characters; (b) a blend form; (c) a form distinct from either parent, often reversionary [resembling a more distant ancestor]" (Bateson, 1902, pp. 20-21). Castle stated what he called the "law of dominance" but then continued by saying that the law is "not of universal applicability." He concluded, "The hybrid often possesses a character of its own, instead of the pure character of one parent, as is true in cases of complete dominance" (Castle, 1903, p. 397).

Thus, it is debatable whether, even as early as 1900-1903, dominance-recessiveness should be included as an empirical generalization in the domain (Item 1 of Table 5-1) and included as a theoretical component of Mendelism (Component 4 in Table 5-2). Certainly Component 4.1 was not an essential theoretical component of Mendelism; also, exceptions to Item 1, the empirical generalization about dominance, were not seen as disconfirming evidence against theoretical components other than the dominance component (C4.1). Because many empirical exceptions were known, the lack of an explanation of dominance was not a serious problem for the theory. Nonetheless, attempts to explain dominance-recessiveness were made, as will be discussed in Chapter 6.

5.6 Component 5. Segregation

C5.1 In the formation of germ cells in a hybrid produced by crossing parents that differ in a single pair of characters, the parental characters segregate or separate, so that the germ cells are of one or the other of the pure parental types. (Each germ cell has either *A* or *a* but not both; called "purity of the gametes.")
C5.2 The two different types of germ cells are formed in approximately equal numbers.
C5.3 When two hybrids are fertilized (or self-fertilization occurs), the differing types of germ cells combine randomly.

$[(A + a) (A + a) = AA + 2Aa + aa$; appearance $3A:1a]$

Segregation was often referred to as the essential discovery of Mendelism. It explained the 3:1 ratios. No single biologist stated the discovery as explicitly as my characterization in Components 5.1 to 5.3. At the end of Chapter 3, I stated more than three separable claims in a hypothetical path for constructing an explanation for the 3:1 ratios. For the actual historical reconstruction of the 1900-1903 claims, I here delineate only three separable assumptions: separation, equal numbers, and random combination. We will see how additional assumptions were uncovered after 1903, that is, implicit assumptions in the 1900-1903 statements were explicitly stated at later stages of theory development. For the early geneticists, separation or segregation (Component 5.1) was the primary focus of discussion. Component 5.2, equal numbers, and Component 5.3, random combination, only sometimes were explicitly mentioned in any given account. Often they were depicted implicitly in diagrams and symbols in the published paper. Almost without exception, the level of the discussion of segregation was at the level of the germ cells, not at the level of units within germ cells.

As discussed in Chapter 4, in his hypothesis of pangenesis of 1868, Darwin proposed that hybrids not only carried parental gemmules latently but also made new hybrid units (Darwin,

1868, V.2, p. 479). In other words, a pink flowered hybrid produced by a cross of red and white parentals would, he claimed, have dormant gemmules for red and white, as well as producing new "pink" gemmules. As mentioned earlier, de Vries explicitly altered and extended Darwin's 1868 hypothesis in his 1889 theory of intracellular pangenesis. De Vries too was interested in the latent pangens carried by hybrids. He agreed with Darwin that evidence supported an assumption of latent hereditary units in hybrids: characters in the offspring of hybrids often showed reversion to characters found in ancestors more distant than the parents. After publishing his theory in 1889, de Vries began hybridization experiments, in part, to find evidence about the nature of latent pangens carried by hybrids (de Vries, 1889, p. 91; Darden, 1976, pp. 152-153). He crossed varieties, raised large numbers of offspring, and followed the results through several generations. He collected data about the percentage of characters that showed reversion to a grandparental character, and then noted the lack of new characters (that differed from the parental characters). He hoped to be able to make inferences about the latent pangens that were carried by, but were not observable in, the first generation of hybrids (Darden, 1976).

De Vries's 1900 rediscovery paper reported on his hybridization experiments and the conclusions that he had reached. Although in his 1900 paper de Vries did not mention Darwin's view of hybrid units, one of the ways he stated the law of segregation makes a strong case for the claim that he believed that he was correcting Darwin's mistaken view that hybrids produced new hybrid gemmules: "*The pollen grains and ovules of monohybrids* [hybrids formed by crossing parents differing in one character] *are not hybrids* but belong exclusively to one or the other of the two parental types" (de Vries, 1900, p. 112). In another formulation of segregation, de Vries referred to the characters separating: "*In the formation of pollen and ovules the two antagonistic characteristics separate*, following for the most part simple laws of probability" (de Vries, 1900, p. 110).

One might have expected de Vries to express the law in terms of the underlying units, the pangens. After all, in 1889, he had proposed pangens as the material bearers of the characters. Nonetheless, both of de Vries's formulations of segregation, as most of the others in 1900-1903, were in terms of characters and germ cells, not units carried by the germ cells.

Another example is Castle's statement of the law of segregation: "The great discovery of Mendel is this: *The hybrid, whatever its own character, produces ripe germ-cells which bear only the pure character of one parent or the other* . . . . This perfectly simple principle is known as the law of '*segregation*,' or the law of the 'purity of the germ-cells' " (Castle, 1903, p. 398).

Yet another example is found as the first element in Bateson's list of the important new conceptions of Mendelism: "The purity of the gametes in regard to certain characters" (Bateson, 1902, p. 21). In the early stages of Mendelism and throughout most of his publications in the later developments in the field, Bateson focused at the level of the germ cell. He even suggested that there might be a *visually observable* difference between the different types of germ cells:

> Both Mendel and those who have followed him . . . assert that these facts of crossing prove that each egg-cell and each pollen grain is pure in respect of each character to which the law applies. It is highly desirable that varieties differing in the form of their pollen should be made the subject of these experiments, for it is quite possible that in such a case strong confirmation of this deduction might be obtained. (Preliminary trials made with reference to this point have so far given negative results. Remembering that a pollen grain is not a germ-cell, but only a bearer of a germ-cell, the hope of seeing pollen grains differentiated according to the characters they bear is probably remote. Better hopes may perhaps be entertained in regard to spermatozoa, or possibly female cells.) (Bateson, 1902, p. 14)

Correns, in a footnote, claimed that such a case had been observed:

> If the pollen grains of the two parental strains differ externally, one may, if Mendel's Law holds, expect the hybrid to form two externally different types of pollen grains. That this is true was first observed by Focke. (Correns, 1900, p. 131, note 15)

Correns explicitly distinguished between the character and the anlage of the character:

> In order to *explain* the facts, one must assume (as did Mendel) that following the fusion of the reproductive nuclei* the "anlage" for one trait, the "recessive" one (*green* in our case), is suppressed by the other trait, the "dominating" one; therefore all embryos are *yellow*. The "anlage," however, although "latent" is preserved, and prior to *the definitive formation of the reproductive nuclei a complete separation of the two anlagen occurs, so that one half of the reproductive nuclei receive the anlage for the recessive trait, i.e. green, and the other half the anlage for the dominating trait, i.e. yellow*.
>
> *[Correns included the following note:] Mendel, of course, does not mention nuclei, but only "germinal cells" and "pollen cells." (Correns, 1900, pp. 125-126)

In addition to being clearer than others about the necessity of discussing a cause of characters in their germ cells, rather than the characters themselves, Correns also pushed the theoretical level lower than the germ cells to the nuclei within germ cells. He thus explicitly made use of knowledge from the field of cytology that fertilization results from a fusion of nuclei of the sperm and egg.

Despite the minor differences between these statements of segregation, there was general agreement about the importance of the new idea of the "purity of the germ cells." It corrected Darwin's erroneous view that hybrids produced hybrid hereditary units, and it represented a new claim about (possibly observable) differences between types of germ cells in the hybrid.

In order for two pure types of germ cells to produce a 3:1 ratio of dominant to recessive characters in the F_2 generation, at least two additional conditions besides purity of the gametes are necessary. These assumptions are stated explicitly in Component 5.2, that the types of germ cells are formed in equal numbers, and in Component 5.3, that they combine randomly in fertilization. The early geneticists differed as to how explicitly they stated these conditions. As we will see, these assumptions became clearer as anomalies forced a more careful delineation of the separable theoretical claims.

De Vries stated: "The combinations take place according to the probability calculus" (de Vries, 1900, p. 112). This implies at least random combinations and possibly also equal numbers. His symbolic representation clearly shows the idea of equal numbers of the different types:

50% dom. + 50% rec. pollen grains, and

50% dom. + 50% rec. ovules.

$(d + r)(d + r) = d^2 + 2dr + r^2$

or

25%d + 50%dr + 25%r   (de Vries, 1900, p. 112).

Correns explicitly stated that the fertilizations were random:

> In the hybrid, reproductive cells are produced in which the anlagen for the individual parental characteristics are contained in all possible combinations, but both anlagen for the same pair of traits are never combined. Each combination occurs with approximately the same frequency. (Correns, 1900, p. 130, emphasis omitted)

Castle, on the other hand, stated that the different types of germ cells were produced in equal numbers, but he did not explicitly state that fertilization occurred randomly (Castle, 1903, p. 398).

Thus, biologists differed as to whether they explicitly stated all the conditions that were necessary if the postulated types of germ cells were indeed to give rise to 3:1 ratios. My statements of the theoretical components of segregation are a little more precise and explicit than the statement of any one biologist of the time, although all the claims can be found implicitly in their verbal statements or in their symbolic representations.

5.7 Component 6. Explanations of Dihybrid Crosses

C6.1 In crosses involving parents that differ in two pairs of differentiating characters, the pairs behave independently. Symbolically, $(AB + Ab + aB + ab)$ $(AB + Ab + aB + ab)$ = complicated array that in appearance reduces to $9AB$:$3Ab$:$3aB$:$1ab$.

In the period from 1900 to 1903, no one separated the law of segregation from what was later called the "law of independent assortment." Explanation of the 9:3:3:1 ratios appeared to be a simple extension of segregation to another pair of characters. Some statements of "Mendel's law" included reference to two pairs of characters; yet others did not mention the two pair cases in the statement of the law. De Vries said, "The *same law* holds also, as already mentioned, if one investigates dihybrids [hybrids whose parents differed in two pairs of characters] or studies two pairs of antagonistic traits in polyhybrids" (de Vries, 1900, p. 115, my emphasis).

In discussing what he called "Mendel's law," Correns included di- and trihybrid crosses:

> If the parental strains differ in only *one* pair of traits (2 traits: A,a) the hybrid will form only *two types* of reproductive nuclei (A,a) which are like those of the parents. Each type is 50 percent of the total. If the parents differ in *two* pairs of traits (4 traits: $A,a;B,b$) four types of reproductive nuclei will be formed (AB,Ab,aB,ab), and 25% of the total will be of each type. (Correns, 1900, p. 131)

Correns continued the discussion for crosses in which three pairs of traits were followed.

Bateson did not explicitly discuss dihybrid crosses in his list of the important new conceptions of Mendelism; however, as he discussed Mendel's paper, he included such crosses. He, as did de Vries, saw the case as simply an extension of the law of segregation:

> Mendel made further experiments with *Pisum sativum*, crossing pairs of varieties which differed from each other in *two* characters, and the results, though necessarily much more complex, showed that the *law* [my emphasis] exhibited in the simpler case of pairs differing in respect of one character operated here also. (Bateson, 1902, p. 10)

Castle extended the discussion to crosses with two different pairs of characters. He stressed that later hybrid generations have "individuals possessing *new combinations* of the characters found in the parents; indeed, *all possible combinations* of those characters will be formed, and in the proportions demanded by chance" (Castle, 1903, p. 400). In another paper the same year, however, in listing "the Mendelian principles of heredity," Castle did not mention the case of dihybrid crosses (Castle and Allen, 1903a, pp. 381-382).

As Chapter 9 will discuss, independent assortment came to be enunciated as a separate "law" in the light of anomalies to the 9:3:3:1 ratios. Anomalies played the role of bringing to the surface implicit assumptions and forcing distinctions to be made for which there was no need at an earlier stage. It is a mark of this early stage of theory formation that much was not stated explicitly, that much within the theoretical components was vague and unclear.

5.8 Additional Claims

My list of theoretical components (Table 5-2) omits a few additional claims that were made by a few of the early Mendelians. Bateson, for example, came to Mendelism with an interest in studying the nature of variation; he, thus, included a discussion of different types of variations in his statement of the Mendelian conceptions. He distinguished "analytical" from "synthetical" variations. Analytical variations appeared, he claimed, when what had previously behaved as one character "resolved into integral constituent-characters, each separately transmissible" (Bateson, 1902, p. 23). Synthetical variations, on the other hand, occurred "not by the separation of preexisting constituent characters but by the addition of new characters" (Bateson, 1902, p. 24). The problem of determining what constituted a single unit-character was sometimes difficult, as Bateson's definition of analytical variations indicated.

Similarly, Castle (1903) discussed what he called "coupled" versus "disintegrating" characters. More specifically, Correns (1900, p. 130) associated correlated traits (now called "completely linked") with exceptions to independent assortment. Thus, we see the early geneticists struggling with the issues of how separate unit-characters were to be distinguished, the origin of new characters, and the existence of correlated characters. All these issues were important in later developments; however, I have not included any of these claims in the lists of domain items or theoretical components: their inclusion is not warranted because none of the early discussions stated these views sufficiently clearly nor claimed sufficient generality for them.

Another claim that I have not included was stated by Castle: "Every gamete (egg or spermatozoon) bears the determinants of a complete set of somatic characters of the species. When two gametes meet in fertilization, there are accordingly present in the fertilized egg the representatives of two sets of somatic characters, which may or may not be the same" (Castle and Allen, 1903a, p. 381). In 1903, little evidence supported this sweeping general claim about all gametes. It was later called the "haploid" nature of the germ cells and the "diploid" nature of both the zygote (the fertilized egg) and the somatic cells (all nongerm cells). By 1903, however, too few hybrid crosses had been made, too few pairs of characters had been examined, in order to support general claims about all characters of all species. Unit-characters were individuated by whether they could vary independently. For the many characters for which variants had not been found, it was questionable whether they should be considered separable unit-characters in the Mendelian sense. Although Castle's assumption was a reasonable one, given the evidence of inheritance of characters from both parents, it was not a commonly made claim. Neither de Vries, nor Correns, nor Bateson made such a sweeping claim during this early period. Later, debates occurred as to whether Mendelism applied to all hereditary characters or whether it applied only to superficial differences between varieties. In other words, Mendelians had to defend their view that studying characters, such as flower color in plants or eye color in fruit flies, could produce a theory applicable to all hereditary characters.

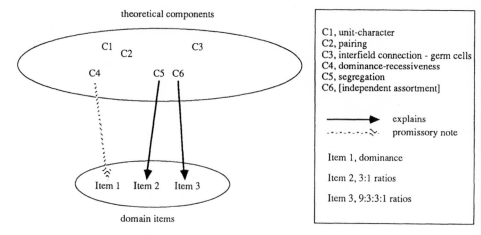

Fig. 5-1. Relations between domain and theoretical components.

The early papers lack clear statements that postulate the existence of a new theoretical entity, now called the gene. Correns came closest with his discussion of anlagen in the germ cells. Most discussions, however, were in terms of characters, not their underlying determinants. Some unknown difference in the germ cells was postulated that could cause (or was associated with) different characters. Thus, the field began with a vague idea that may be labeled a "promissory note" or a "black box." It called immediately for further work.

Bateson stressed how little was known:

> We can study the processes of fertilisation and development in the finest detail which the microscope manifests to us, and we may fairly say that we have now a considerable grasp of the visible phenomena; but of the nature of the physical basis of heredity we have no conception at all . . . . We do not know what is the essential agent in the transmission of parental characters, not even whether it is a material agent or not . . . . We are in the state in which the students of physical science were, in the period when it was open to anyone to believe that heat was a material substance or not, as he chose. . . . But apart from any conception of the essential modes of transmission of characters, we *can* study the outward facts of the transmission. (Bateson, 1902, p. 3)

5.9 Relations between Domain and Theoretical Components

Figure 5-1 illustrates some of the relations between the theoretical components and the domain items. The upper oval represents the theoretical components; numbers, such as C4, refer to the theoretical components listed in Table 5-2. The lower oval represents the domain to be explained; the numbered items refer to the empirical generalizations listed in Table 5-1. Ideally, the theoretical components explain all the domain items, and no theoretical components unnecessary for explaining domain items are included. Some components are more "directly" involved in the explanation of specific items than others, in the sense that, if the domain item is deleted, the component is unnecessary. Components 4, 5, and 6 can be regarded as directly tied to Items 1,

2 and 3, respectively. Components 1, 2, and 3 are more general and less directly tied to the domain items. Component 5, segregation, explains Item 2, the 3:1 ratios. Component 6 (later called "independent assortment") explains Item 3, the 9:3:3:1 ratios. The relation between Component 4, dominance, and Item 1, dominance, was not explanatory in the period 1900-1903. As we discussed, C4 was a promissory note: some difference would be found to explain why one character dominated over another.

This representation shows the "direct" relations between Components 4 to 6 and the domain items. Such direct relations will be seen to be important as anomalies arise and changes are made to both the domain and the theory. The more "indirect" relations between Components 1 to 3 and the domain items are not represented by any specific links.

Relations among the theoretical components themselves are complicated. Those relations are somewhat modular and somewhat systematically interconnected. Because of their complexity, the relations among components are not represented in Figure 5-1 by explicit lines; nonetheless, some relations among the components are suggested by the positions of the components relative to one another. Component 1, unit-characters, and Component 2, pairing, are both used in stating Components 4, 5, and 6 and are located at the top of the components circle. Component 3, germ cells, is not used in C4, dominance, but is used by C5, segregation, and C6, independent assortment. Thus, C3 is located closer to C5 and C6. C4, dominance, is independent of both C5, segregation, and C6, independent assortment, as we will discuss in Chapter 6. Independent assortment, however, is not completely separate from segregation, so C6 is close to and slightly below C5. As we will see in Chapter 9, segregation can occur without independent assortment also occurring, but not vice versa. Figure 5-1 provides a representation that will be useful for depicting changes that occur, both to the domain items and to the theoretical components in the development of the theory of the gene.

This form of representation was developed to depict medical diagnostic reasoning. Reggia, Nau, and their colleagues use a lower oval to represent symptoms; the higher oval depicts diseases that explain the symptoms. The upper circle is said to be a set of diseases that "covers" or explains all the items in the lower circle; hence, their term for this form of explanation is the "set covering model" (Reggia, Nau and Young, 1983). Similarly, a scientific theory may be considered a set of components that must "cover," or explain, all the items in its domain.

5.10 Conclusion

Hindsight shows that the new field began with much that was vague or implicit or promissory in its theoretical components. Nonetheless, the field did have a clear research program: make predictions by manipulating symbolic generalizations, then test those predictions by doing hybrid crosses. Experimental breeding provided an exact experimental technique to extend the scope of the domain and to test theoretical predictions. The early successes of test crosses provided impressive evidence for the theoretical components of Mendelism; however, very soon anomalies began to arise that required changes in the theoretical components. No theoretical component was immune from challenges, although a few remained unchanged by 1926, as we will see.

In its early days, the scope of the domain of Mendelism was limited to data about hybrid crosses involving "alternative" characters. Much was outside this domain. For those who accepted the belief that a cross must show dominance for Mendelian explanations to be applicable, the domain did not include crosses that showed a blend form in the F_1 generation. If plants with red and white flowers, for example, were crossed and produced pink in the F_1, then such a case of blending was outside the domain of Mendelism. Similarly, characters, such as height in humans, showing "continuous" (quantitative), rather than "discontinuous," variation were also considered outside its scope. Furthermore, whether characters that had not yet shown any variation could be subsumed under the unit-character concept was an open question. In addition, some hybrids did not show segregation and were thought to be outside the scope of Mendelism. Given so many types of hereditary phenomena outside its scope, few of the early Mendelians saw the promise of Mendelism as a general theory to explain all hereditary phenomena. Bateson and Castle seem to have made the most general claims for it. Thus, the field began with a domain of limited scope.

In a theory's early development, it may be an open question what is inside the domain and so is an anomaly for a theory and what is outside the theory's scope (and thus requires some other theory for its explanation). Such was the case for some hereditary phenomena, as we will see. Furthermore, debates occurred as to whether a general theory of heredity also had to be a theory of embryological development, sex determination, and the origin of new species. In nineteenth-century theories of heredity, the problem of the transmission of hereditary characters had not been clearly separated from these other problems. That the problem of the transmission of hereditary characters, in contrast to the embryological and evolutionary development of those characters, could be considered a separate problem was, to some extent, discovered in the process of solving it. Mendelism became an adequate theory for transmission genetics but it did not solve these other problems (as will be discussed more fully in Chapter 13). We now turn to the development of the 1900 theoretical components of Mendelism into the 1926 theory of the gene.

CHAPTER 6

Unit-Characters, Pairs, and Dominance

6.1 Introduction

The task in Chapters 6 to 13 is to trace the changes between 1900 and 1926 in the theoretical components of Mendelism. Furthermore, strategies for producing the changes will be described. Each chapter will discuss challenges to one or more components, the way the components changed in the light of the challenges, and one or more strategies exemplified in changes. As will be seen, changes were driven by the need to explain anomalies, to expand the scope of the domain of Mendelism, and to clarify the explanatory concepts.

Sections 6.2 and 6.3 discuss changes to Component 1, the unit-character concept, to Component 2, the pairing of characters, and to Component 4, dominance-recessiveness (see Table 5-2 for a complete statement of all the theoretical components as of 1900-1903). Those changes produced several theoretical alterations. Component 1 was changed to clearly distinguish the factor in the germ cells from the character with which it was associated. Changes to Component 2, the pairing of characters, resulted in the development of hypotheses about "multiple factors" and "multiple allelomorphs." The fate of Component 4, the dominance component, was eventual deletion from the set of the theoretical components. Strategies exemplified in those changes illustrate ways of slightly altering theoretical components. Section 6.4 analyzes the strategies of **generalizing, specializing, complicating,** and **deleting.** The changes to the theoretical components resulted in the resolution of empirical anomalies and the expansion of the scope of the domain.

6.2 Changes to Component 1: Unit-characters

Mendelism, 1900-1903:

C1. Unit-characters
C1.1 An organism is to be viewed as composed of separable unit-characters.
Correns (1900) explicitly distinguished the character from the theoretical entity that produced it. He designated the underlying Mendelian units, found in the germ cells, "anlagen." Gradually

65

other geneticists saw the need for distinguishing the underlying "factor," as it came to be called, from the visible character. The distinction between the character and the factor that causes it was needed. Characters are not literally carried by the germ cells; instead some factor that causes the character to develop in the offspring is carried by the parental germ cells and is transmitted to the offspring during fertilization. That distinction was often not made in the 1900-1903 papers, as discussed in Chapter 5. The "factor hypothesis" emerged by about 1906 as the need to discuss numbers and states of factors different from their associated characters emerged (see, e.g., Bateson, Saunders, and Punnett, 1906).

The simplifying assumption that *one* factor produced *one* character was very important. Since the only access to the factors was by inference from the visible characters, this one-to-one assumption allowed inferences to be made about the presence and behavior of factors, based on the ratios of characters. Once the simple cases of 3:1 and 9:3:3:1 were understood, then other anomalous ratios were explained by more complex factor combinations.

East (1910), while studying color inheritance of seeds in maize, found a ratio of 15 yellow to 1 white in the F_2 generation, with the yellow varying from dark yellow to light yellow. He explained this by postulating two pairs of factors that interacted in producing the yellow color. Thus, the 9:3:3:1 ratio could be extended from cases of *two* independent pairs of factors producing *two* independent characters to cases in which *two* pairs of factors interacted in the production of *one* character in a "continuously" varying array in the F_2 generation. The 9:3:3 ratio collapsed to 15.

In 1909, Nilsson-Ehle, independently of East, found a similar case for color characters in wheat. In this case, the ratio of 63 red:1 white was observed, indicating that three factor pairs were interacting in the production of the red color (Dunn, 1965a, p. 100). Thus, anomalous new ratios were included in the domain of Mendelism. Resolving these anomalies drove the conceptual change from one unit-one character to the possibility of multiple factors interacting to produce a character that exhibited the new ratios in crosses.

The new theoretical assumption of multiple factors allowed extension of the scope of the theory. Mendelism in 1900 was applied only to "alternative" characters that had two discontinuous states, for example, red and white, yellow and green. Characters that varied "continuously," such as height in humans, were outside the scope of Mendelism in 1900. As early as 1902, Bateson speculated that four or five Mendelian pairs might interact to produce "continuous" characters (Bateson and Saunders, 1902, p. 60). By 1910 East gave a more exact exposition of Bateson's idea:

> When four such units A1, A2, A3, A4, are crossed with a1, a2, a3, a4 . . . only one pure recessive is expected in 256 individuals. And 256 is a larger number than is usually reported in genetic publications. When a smaller population is considered, it [appearance in the F_2] will appear to be a blend of the two parents with a fluctuating variability on each side of its mode. (East, 1910, p. 72)

Thus, anomalous ratios of 15:1 or 63:1 were explained by "complicating" the original one-to-one relation to many-to-one. This modification of the theory then increased the explanatory power by extending the scope of the kind of data for which the theory could account. Mendelism moved beyond the early simple cases of 3:1 and 9:3:3:1 ratios of the narrow, original domain to include additional ratios. The set of theoretical components changed from the oversimplified unit-character assumption to include multiple factors. As a result, Component 1, the unit-character assumption, was altered to decouple the unit from the character and allow the possibility of either one unit producing one character or multiple factors interacting to produce an array of variations in a character.

Not all geneticists readily accepted either the factor hypothesis or its extension to multiple factors. Morgan, who became one of the major developers of Mendelism, was a severe critic during the period before 1910. (See Allen, 1978, for a thorough discussion of Morgan's objections and his reasons for changing his views.) In 1909, Morgan, in a paper entitled "What are the 'Factors' in Mendelian Explanations?", objected to the rapid rate at which "facts are being transformed into factors" (Morgan, 1909d, p. 365). He was willing to admit the possibility that Mendelism indicated that some difference existed between the germ cells of hybrids, but he was unwilling to concede that the evidence supported the hypothesis that such a difference was to be attributed to different material *units* in the germ cells. In 1909, Morgan wanted an integrated theory of heredity and development that did not include preformed units in the germ cells. He said:

> The egg need not contain the *characters* of the adult, nor need the sperm. Each contains a particular material which *in the course of the development produces* in some unknown way the character of the adult. Tallness, for instance, need not be thought of as represented by that character in the egg, but the material of the egg is such that placed in a favorable medium it continues to develop until a tall plant results. (Morgan, 1909d, p. 367)

Within two years Morgan abandoned his critique of the factor hypothesis and embraced it within the context of developing the chromosome theory (discussed in Chapters 9 and 10). Morgan often proposed working hypotheses and readily abandoned them in the face of counterevidence. He also proved able to embrace a theory he had previously opposed. Morgan is a good example of a critic who understood well the view he opposed, and developed arguments against it and alternatives to it. When his own research supplied evidence against those counterarguments, he rapidly shifted to become an advocate for the hypothesis that he had so vigorously opposed. From this case, the generalization can perhaps be made that critics who take a view seriously enough to argue against it are more likely to shift their position (in the light of new evidence that satisfies their objections) than the critic who merely ignores the opposing view.

Other criticisms and changes to the unit-character concept will be discussed in later chapters. Castle's critiques of multiple (modifying) factors and his eventual acceptance of that hypothesis will be discussed in Chapter 8, in the context of examining Castle's challenges to segregation. The further development of the unit-character concept from factor to gene will be discussed in Chapter 11, after a discussion of the development of the chromosome theory, namely, the theory that genes are parts of chromosomes.

In summary, the multiple factor hypothesis became an additional theoretical component of the theory after 1910. Geneticists used the multiple factor assumption successfully in additional cases to explain new ratios. Test crosses supported those explanations. As a result, the hypothesis of multiple factors ceased to be "hypothetical" and became an accepted theoretical component of Mendelism. Geneticists thus added to their explanatory repertoire: in some cases one factor produced one character, but in more complex cases, multiple factors interacted to affect one character. The actual numerical value of the ratios between characters indicated whether one pair of factors or multiple pairs of factors were present. The set of theoretical components of the theory had expanded, producing a concomitant expansion in the scope of the domain.

6.3 Components 2 and 4: Paired Allelomorphs and Dominance-recessiveness

Mendelism, 1900-1903:

C2 Differentiating pairs of characters

C2.1 In varieties of organisms, the traits by which they differ are antagonistic or differentiating pairs of characters (allelomorphs).

C4 Dominance-recessiveness

C4.1 In a hybrid formed by crossing parents that differ in a single pair of characters, there is some difference such that one character dominates over the other; thus, the character in the hybrid resembles one but not the other of the parents. (Let A symbolize the dominant character which appears; a the recessive which is not apparent.)

Originally the domain of Mendelism contained only data about characters that showed dominance. Just as East's work and the postulation of multiple factors extended the scope to apply to additional kinds of characters in the F_2 generation, the elimination of the claim that dominance was universal extended the scope to blending inheritance in the F_1 generation. An example of blending is the production of a pink flower in the F_1 hybrid from a cross of parentals with red and white flowers. In the F_2, the Mendelian ratio is 1 red: 2 pink: 1 white.

Not only were dominant or intermediate blend forms found in the F_1 hybrid, sometimes a new form different from either parent appeared. The Mendelian symbolism was easily extendable to represent the blending or new forms. The Aa symbolism already contained a distinction at the factor level that dominance obscures among the visible characters. In other words, AA and Aa, which had both been used to symbolize the dominant hybrids, could easily represent the distinction between the one parental form (AA) and the blend or new form (Aa). An already existing, but unused, distinction in the symbolism thus made the representation of the new phenomena an unproblematic addition (see, e.g., Bateson, 1902, p. 19-21).

As noted in the discussion in Chapter 5, the phenomenon of dominance was considered to be universal by some, but not others, of the earliest geneticists. As work proceeded to test the Mendelian claims, the mere lack of dominance in the F_1 generation was no problem for Mendelian theory or its representation in the A and a symbolism.

In 1907, however, Morgan discussed crosses that not only lacked complete dominance in the F_1, but also seemed to lack pure parental forms in the F_2. Such lack of dominance in the F_1, along with the lack of pure AA or aa in the F_2, was a challenge to Mendelian segregation. Morgan said:

> Hurst estimated [for poultry] that incomplete dominance is twice as numerous as complete dominance. It was also observed that in the second generation there is often a mixing of the characters, so that it is difficult or impossible to distinguish the pure forms [AA] from the "dominant-recessives" [the hybrid Aa]. Such results are difficult to account for on the basis of "pure" gametes, although a tendency towards segregation may be distinctly recognized. (Morgan, 1907, p. 638)

The lack of dominance in the F_1 alone was not a problem for Mendelism, but lack of the reemergence of pure parental forms in the F_2 would have been a challenge to the more central Mendelian claim in Component 5 (segregation), which will be the subject of discussion in

Chapter 8. If three or more factors were involved, then the numbers of pure parental forms were small and might not be found in any one cross, as noted in the multiple factor discussion earlier. Usage of the term "incomplete dominance" did not always clearly distinguish these two possibilities of (1) a blend form in the F_1, and (2) cases attributable to multiple factor interactions, resulting in many forms in the F_2 generation. Morgan's 1907 discussion shows the tendency to use this term for both (e.g., Morgan, 1907, p. 465). Eventually these different cases were sorted out (Morgan et al., 1915, pp. 27-32). Lack of dominance in the F_1 was not a challenge to Mendelian segregation; *actual* challenges to segregation will be discussed in Chapter 8.

Component 4, the theoretical component of dominance, is not explanatory. It contains no explanation of the difference between A and a that causes one to dominate over the other. Some attempts were made to explain, or partially account for, why one character dominated over another, but they all failed. The explanation of dominance was left as a promissory note, one that was never cashed by Mendelian genetics, but whose delinquency became insignificant.

De Vries (1900, p. 111) speculated that the dominant character was older in evolutionary ancestry than the recessive; however, he did not explain why being older would cause dominance. The idea that the evolutionarily older character was prepotent over newer ones was not a new idea; it was discussed by Darwin (Olby, 1985, p. 42). Such a claim might have made predicting dominance possible, if the evolutionary history of the characters were known; however, it did not explain why the older character dominated. Even when the evolutionary order was agreed upon, later work did not confirm the prediction that the older character was dominant (Bateson and Saunders, 1902, p. 42; Morgan, 1907, p. 466).

In 1905, Bateson and Punnett developed an idea that had been suggested by several earlier Mendelians (Carlson, 1966, p. 58) to provide an explanation, or partial explanation, for dominance and recessiveness—the presence and absence hypothesis. They claimed that the presence of an allelomorph somehow caused the dominant character, while its absence caused the recessive. Bateson had originally introduced the term "allelomorph" to refer to alternative characters, such as yellow versus green peas (Bateson, 1902, p. 22). He, along with other geneticists, however, came to see the need to discuss underlying factors. The development of the presence and absence hypothesis was one of the developments requiring such a distinction. Gradually "allelomorphs" or "allelomorphic factors" came to refer to the underlying, paired, alternative factors, not the pairs of characters themselves. According to the presence and absence hypothesis, the recessive character was due, not to a recessive allelomorph, but to the absence of the dominant allelomorph.

Bateson and Punnett originally suggested this explanation to account for a complex case of the inheritance of combs in fowls. Earlier, they had suggested that more than two alternative allelomorphs might occur in a hybrid fowl to account for the pattern of inheritance (Bateson, 1903). The idea of three alternative allelomorphs segregating into three different types of germ cells in one hybrid organism was not an unreasonable extension of Mendelism at the time. (It became less plausible with the proposal that allelomorphs were associated with *paired* ho-mologous chromosomes, as Chapter 7 will discuss.) In 1905, however, Bateson and Punnett suggested a "much simpler and probably more correct" account in terms of two pairs of allelomorphs (four allelomorphic factors). The observable character pairs, based on the appear-ances of the combs in the fowls, were called "rose + no-rose" and "pea + no-pea." The allelomorphs were then designated as "rose and absence of rose" and "pea and absence of pea" (Bateson and Punnett, 1905, p. 136). Swinburne (1962), in his discussion of Bateson and Punnett's hypothesis, plausibly argued that the nomenclature within the hypothesis arose from a consideration of the phenotypic characters; for example, the no-pea character easily suggested the theoretical idea of the absence of the pea allelomorph. An assumption of strict parallelism between phenotype and genotype (which had not been clearly distinguished at this stage of the development of Mendelism) led Bateson and Punnett to assume that they had explained the lack of a phenotypic character by proposing the absence of its allelomorph (Swinburne, 1962, pp. 132-133).

By 1907, Bateson extended the presence and absence hypothesis beyond an explanation for combs in fowls to a general hypothesis to explain all cases of dominance. The appearance of a dominant character was explained by the presence of a dominant allelomorphic factor, while the loss of that factor resulted in the recessive character (Bateson, 1907). New symbolism came into usage: instead of Aa, the heterozygote was represented by AO to indicate the presence of an allelomorphic factor for the dominant character and its absence in the recessive.

Problems arose, however, as others began testing the applicability of the presence and absence hypothesis for Mendelian characters in general. Sometimes the lack of a character, such as hornlessness in sheep or lack of pigmentation in fowl, was found to be dominant to the presence of the character. To explain such cases, proponents of the hypothesis introduced the idea of inhibiting factors and claimed that the presence of inhibiting factors caused the loss of characters (Bateson and Punnett, 1911c). Morgan criticized this as an assumption that opponents of the presence and absence theory did not need to make (Morgan, et al., 1915, p. 222).

A further problem was presented by new data. Presence and absence is a hypothesis that explains two and only two states of a character. One might imagine a strategy for constructing such a hypothesis: to explain two different states, postulate an underlying phenomenon with two states, such as present-absent, on-off, or active-dormant. The early data were gathered for varieties with alternate characters, such as yellow and green peas. The presence and absence theory illustrates the relation between Component 2, the claim that characters occur in pairs, and Component 4, the claim of dominance. The relation between these different components of Mendelism only became evident as the theory developed. If characters (or more precisely, character states, such as yellow and green) existed only in pairs, then the range of possible explanations for dominance was different from the range of explanations for characters with multiple states.

As Mendelian crosses were made testing more and more characters, varieties with more than two possible states of a given character were found; for example, Cuénot (1904) found three coat colors in mice: yellow, agouti (French *gris*), and black. Additional cases of multiple allelo-morphic characters began to turn up in other species (Dunn, 1965a, p. 101; Provine, 1986, p. 87). Sturtevant (1913a) found that eye color in groups of fruit flies, *Drosophila*, exhibited "multiple allelomorphs" for red, white, or eosin. In any given organism, the allelomorphs occurred only in pairs, but in a stock of flies more than two allelomorphs for eye color were found. A given heterozygous fly, for example, might carry allelomorphs for red eye and white eye, or eosin and white, or red and eosin. Yet, in the stock, all three allelomorphs for that character were found.

The reader should clearly distinguish the multiple allelomorph hypothesis from the multiple factor hypothesis discussed in the last section. In contrast, multiple factors are two or more pairs of allelomorphs that affect the same character in a given organism; for example, East proposed two pairs of allelomorphic factors to explain the 15:1 ratios for seed color in corn. When instances of multiple factors occur, more than one pair of factors affecting one character are found in one organism. Multiple allelomorphs are postulated when instances of more than a pair of character states are found in a population of organisms. In a stock of fruit flies, for example, three different alternative eye colors were observed, red, white, and eosin. Three different allelomorphs for eye color were postulated; however, any one fruit fly had at most two of those three different allelomorphs. The possibility of more than two alternative allelomorphs in any one organism was ruled out by the chromosome theory, in which allelomorphs were proposed to be parts of paired, homologous chromosomes, as will be discussed in Chapter 7. (In modern genetic terminology, the difference is between, on the one hand, multiple alleles, and, on the other hand, epistatic genic interactions between two or more different pairs of alleles.) Multiple alleles are alternative states of the same locus in a linkage group; many different alleles may be present in a population, such as human blood types. Because any given diploid organism has pairs of homologous chromosomes, at most two different alleles can be found at the two corresponding loci on the genetic maps of those chromosomes (with a few exceptions involving somatic mutations). On the other hand, multiple factors are two or more pairs of alleles located at different loci along the same or different chromosomes that somehow interact during the development of a character in one organism.

Morgan (1913a) and Sturtevant (1913a) interpreted the finding of multiple allelomorphs as evidence against the presence and absence hypothesis. As discussed earlier, Bateson himself, in 1903, had suggested explaining a complex case of inheritance as a possible instance of more than two alternative allelomorphs occurring *in one hybrid organism*. (In hindsight, the idea of more than two alleles in a single organism is a very non-Mendelian idea.) He abandoned that idea, however, when he developed the presence and absence hypothesis. Bateson saved the assumption of pairing in any given organism and in any population by suggesting that each allelomorph was paired with its absence (Bateson and Punnett, 1905).

Although Morgan and his students argued strongly in favor of pairing of allelomorphs in any given organism (in the light of paired chromosomes), they were not committed to saving the assumption of pairing within populations. They interpreted the case of eye color allelomorphs in *Drosophila* as a genuine case of multiple, alternative allelomorphs found in a given stock of flies. Additional evidence helped to support their argument against the presence and absence hypothesis. "Back" mutation (spontaneous changes in allelomorphs) occurred in *Drosophila* stocks: in some cases, white eyes mutated to eosin, and, in other cases, eosin mutated to white. If the difference had been due to the loss of an allelomorphic factor, then somehow the factor that had been lost would have been restored. There was no obvious mechanism for restoring absent factors (Sturtevant, 1913a). Although Bateson and Punnett continued to advocate the presence and absence theory, others abandoned it in the light of the evidence for multiple allelomorphs and back mutations (Punnett, 1927; Swinburne, 1962).

Multiple allelomorphs resulted in a change in the theoretical components of Mendelism. A new qualifying condition had to be added to the general claim about pairs of allelomorphic factors, namely whether the number of allelomorphs was being counted *in any one organism* or *in a population of organisms*. Thus, Component 2, the claim that all characters occurred only in pairs, was refined. It was replaced by Component 2.1', the claim that *in any one organism*, allelomorphs for a given character occur in pairs. Moreover, a new claim was added. Component 2.2' claims that *in a population*, multiple allelomorphs for a character may occur. (Detailed discussion of these components is found with discussion of Figures 6-1 to 6-3.)

The discovery of multiple allelomorphs presented problems of finding adequate symbolism for the theory. Both Mendel's symbolism of capital and small letters and Bateson's of a capital letter and a zero were rendered inadequate. More than two possible allelomorphs for the same character needed to be represented and clearly shown to be related. Cuénot's early use of three different letters for the three alternative mouse coat colors initially prevented Sturtevant from recognizing that the three were actually three alternative allelomorphs (Sturtevant, 1965, p. 53). The Morgan group developed an alternative symbolism with + for the character found in wild populations and small letters, such as *w* for white eyes and *e* for eosin eyes, for mutants (Sturtevant, 1965, p. 53). This symbolism could easily represent any number of alternative allelomorphs, but it had the disadvantage of not clearly indicating that they were allelomorphs for the same character in the way capital and small letters did. Thus, theoretical changes necessitated changes in the symbolic notation, with benefits and drawbacks as to what could easily be represented by the new notation.

Yet this expansion of Mendelism to include multiple allelomorphs did not solve the problem of explaining dominance and recessiveness. The presence and absence theory was abandoned; however, nothing replaced it. In fact, dominance was never explained by Mendelian genetics. The development of the chromosome theory of Mendelian heredity, which is the subject of the next chapter, helped to explain why allelomorphs occurred in pairs in an individual organism: because they are located on paired chromosomes. The chromosome theory, however, did not help in explaining dominance: normal, paired chromosomes (except for the sex chromosomes) are identical in appearance when viewed with the light microscope. Although localizing the genes in the chromosomes aided the development of other aspects of the theory of the gene (such as linkage, to be discussed in Chapter 9), it was of no help with dominance. Such a similarity of chromosomes that carried different allelomorphs could be taken either as evidence against the chromosome theory or as an indication that dominance might require lower levels of organization to explain it. Bateson argued for the former position; one of his many arguments against the chromosome theory was that no difference in the chromosomes was observed to correspond to a character difference (except for sex chromosomes) (Bateson, 1907, p. 166). Sturtevant, on the other hand, speculated that dominance and recessiveness might be explained by a change in a chemically complex molecule (Sturtevant, 1913a, p. 18). Expositions of Mendelism stressed dominance less and less as the field developed. In his 1906 textbook account of Mendelism, Lock stated that dominance was not universal (Lock, 1906, p. 180). Morgan and his students, in their seminal book of 1915, entitled *The Mechanism of Mendelian Heredity*, listed dominance in the table of contents as a section heading within a chapter on "Types of Mendelian Heredity," but noted: "Whether a character is completely dominant or not appears to be a matter of no special significance. In fact, the failure of many characters to show complete dominance raises a doubt as to whether there is such a condition as complete dominance" (Morgan et al., 1915, p. 31).

By 1926, in *The Theory of the Gene*, Morgan did not list dominance in the contents or in the index or in his statement of the theory of the gene. After presenting Mendel's pea crosses, he discussed crosses with four-o'clocks in which the hybrid did not show dominance, but was a blend form. Mendelism came to encompass blending seamlessly.

The evolutionary origin of dominance was taken up in the 1920s and 1930s, as relations between the findings of genetics and issues of evolution were explored (Dobzhansky, 1937, pp. 171-176). As molecular biology developed, explaining dominance was not a major concern. Today, it is still receiving some attention as biologists try to unravel the complex relations between mutant alleles and enzymes (Kacser and Burns, 1981); but it is not a major research topic in genetics. According to Joshua Lederberg (personal communication), sometime in the 1950s biologists stopped talking about dominance as a property of a gene and shifted to the more general framework of gene dosage effects; phenotypes are judged in terms of the quantitative level of a gene product. There is still much research being done to unravel the regulation of such effects.

6.4 Strategies: Complicate, Specialize, Add, Delete

Mendelism began with a small domain. It applied to a few types of ratios produced by hybrid crosses between varieties that had alternative characters. Much in the early theoretical components was vague and unclear. Numerous changes in and refinements to the theoretical components occurred as Mendelism was applied to additional phenomena. This chapter has discussed changes that produced two refined theoretical components, namely, the addition of multiple factors and multiple allelomorphs. The chapter has also traced the decrease in importance of dominance as a general theoretical claim of Mendelism. Collectively, these theoretical changes resulted in the removal of anomalies and the expansion of the scope of the domain of Mendelism. We are now able (after all the preliminaries in this and prior chapters) to begin the central task of this analysis—to find general strategies of theory change that are exemplified in these specific historical changes.

The unit-character concept developed into the theoretical claim that one factor caused (or, more loosely, was somehow associated with) one character. This change allowed inferences to underlying, unobservable factors, based on observable, countable characters. This important change occurred gradually as Mendelism was applied to additional cases and as anomalies required that the underlying factor be distinguished from the character.

The one factor-one character assumption was an oversimplification. It was important in allowing inferences from characters to factors. Gradually, as geneticists came to understand the simple Mendelian cases of complete dominance in the F_1 as well as 3:1 or 9:3:3:1 ratios in the F_2, they were able to "complicate" the oversimplification. When a 15:1 ratio related to one character appeared, it was an obvious complication to say that two factor pairs were affecting one character. Prior to the multiple factor hypothesis assumption, the 9:3:3:1 ratios had been expected only for two independent pairs that produced two different characters. However, complicating the one factor-one character assumption to allow two factors to affect one character explained the anomalous 15:1 ratios (the 9:3:3 collapsed to 15). This new theoretical component also was further complicated to predict 63:1 ratios (for three interacting factor pairs) and 255:1 ratios (for four pairs).

Thus, the addition of multiple factors can be seen as exemplifying the strategy of **complicating a simplifying assumption**. Only some characters are produced by one factor; other characters are affected by multiple factors. The addition of the more complicated component resulted in an expansion of the domain. Anomalous ratios such as 15:1, 63:1, and 255:1 were incorporated into the domain that Mendelian theory explained.

If a simplifying assumption is made at the beginning of theoretical development, then it might be expected that **complication** will be a strategy to employ in the light of anomalies. A question to ask is—what was neglected in forming the simplification? The answer to that question can guide the formation of the next, more complicated theoretical component. In this case, the possibility of factors interacting to produce one character had been neglected in the one factor-one character simplification. Consequently, the hypothesis of multiple factor interactions was an obvious complication to make in the light of certain anomalous ratios. From the change from the one factor-one character assumption to the multiple factor hypothesis we get the strategy of **simplify, work with the simple case until it is well understood, then gradually complicate**.

Figure 6-1 illustrates the addition of the new, more complicated theoretical component of multiple factors and the concomitant expansion of the domain. Component 1, the unit-character concept, was complicated by the addition of the component of multiple factors, represented by

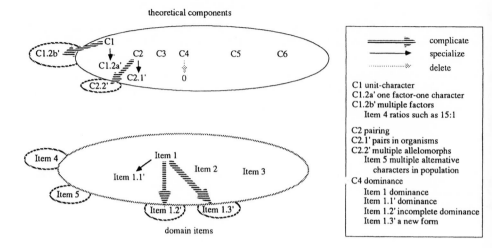

Fig. 6-1. Changes in theoretical components (C1, C2, C4) and domain.

adding C1.2b'. (The reason for this numbering is to coincide with the component list that we are building for the theory in 1926 in Table 14-1.) The figure graphically represents *the complication* of C1, the unit-character concept, to include multiple factors. Furthermore, Figure 6-1 illustrates the expansion of the scope of the domain by the addition of Item 4, which represents ratios such as 15:1. Item 4 was explained by multiple factor interactions. The figure also shows that, in this case, addition of a component by complication resulted in an expansion of the domain.

C1 was the general claim that all characters are produced by one factor. The addition of the multiple factor component resulted in the specialization of that claim. C1 thus was specialized to the less general claim (C1.2a') that *some*, but not all, characters are produced by one factor. When 3:1 ratios were found, then the one factor-one character concept still applied to those cases; however, other ratios, such as 15:1, indicated multiple factors (C1.2b'). The development of the multiple factors component thus resulted in two changes to C1: the specialization of an old component (C1 to C1.2a') and the addition of a new component (C1.2b') by complication, as shown in Figure 6-2. Characters included within the domain of Mendelism thus became grouped into two sets: the ones produced by one factor and the others produced by multiple factors, as indicated in Figure 6-2.

The changes to Component 2 (all allelomorphs occur in pairs) also exemplify the strategies of specialization and complication. Multiple allelomorphic characters were discovered in populations and the theoretical component of multiple allelomorphic factors (in short, multiple allelomorphs) was added to the theory. Within one organism, however, only pairs of alternative characters were found. Thus, the pairing of allelomorphic factors within one organism became a specialized version of the general claim in Component 2. Figure 6-3 illustrates these changes. A new condition—in one organism or in a population—was discovered to be an important qualification. The universal generalization (C2) that "all allelomorphs occur in pairs" was specialized by the modification "*in any one organism*, all allelomorphs occur in pairs." It was also specialized by adding "*in a population*, only some allelomorphs occur in pairs." In the original

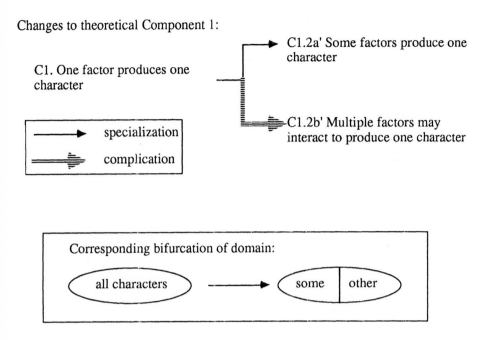

Fig. 6-2. Changes to Component 1 and domain items.

formation of the pairing component, the qualification "in one organism" or "in a population" was not known to be important. As cases were found, however, in which pairing occurred in one condition but not the other, the restrictive condition had to be added to the pairing component. The strategy exemplified by this change, **adding a condition**, is one way of **specializing a generalization**.

Another change to C2 illustrates the strategy of **complication**. C2 was complicated to yield the new theoretical component C2.2b': in a population, multiple allelomorphs may occur. Once again, refinements to the domain occurred, as the ovals in Figure 6-3 show. The addition of multiple allelomorphs (C2.2b') to the theoretical components of Mendelism and the concomitant expansion of the domain to instances of multiple alternative (allelomorphic) characters in a population are illustrated in Figure 6-1 by the additions of C2.2' and Item 5. The domain thus expanded in scope to include the new cases of multiple alternative characters, such as the eye color cases in fruit flies and the coat color cases in mice. Once again, addition of a new theoretical component constructed by complicating an old one removed anomalies and expanded the scope of the domain.

The changes to multiple factors and multiple alleles exemplify the strategies of **specializing one component and then adding another, related component**; these strategies are ways of expanding the theory to take account of anomalies. Ways of adding new ideas in an incremental way include complicating an old idea, such as moving from one-to-one relations to many-to-one or moving from only pairs to both pairs and multiples. Another way of adding a new component is by adding a restrictive condition that serves as a qualification that had previously not been

Changes to theoretical Component 2:

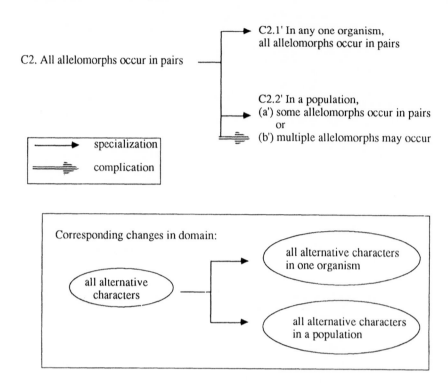

Fig. 6-3. Changes to Component 2 and domain items.

known to be important. The conditions of "in one organism" and "in a population" are examples of such restrictive conditions. Once two different components were formed in this way, they could be changed independently, as was done with the claim that "multiple allelomorphs may occur in a population." Figure 6-3 illustrates the effect of adding a restrictive condition. The strategy of **adding a restrictive condition** might be considered to be a way of specializing a general claim or as a way of complicating a simple claim. The strategies of **simplifying, generalizing**, and **adding a restrictive condition** are summarized in Figure 6-4.

Yet another strategy emerges from the elimination of the theoretical component of dominance-recessiveness—**deletion**. Somewhat confusingly, but reflecting historical usage, I used the term "dominance" to name both the dominance item of the domain (Item 1 in Table 5-1) and the theoretical component corresponding to it (Component 4). The name of the domain item was also the name of the theoretical component, a component that was only a promissory note for an explanation. The lack of an adequate theoretical component to explain the dominance, especially after the failure of the presence and absence theory, might have been seen as a major inadequacy for the entire theory of Mendelism. Yet, such a failure was not viewed that way historically. There are several possible reasons that may help to explain why geneticists adopted this attitude toward the problem of dominance.

First, many exceptions were found to the claim in Item 1 that in all hybrid crosses between

1. simplify--complicate
    e.g. one--many
        pairs--multiple

2. generalize--specialize
    e.g. all--some

3. add a condition

    (may be viewed as either a way of specializing
    or a way of complicating)

    e.g. all a are b ⟶
        some a are b & some a are c

    e.g. all a are b ⟶
        all a are b only in E1
        & all a are c only in E2

Fig. 6-4. Strategies: simplify, generalize, add condition.

varieties with two alternative characters, one character dominates over the other. Sometimes blended or new characters were found. Thus, the domain item was not a universal, empirical generalization. It was specialized in the light of many exceptions. Second, whether the heterozygous hybrid showed (1) dominance, (2) a blend form, or (3) some new form, had to be determined empirically for each pair of alternative characters. No theoretical components were successful at predicting whether dominance would occur, and, if so, which character would dominate. Third, even after the cross was done and the dominant character identified, the theoretical component did not explain why it dominated. The explanation of the development of a dominant character was part of the entire problem of embryological development and gene expression, which Mendelism never solved. Fourth, other theoretical components of Mendelism, as we will see, had numerous successes. The lack of an explanation for dominance did not hinder development of other parts of the theory. Finally, geneticists speculated that dominance was due to some change in chemical molecules; if correct, such speculations would entail that the problem of dominance should be solved at a lower level of organization than the levels of either Mendelism or the chromosome theory. The explanation of dominance would thus become part of the domain of some other theory.

All these reasons help to explain the deleting of the dominance component (C4) from the set of theoretical components and the demoting of the phenomenon of dominance (Item 1) to a more specialized empirical generalization not explained by the theory. Deleting dominance from the set of theoretical components allowed Mendelian segregation (i.e., other theoretical components) to be extended to include characters that blended or showed new forms in heterozygous hybrids. Figure 6-1 shows the elimination of C4, dominance, as a theoretical component and the specialization of Item 1, dominance, into Item 1.1', dominance, Item 1.2', blending, and Item 1.3', a new form. Since these items were not explained by Mendelism and were not considered to be unexplained phenomena that Mendelism should be actively working to explain, perhaps they should not be depicted as still part of the domain (i.e., items to be explained by the theory) of

1. simplify--complicate
   e.g. one--many
        pairs--multiple

2. generalize--specialize

   e.g. all--some

3. add a condition
   e.g. all a are b ⟶
   all a are b only in E1

4. delete

   e.g. all a are b ⟶
   [nothing]

Fig. 6-5. Strategies: complicate, specialize, delete.

Mendelism after some point. I don't have any firm opinion as to what date I would claim that they ceased to be part of the domain.

As we saw for Mendelism in its earliest days, issuing promissory notes for explanations is a method of beginning a theory with some incompleteness. In general, at what stage a failure to cash those notes becomes a serious failure of a theory may be difficult to decide. Deleting a component of a theory and leaving even a more restricted domain item unexplained might sometimes be viewed as a serious theoretical inadequacy. The dominance case suggests some possible guidelines for using the strategy of **deletion of a theoretical component** to improve the theory. First, if the theoretical component accounts only for a single domain item and if that item is deleted, then delete the theoretical component. Or, if the domain item is an empirical generalization and many exceptions to it are found, then consider deleting the theoretical component. Second, if the theoretical component is a promissory note for an explanation and it has not been paid and no prospects for paying it are evident, then delete the component. Third, if an unsolved problem can be reasonably assumed to require another, as yet unavailable, level of organization to solve it, then shelve the problem at this point. Such a move serves to delete the problem from the domain of the current theory, but relegates it to another theory.

Deleting a theoretical component (or a promissory note for one) and providing nothing to take its place when phenomena (even with many exceptions) remain unexplained sounds like a move that would often result in a worse, rather than better, theory. It may only be justified when other components of the theory are very successful. Furthermore, the deletion must be able to be made without affecting the other parts, that is, the deleted component must be a relatively independent module of the theory. Such was the case for the dominance component.

In summary, changes to Components 1, 2, and 4 resulted in anomalies being removed, the scope of the theory being expanded, several theoretical components of the theory becoming more complicated, and dominance ceasing to be a theoretical component. The strategies of **complicate, specialize, add,** and **delete** are strategies for anomaly resolution and expansion of scope

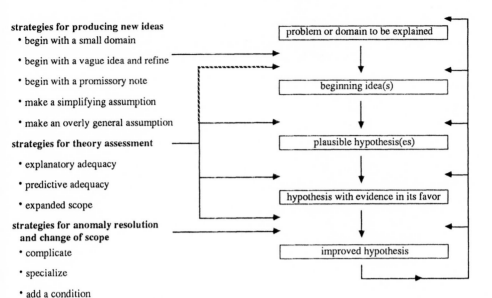

strategies for producing new ideas
- begin with a small domain
- begin with a vague idea and refine
- begin with a promissory note
- make a simplifying assumption
- make an overly general assumption

strategies for theory assessment
- explanatory adequacy
- predictive adequacy
- expanded scope

strategies for anomaly resolution and change of scope
- complicate
- specialize
- add a condition
- delete

Fig. 6-6. Refined stages and strategies, III

that are exemplified in these changes to the theory. They are summarized in Figures 6-4 and 6-5. Figure 6-6 places the strategies discussed in this chapter in a refined diagram of stages and strategies of theory development.

CHAPTER 7

Boveri-Sutton
Chromosome Theory

7.1 Introduction

This chapter traces the early development of the chromosome theory of Mendelian heredity proposed by Sutton and de Vries in 1903 and Boveri in 1904. After a brief discussion of the nineteenth-century background in Section 7.2, the Boveri-Sutton theory (as it came to be called) is discussed in Section 7.3. The extremely fruitful interaction between cytology and Mendelism produced new hypotheses and predictions for both fields. Cytological work on the nature of the sex chromosomes provided additional evidence for a relation between specific chromosomes and specific characters in organisms, as Section 7.4 describes; however, opposition to the claim of the importance of the chromosomes in heredity was common. Objections by Bateson in 1907, Morgan through early 1910, and others show the problems the theory faced. Section 7.5 samples a range of assessments of the chromosome theory of Mendelian heredity, prior to 1910. The further development of the chromosome theory after 1910 will be the subject of Chapters 9 and 10. Section 7.6 begins the analysis of the strategy of **making interrelations between two different bodies of knowledge**. It is a powerful way to form new hypotheses and make connections between diverse scientific fields.

7.2 Weismann and Nineteenth-Century Cytology

The field of cytology developed during the nineteenth century as microscopes improved, staining and fixing techniques developed, and good model organisms were found for microscopic observations. Formulated in the 1820s and 1830s, cell theory claimed that the basic units of organization in animals and plants were cells (Hughes, 1959; Darden, 1978). In the late nineteenth century, steps of normal cell division, called mitosis, were detailed (Wilson, 1896, ch. 2). With hindsight this work seems straightforwardly observational. Nevertheless, cytologists, who were working with fixed material, faced a number of difficulties in determining the temporal order of the various stages of cell division in living material. They had to distinguish actual cell components from artifacts created in the staining and fixing processes. In addition, using two-dimensional data to make inferences about three-dimensional phenomena presented challenges (Baxter, 1974).

These problems were even more pronounced with the attempts to determine the steps in the formation of germ cells. The long history (see Stubbe, 1972; Farley, 1982) of the unraveling of the role of the male and the female in reproduction had, by the late nineteenth century, resulted in an understanding in terms of cell theory. This history included discoveries that the male plays a material role in reproduction, that semen is relevant, and that sperm are not parasites but have a functional role. That work also led to the conclusions that the female menstrual fluid is not female semen and that eggs are cells. Cell theory provided a new level of organization, below that of the easily visible fluids, that was a necessary level of organization for solving the problem of the nature of fertilization. The existence of germ cells, as well as the view that in sexual reproduction the sperm and egg cells combine, were widely accepted ideas by the late nineteenth century, despite some anomalies. In particular, work on parthenogenesis and other examples of asexual reproduction raised questions about the universal claim that both male and female germ cells were necessary to produce the new individual (Baxter, 1974; Farley, 1982). As already noted, in 1865, Mendel had available the level of the germ cells in plants, the "pollen cells" and the "germinal cells," which he used to formulate an explanation for the ratios that he discovered.

By the 1880s and 1890s, cytologists differed in their opinions about the importance of the nucleus in the functioning of the cell and the role of the nucleus in heredity. Also at issue was the nature and role of the chromatin, the darkly staining substance within the nucleus. Immediately before cell division, the tangle of chromatin was observed to form darkly staining threads, called "chromosomes." In normal cell division, mitosis, the chromosomes duplicated and divided. Mitosis thus resulted in daughter cells with the same number of chromosomes as the parent cells. During germ cell formation, however, the behavior of chromosomes differed from their behavior during mitosis. Not until the turn of the century were the details of chromosome behavior during germ cell formation determined and the process labeled "meiosis." The details of the behavior of chromosomes during meiosis and fertilization were active areas of research and debate in the 1890s and early 1900s (Wilson, 1896; Wilson, 1925; Churchill, 1970; Baxter and Farley, 1979; Farley, 1982).

At least three different theories have been called the "chromosome theory of heredity": (1) a purely cytological claim that the chromatin or the chromosomes carried the hereditary material, which was proposed by several cytologists in the late nineteenth century; (2) Weismann's theory of the germplasm, which underwent various developments in the 1880s and 1890s; and (3) the Boveri-Sutton chromosome theory of Mendelian heredity of 1903-1904. The latter is the most important from the perspective of the theory of the gene and will be the theory called "the chromosome theory" in subsequent chapters of this book. The other two, however, provided a background for the development of the Boveri-Sutton theory and will be briefly sketched here.

The "purely cytological" chromosome theory of heredity proposed that the chromosomes were the hereditary material. The first step to this theory was the recognition of the role of the nucleus in heredity (Coleman, 1965). Wilson's important book *The Cell in Development and Inheritance* (Wilson, 1896) provided an excellent review of cytology in the 1890s. According to Wilson, before the chromosome theory could be developed, it was necessary to have the view that heredity was to be explained by a definite material passed from one generation to the next. Wilson attributed this idea of a physical basis of heredity to Naegeli, who labeled the hereditary material the "idioplasm." Although Naegeli did not locate the idioplasm within a particular part of the cell, others suggested the chromatin as its location (Wilson, 1896, p. 301). The study of fertilization by Hertwig, Strasburger, and Van Beneden proved that in the sexual reproduction of both plants and animals the "nucleus of the germ is equally derived from both sexes" (Wilson,

1896, p. 301). Wilson stated the cytological version of the theory: "Hertwig, Strasburger, Kolliker, and Weismann independently and almost simultaneously [were led] to the conclusion that *the nucleus contains the physical basis of inheritance, and that chromatin, its essential constituent, is the idioplasm postulated in Naegeli's theory* " (Wilson, 1896, p. 302).

In the late 1880s, Weismann developed this cytological claim into his theory of the germplasm, which was more theoretical and less empirical than the purely cytological theory (Weismann, 1891). He argued for a distinction between the germplasm, from which the germ cells were formed, and the somaplasm, the rest of the body. The germplasm, he argued, was isolated from the soma early in development. Because of this isolation, Weismann believed that influences of the soma on the germ were impossible; this claim led him to reject the widely held view of the inheritance of acquired characters (often misnamed "Lamarckism").

Weismann claimed that the chromosomes, carrying the hereditary material, behaved differently in the somaplasm and the germplasm. To explain the differences among the differentiated cells of the somaplasm (the rest of the body other than the germ cells), Weismann claimed that the chromosomes qualitatively divided during embryological development. Individual chromosomes split into parts and different parts were distributed to the various cells of the body during development, thus causing differentiation of somatic cells into different types. Weismann's theory thus predicted a difference in appearance of the chromosomes in different types of somatic cells. Cytological observations of somatic cells did not support this prediction; the chromosomes appeared to stay the same, not be split into parts, and to remain remarkably similar from cell to cell (Wilson, 1896, p. 306).

According to Weismann, entire ancestral germplasms were carried on the chromosomes destined for the germ cells. In order not to double the number of chromosomes with each fertilization, Weismann predicted a "reducing division." During the reducing division, one-half of the total number of chromosomes of the adult were passed to each germ cell. Consequently, when fertilization occurred with the union of two germ cells, the adult number of chromosomes would be restored and thereby maintained from generation to generation (Weismann, 1891). Of this prediction, Wilson said: "The fulfillment of Weismann's prediction is one of the most interesting results of cytological research" (Wilson, 1896, p. 185).

After being influenced by Hugo de Vries's (1889) claim that material hereditary particles existed, Weismann further developed his theory in order to include the idea of hereditary particles making up the chromosomes and organized into a hierarchy. At the lowest level were the "biophors," which corresponded to de Vries's pangens (Weismann, 1892, p. 42). Each biophor produced one independently variable hereditary character. Biophors were organized into determinants, which corresponded to anatomically related groups of characters within the body. A complete set of determinants for a whole body was called an "id." The chromosomes, he claimed, were composed of groups of ancestral ids. Thus a single chromosome, according to Weismann, contained more than one set of all hereditary units for producing an entire organism (Weismann, 1892).

During germ cell formation and the reduction in the number of chromosomes, Weismann claimed that chromosomal divisions resulted in different germ cells having different hereditary material. These differences in the ancestral ids parceled out to germ cells contributed to variability produced by sexual crossing. Such variability was important as raw material for Darwinian natural selection (Weismann, 1891; 1892).

Although cytologists observed that germ cells contained half the total number of chromosomes of non-germ cells, the details of germ cell formation were debated. Especially at issue were details about the nature of the reducing division (Churchill, 1970; Baxter and Farley, 1979; Farley, 1982; Churchill, 1987). Some cytologists objected to the idea that different germ cells carried different ancestral hereditary units; instead, they maintained that all germ cells of an organism were alike and that during fertilization the chromosomes fused so as to "maintain the permanence of the species as a whole" (Baxter and Farley, 1979, p. 149). In addition, Weismann's claim that each chromosome carried all the hereditary units necessary for the formation of a complete individual was subjected to test by Boveri. Boveri did not confirm Weismann's view.

7.3 Boveri-Sutton Chromosome Theory, 1903-1904

Boveri's work in the 1890s, along with that of other cytologists, aimed at understanding the role of the chromosomes in germ cell formation and in ontological development. In 1901, Montgomery did morphological studies of chromosomes and concluded that they occurred in homologous pairs, one of maternal origin and the other paternal. During germ cell formation, he claimed, the paternal and maternal chromosomes aligned (conjugation) and then separated; consequently, any given germ cell contained a reduced number that were all of either maternal or paternal origin (Montgomery, 1901).

Baxter and Farley (1979) discussed the context in which Montgomery discovered chromosome pairing: Montgomery believed that the conjugation of chromosome pairs was the lower level manifestation of the rejuvenating effect of sex. A common view in the late nineteenth century was that sex had a rejuvenating effect, with an egg being rejuvenated by the passage of sperm into it. Baxter and Farley, interestingly, interpreted Montgomery's discovery of maternal and paternal pairing as an example of a correct observation that followed from an incorrect theory. They argued that his morphological, observational evidence was weak:

If one rejected his idea that the maternal and paternal elements conjugated for rejuvenation, it was easy to reject his other reasons [for maternal and paternal pairing]. As the debate over maturation division [of the germ cells] clearly indicates, theories based solely on observations of chromosomes were weak and open to dispute. One could never be sure if one's observations reflected too much stain or the wrong angle. What was needed was more information about chromosomes based on experiment and not simply observations. (Baxter and Farley, 1979, p. 162)

Boveri's sea urchin experiments provided such evidence. Boveri tested Weismann's claim that a single chromosome carried all characters by examining abnormal sea urchin embryos with some chromosomes missing. The result was that certain body parts, such as parts of the skeleton, were missing. Boveri concluded that Weismann was wrong—the chromosomes were not all equivalent; instead, different chromosomes affected different characters. Boveri's developmental experiments were unable to provide more detailed evidence to allow a given hereditary character to be associated with a specific chromosome. All he could show was that abnormal chromosome numbers produced large-scale developmental irregularities (Boveri, 1902).

Inspired by Montgomery's observations and Boveri's sea urchin experiments, Sutton examined the chromosomes of grasshoppers to see if chromosomes could be shown to be distinct individuals morphologically. Because chromosomes seem to dissolve into a large tangled thread between cell divisions, debate arose as to whether the same individuals as dissolved emerged from that tangled mass. Sutton concluded:

Although after each division there is a brief interval, during which chromosomic boundaries can no longer be traced, the regular correspondence, unit for unit, of the mother series with the daughter series establishes a high probability that we are dealing with morphologically distinct individuals, each of which bears to its mother element a genetic relation comparable to that existing between mother- and daughter-cells. (Sutton, 1902, p. 35)

By 1902, Boveri's and Sutton's work showed the chromosomes to be *individuals* in both a morphological and a functional sense: first, they persisted as distinct individuals through the resting stage from one cell division to the next, and, second, they carried different hereditary qualities. The cytological and developmental studies, however, did not provide a means of analyzing the chromosomes down to the level of individual hereditary characters. Boveri had reached a limit of his techniques, which showed that missing chromosomes resulted in many character abnormalities. Furthermore, by 1903, questions remained about the nature of the germ cells. Were all germ cells alike in their chromosome complement? Alternatively, did each germ cell receive a set of either all maternal or all paternal chromosomes when it formed? Or, did germ cells have some other possible chromosome assortment? When two germ cells united at fertilization, what happened to the chromosomes? Did the maternal and paternal chromosomes fuse or did they remain distinct in both the soma and germplasm?

Despite these remaining questions, by 1903, cytologists, working at the microscopic level and examining the chromosomes in the cell nucleus, had evidence for the conclusion that the chromosomes were the hereditary material. Moreover, they had found that the chromosomes were pure individuals, that they occurred in pairs (one from the female, the other from the male), and that the reducing division resulted in the germ cells having one-half the normal number of chromosomes. These results were produced by cytological observations with the microscope of fixed specimens and by fertilization experiments with observations of the resulting developmental features. Furthermore, Weismann's theory had raised the questions whether hereditary characters were caused by units carried by the chromosomes, and if so, what was the relation between specific chromosomes and the units they carried.

In comparison, by 1903, Mendelism, which used the technique of artificial breeding and the counting of characters, had reached a number of conclusions about hereditary characters that were similar to results about chromosomes. Characters of varieties occurred in antagonistic, alternative, allelomorphic pairs. These characters were considered to be individuals, in that they did not contaminate each other by being together in a hybrid; pure characters re-emerged in subsequent generations (purity of the gametes). Segregation resulted in one or the other but not both characters being represented in a given germ cell (gamete). Finally, different pairs of characters assorted independently so that maternal and paternal characters were mixed in the offspring. Questions remained, however. What were the differences between the different types of germ cells? Did allelomorphic factors really not influence each other while they were together in the hybrid? In other words, would pure breeding forms really emerge after a hybrid cross, never again to show the influence of their mongrel ancestry? What was the "physical basis" for Mendelian heredity? Table 7-1 summarizes my analysis of the mappings between similar features of chromosomes and Mendelian allelomorphs.

Various of these similarities were noted by Cannon (1902), Boveri (1904), Sutton (1903), and de Vries (1903), each of whom independently proposed that the Mendelian anlagen, allelomorphs, or pangens were parts of the chromosomes.

Chromosomes	Allelomorphs
1. Pure individuals (remain distinct, do not join)	1. Pure individuals (remain distinct, no hybrids)
2. Found in pairs (in diploid organisms prior to gametogenesis and after fertilization)	2. Found in pairs (in diploid organisms prior to segregation and after fertilization)
3. The reducing division results in one-half to gametes	3. Segregation results in one-half to gametes
4. Prediction: random distribution of maternal and paternal chromosomes in formation of gametes	4. Characters from maternal and paternal lines found mixed in one individual offspring; independent assortment of allelomorphs
5. Homologous chromosomes A and a combine to give: $AA:2Aa:aa$.	5. Mendelian ratios of allelomorphs A and a in F_2 give: $AA:2Aa:aa$.
6. Prediction: abnormalities in chromosomes (e.g., abnormal sex chromosomes found by Bridges)	6. Prediction: corresponding abnormalities of characters (e.g., abnormal sex-linked characters)
7. Paired homologous chromosomes in normal, purebred organisms	7. Prediction: dual basis of each character in normals (two alleles for each gene in normals as well as hybrid)
8. Chromosome number smaller than character number	8. Prediction: more than one allelomorph on a chromosome; their corresponding characters linked in inheritance
9. Problem: no visible difference between homologous chromosomes	9. Dominant allelomorphs may be associated with a corresponding (but somehow different) recessive
10. Prediction: recombination of parts of the chromosome or of the allelomorphs on the chromosome	10. Prediction: more combinations of linked allelomorphs than number of chromosomes

Table 7-1. Relations between chromosomes and allelomorphs, 1903-1904.

Boveri, in a 1904 book expanding a brief discussion in 1903, remarked on the similar conclusions reached in the two fields and postulated the chromosome theory of Mendelian heredity:

> Thus we see here two fields of research which had developed entirely independently of one another reach results which agree so exactly [that it appears] that one was derived theoretically from the other; and when we realize what we have concluded about the significance of the chromosomes from their inheritance, the probability that the transmitted characters in the Mendelian investigations are actually bound to definite chromosomes is unusually large. (Boveri, 1904, p. 117, my translation)

Sutton too was struck by the similarity of the findings in Mendelism and cytology:

> Thus the phenomena of germ-cell division and of heredity are seen to have the same essential features, viz., purity of units (chromosomes, characters) and the independent transmission of the same; while as a corollary, it follows in each case that each of the two

antagonistic units (chromosomes, characters) is contained in exactly half the gametes produced. (Sutton, 1903, p. 32, italics omitted)

De Vries discussed the two fields that were contributing to the theory of heredity: morphological, microscopic studies of fertilization and "physiological" studies of relations of offspring and parents, especially in hybrids, investigated by artificial breeding:

> Few investigators master both provinces; their extent is much too great for that. And especially has the study of hybrids so greatly advanced in recent years, that even here a division of labor will soon be necessary. Both lines of work have therefore developed more or less independently of each other. In both the main features of the problem begin gradually to arise out of the abundance of individual phenomena. And thereby there is disclosed, one might almost say, beyond all expectation, an agreement in the results of both lines of investigation, which is so great, that almost everywhere the physiological processes are reflected in the microscopically visible changes. (de Vries, 1903, p. 220)

De Vries's discussion was much less precise than Boveri's and Sutton's in laying out various similarities between Mendelism and cytology. Thus, the table is based primarily on Boveri's and Sutton's claims and predictions.

One feature of Mendelism with no cytological counterpart (see Table 7-1, no. 9) was dominance. Sutton admitted: "Dominance is not a conception which grows out of purely cytological consideration" (Sutton, 1903, p. 35). The chromosomes of a pair looked the same, so cytology provided no means of explaining dominance and recessiveness. Sutton did not consider this a serious problem, because, as he noted, dominance was not a universal phenomenon and other relations, such as blending or new forms, might appear in the hybrids. He concentrated on other aspects of Mendelism in his arguments for the claim that cytology provided the "physical basis of the Mendelian law of heredity" (Sutton, 1902, p. 39).

Sutton discussed the relation of maternal and paternal chromosomes in the light of the connection to Mendelian phenomena. Cytologists had, he said, mistakenly claimed that the set of chromosomes from the mother and the set from the father would remain together when the germ cells formed. If chromosomes carry hereditary characters and offspring can inherit some characters from the mother and others from the father (or from any grandparent), then the maternal and paternal chromosomes must be mixed in the formation of germ cells (Sutton, 1903, p. 29). Thus, Sutton made a prediction for cytology that the maternal and paternal chromosomes would be found to mix in the formation of germ cells, in contrast to Montgomery's view.

De Vries too saw the conflict between the fact that an organism inherited a random assortment of paternal, maternal, grandpaternal, and grandmaternal characters (and therefore their pangens) and the claim that the maternal and paternal sets of chromosomes stayed together when germ cells were formed. Yet, he proposed a different hypothesis from Sutton's. De Vries suggested that, during the period in germ cell formation when the pairs of homologous chromosomes lined up, some of the pangens were exchanged from one chromosome to the other (de Vries, 1903, p. 243). It is unclear from his discussion which of two possibilities de Vries was proposing. On the one hand, he may have believed that the pangens were to be identified with small parts of the chromosomes, and thus he believed that actual parts of the chromosomes switched places. Alternatively, he may have believed that the pangens were carried by the chromosomes and the chromosomes remained intact as a substratum (he called them "nuclear threads") on which pangens jumped back and forth. No visible evidence existed for either possibility, but de Vries was never reluctant to postulate processes below the level of current microscopic observation. At any rate, Sutton's view that maternal and paternal chromosomes assorted randomly in the formation of germ cells was subsequently confirmed by cytologists. Such confirmation was dependent on the availability of abnormal chromosomes so that the usually identical, homologous chromosomes could be distinguished (Carothers, 1913).

Both Sutton and Boveri predicted that characters whose allelomorphs were carried by the same chromosome would be inherited together. Using Bateson's term, "allelomorph," to refer to the factors in the germ cells, Sutton asked whether "an entire chromosome or only a part of one is to be regarded as the basis of a single allelomorph," and concluded that an allelomorph must be a part of a chromosome. He reasoned that, because the Mendelian character number was greater than the chromosome number, numerous characters would be associated with a single chromosome. In an insightful comment, Sutton said, "If then, the chromosomes permanently retain their individuality, it follows that all the allelomorphs represented by any one chromosome must be inherited together" (Sutton, 1903, p. 34). Because Boveri and Sutton (1902) had only recently shown that the chromosomes were individuals, Sutton probably added his qualification about individuality because his contemporaries might doubt that chromosomes were definite structures. It is unlikely that Sutton foresaw that a phenomenon such as the switching of pieces (crossing-over) between paired chromosomes might occur (the discovery of crossing-over is discussed in Chapter 9).

Sutton mentioned some examples of correlated characters that had already been noted; however, most of the different characters studied by Mendelians at the time (such as Mendel's pea color and shape) had been observed to assort independently, according to what would later be called Mendel's second law. This lack of evidence of "coupling" or "linkage" of characters was argued by some to be evidence against the chromosome theory of Mendelian heredity, as we will see. De Vries's view did not predict linkage of characters produced by pangens carried by the same chromosome, because he believed the pangens were exchanged between chromosomes.

Boveri also predicted:

> if a hybridization experiment included numerous characters and if it should be found in successive breeding that the number of combinations in which the separate traits can occur is greater than would correspond to the possibilities of recombination of the chromosomes present, then it would have to be concluded that the traits localized in a chromosome can go independently of each other into one or the other daughter cell which would point to an exchange of parts between the homologous chromosomes. (Boveri, 1904, p. 118, my translation)

Although Boveri has been credited with the prediction of the crossing-over that T. H. Morgan later found (Stern, 1950), we must be wary of drawing that conclusion. In a footnote to the above passage, Boveri said: "In a just published work of de Vries (1903) a very clear presentation is given of this point" (Boveri, 1904, p. 118); however, de Vries's view might not have been that lengths of the paired, homologous chromosomes were exchanged (as Morgan later claimed). Instead, de Vries was focused at the level of the separate pangens, some of which were individually exchanged between chromosomes. The question remains whether Boveri predicted that the anlagen on the chromosomes shifted individually or whether exchanges occurred between actual lengths of chromosomes. At this early period, instances of correlated characters were rare; distinguishing these various possibilities of complete linkage, partial linkage, and crossing-over was seven years away. The role of the chromosome theory in dealing with these anomalies will be discussed in Chapters 9 and 10. As of 1903, no data yet suggested the level of organization between individual characters (or their allelomorphs) and the characters (or their allelomorphs) carried by an entire chromosome. The "chunking" together of factors into linkage groups was an important level of organization discovered after 1903. No methods existed in 1903 for associating any specific Mendelian character with any specific chromosome. Work on sex determination proved crucial to that, as will be discussed in Section 7.4 and in Chapter 9.

The Boveri-Sutton version of the chromosome theory of Mendelian heredity provided explanations and predictions for both genetics and cytology. It provided support for the claim that all hereditary characters were caused by paired allelomorphic factors for organisms with a diploid (dual) chromosome number. If factors were parts of chromosomes, and chromosomes were paired, one would expect all factors to be paired also. Although Castle had made such a sweeping claim for all hereditary characters (see Chapter 5, Section 5.7), other Mendelians had been more careful about making claims about characters for which no variant was known. A chromosomal basis for characters provided new evidence for the inference that all allelomorphic characters were paired.

Furthermore, the chromosome theory and the reducing division in the formation of germ cells explained why Mendelian factors segregated: factors (allelomorphs, anlagen) were part of the paired homologous chromosomes that separated during germ cell formation. The different types of germ cells postulated to explain Mendelian ratios were shown to be different because they contained different chromosomes. The smaller number of chromosomes than characters allowed the prediction that characters would be correlated in inheritance, that is, that not all character pairs would assort independently as did yellow-green and tall-short characters in peas.

As Baxter and Farley (1979) have argued, cytology also benefited from its connections to Mendelism. The confusing observational evidence acquired a theoretical framework once Mendelian factors were claimed to be parts of chromosomes. Moreover, the prediction that maternal and paternal chromosomes were randomly distributed during germ cell formation was confirmed (Carothers, 1913).

Despite these benefits provided by the chromosome theory that are so readily apparent with hindsight, the theory did not win immediate universal acceptance. Arguments for and against the chromosome theory will be discussed in Section 7.5 and in Chapter 10. Reasons for the lack of response to the chromosome theory were not all of an evidential nature, however. Its proposers were occupied with other matters. After postulating the connections between Mendelism and cytology, Boveri continued his cytological and developmental studies and did not pursue the chromosome theory of Mendelian heredity (Baltzer, 1967). Sutton was a graduate student; he left cytological research to become a physician (McKusick, 1960). De Vries continued his studies of mutations and did not pursue the chromosome theory or Mendelism as primary areas of research, although he followed their developments and continued to develop his idea that the pangens were carried by the chromosomes (e.g., de Vries, 1924b).

Interdisciplinary endeavors often have difficulty getting established for a number of reasons, not the least of which is that in their early stages of development no practitioners may have training in both fields. Boveri did not do breeding experiments and de Vries was not a cytologist. Graduate students trained in both have to have opportunities open to them to continue their bridging work. (For further discussion of interdisciplinary work, for example, see Bechtel, 1986.) As we will discuss in Chapter 9, what came to be called the Boveri-Sutton theory—that genes are parts of the chromosomes—was developed by Morgan and especially by his graduate students after 1910. The role of the interfield interactions between genetics and cytology will be shown to have played an important role in developing a major new set of theoretical components for Mendelism after instances of anomalous correlated ("partially linked") characters were found. As of 1903, the lack of evidence in favor of correlated characters (characters that did not independently assort) provided evidence against Sutton's version of the theory.

7.4 Sex Chromosomes

Between 1900 and 1910, cytologists and embryologists were far from agreement about the importance of the nucleus and chromosomes in heredity. One issue in these debates was the cause of sex determination. Theories of sex determination that had been proposed over the centuries can be grouped in several general types. First, they may be characterized as postulating *external*, environmental causes versus *internal* causes. Internal theories can be further classified as to whether they appealed to either *preformed* internal determinants or some sort of "*epigenetic*" control that became apparent during the development of the embryo. Finally, theories also differed as to whether they claimed *nuclear* versus *cytoplasmic* localization for either the preformed parts or the epigenetic controlling elements (see, e.g., Baxter, 1974; Gilbert, 1978; Farley, 1982). More specific theories within these general types appeared to have evidence in their favor at certain times. For example, in the second edition to his book *The Cell in Development and Inheritance* in 1900, Wilson discussed various studies that he claimed showed the effects of nutrition in determining whether an egg or an embryo developed into a male or female, after which he concluded:

> The observations cited above, as well as a multitude of others that cannot here be reviewed, render it certain, however, that sex as such is not inherited. What is inherited is the capacity to develop into either male or female, the actual result being determined by the combined effect of conditions external to the primordial germ-cell. (Wilson, 1900, p. 145)

By 1910, Wilson's own work, along with that of others, rendered this "certain" conclusion doubtful.

Between 1900 and 1910, various Mendelian explanations of sex were attempted using the Mendelian concepts of dominant and recessive or homozygous and heterozygous, but these Mendelian attempts to account for sex foundered on several problems. The Mendelian ratio of 1:2:1 does not correspond to the sex ratio of 1:1 males to females. Several ideas were advanced to explain the 1:1 sex ratios by assuming that one sex was a heterozygote and the other a homozygote, but they met with various difficulties (discussed in more detail in Chapter 9). Further complications were posed in attempts to explain how some insects produced parthenogenetically, without fertilization. None of these early attempts by Mendelians to account for sex determination localized the Mendelian factors on chromosomes (see Gilbert, 1978, pp. 327-328; Allen, 1978, pp. 125-129; Farley, 1982, pp. 218-221 for further discussion).

Cytological work, rather than Mendelian, provided a vital piece of evidence. The recent acceptance of the existence of paired, homologous chromosomes, one of maternal and the other of paternal origin, focused attention on their numbers, their pairing, and their reduction in number during germ cell formation. A bit of unpaired chromatin in sperm had been described in 1891 and labeled *X* because it was puzzling at the time (Wilson, 1925, p. 748). In 1902, McClung called attention to this earlier finding and suggested that this unpaired "accessory" chromosome was associated with sex determination because during the reduction division one-half of the sperm received it, while the other half did not. He further suggested that the presence or absence of this accessory chromosome in the sperm made the sperm responsible for sex determination: eggs fertilized by sperm carrying the accessory, he suggested, became males, while those fertilized with sperm lacking the accessory became females. In order to accommodate evidence that external conditions, especially nutrition, could influence sex ratios, McClung further suggested that the egg could select or "choose" which type of spermatozoan would be allowed to penetrate and effect fertilization. He preferred, he said, to endow the female with the power to select that which is for the best interest of the species (McClung, 1902).

McClung's hypothesis that the accessory or X chromosome was sex determining was confirmed by work by Stevens and Wilson in 1905. Whether Wilson arrived at his conclusions independently of Stevens has been questioned (Brush, 1978). At any rate, they both demonstrated that sperm in numerous insect species did indeed differ, by one having the X chromosome and either the other having nothing that paired with it, or the other having a smaller Y chromosome (to use Wilson's later symbols of X and Y for representing the different types of sex chromosomes). They both corrected McClung's speculation that the X chromosome was male determining; a more careful examination of the chromosome complements of males and females in insects showed that the female had two X chromosomes and males had either one X or an XY constitution (Stevens, 1905-6; Wilson, 1905; Gilbert, 1978). Wilson concluded: "The foregoing facts irresistibly lead to the conclusion that a causal connection of some kind exists between the chromosomes and the determination of sex" (Wilson, 1905, p. 105). The nature of that causal connection was still a matter of debate, however, with Wilson suggesting it was due to differences in degree of activity of the chromosomes and Stevens questioning whether any evidence for difference in activity existed (Wilson, 1905; Stevens, 1905-6).

Thus, sex determination and sexual characters were associated with a specific chromosome by the cytological work. Specific relations between Mendelian breeding ratios and sexual characters or sex chromosomes were yet to be established. Wilson was more cautious about accepting the chromosome theory of Mendelian heredity than his students Sutton and Cannon had been, but by 1910, Wilson was probably a supporter (Wilson, 1902; Baxter, 1974; Gilbert, 1978).

7.5 Assessments of the Chromosome Theory, 1906-1910

Reactions to the chromosome theory between 1906 and 1910 were diverse. As a sampling of the range of these opinions, five different views will be discussed. Two come from textbooks: Lock's *Recent Progress in the Study of Variation, Heredity and Evolution* of 1906, which was used by Wilson in his courses and studied by Muller and other of Morgan's students (Allen, 1978, p. 308; Carlson, 1981, p. 27); another from Thomson's *Heredity* (1908). Thomson was much more skeptical than Lock about the conclusions to be drawn about the role of sex chromosomes. A third position is that of Guyer, who enumerated many deficiencies of the chromosome theory in a paper published in 1909; he pleaded for other working hypotheses to be maintained. Bateson's objections in 1907 show that Bateson, although an advocate of Mendelism, opposed the chromosome theory. Morgan, in contrast, was an early critic of both Mendelism and the chromosome theory between 1900 and 1910, after which he became a proponent and major developer of both. These views show the diverse assessments of the chromosome theory between 1906 and 1910.

Lock (1906) was a proponent of both Mendelism and the chromosome theory. After enumerating the many similarities between Mendelian and cytological phenomena, he concluded that "there must be some real connection between the behavior of chromosomes as seen microscopically on the one hand, and the behavior of allelomorphic characters as deduced from the results of experiment on the other" (Lock, 1906, p. 250). He called this the "chromosome-allelomorph" view and endorsed de Vries's claim that during pairing before the reduction division, allelomorphs were exchanged between homologous chromosomes. He suggested that for the few known instances of coupled characters "some mechanism" existed for keeping them together while the others are "being reassorted between the chromosomes" (Lock, 1906, p. 252). Further, he considered Wilson's discovery of sex chromoses to be strong evidence for association of particular characters with particular chromosomes (Lock, 1906, p. 253).

Thomson (1908), in contrast, was a strong proponent of Mendelism, but he gave only cautious support to the cytological view of the nucleus and the chromatin as providing the "physical basis of inheritance" (Thomson, 1908, ch. 2). He provided only a brief mention of Sutton's hypothesis that the "chromosomes are the vehicles of hereditary qualities" (Thomson, 1908, p. 348). Earlier Thomson had developed a physiological theory of sex determination, with external factors as primary agents (Gilbert, 1978, pp. 323-325). By 1908, he, like Wilson, found the evidence for external factors in sex determination less convincing. In contrast to Wilson, however, he did not accept the evidence that different chromosomes determined the sex of the embryo; instead he believed that each fertilized egg usually had the potential to develop into either a male or a female, with the *internal* determining factor still unknown (Thomson, 1908, ch. 13).

In 1909, Guyer published "Deficiencies of the Chromosome Theory of Heredity," which attacked the view that the chromosomes carried the hereditary material. He only briefly raised questions about whether the Mendelian claim of purity of the germ cells with respect to a given character was to be accepted. Against Mendelian segregation, he mentioned doubts cast by various cases of non-Mendelian inheritance, such as blending and purported evidence of "lingering" "influences" instead of purity (Guyer, 1909, p. 11). Guyer was an advocate of the view that the egg cytoplasm played a central role in heredity of the generic characters of a species. He objected to the unit-character view of transmission of hereditary traits by the chromosomes as insufficient to account for the "fundamental physiological unity of the entire organism" (Guyer, 1909, p. 13). Sapp (1987) discussed Guyer's and other cytological theories of development, as well as the opposition that proponents of such theories had to the chromosome theory throughout the period from 1909 to well into the 1930s. Geneticists and proponents of the chromosome theory continuously had to argue that they were not merely studying superficial variable characters that distinguished varieties from one another, but that their theories applied to all hereditary characters, including "fundamental" ones, such as the nature of vital organs and general body organization.

Bateson was a major proponent of Mendelism and has been rightly called the founder of genetics. Nevertheless, he was an opponent of the chromosome theory and in 1907 raised a number of specific objections against it. The visible appearance and numbers of chromosomes did not have features that Bateson said would be expected if they were the chief agents in the production of characters. First, more differences between the chromosomes, corresponding to different characters, would be expected, but the shapes showed much uniformity (Bateson, 1907, p. 165). Bateson reiterated his support of his presence and absence theory to explain dominance and recessiveness (discussed in Chapter 6). Although he did not specifically mention the chromosomes in relation to this theory, explaining recessive characters as due to an absence of the dominant factor would have had implications for the physical basis of the factors. Bateson would not have expected paired entities, such as the homologous chromosomes, to account for paired allelomorphs. Instead, he would have expected something that was present in the dominant form and missing in the recessive. Chromosomes did not have gaps along their lengths; nor did the paired, homologous chromosomes show differences in their over-all lengths.

In addition, Weismann's claim that only parts of chromosomes were distributed throughout the body to account for the differentiation of cells into various types was not supported by recent cytological examination of various tissues. Bateson cited evidence that no distinctions between chromosomes in the nuclei in various tissues throughout the body had been found. Furthermore, when the chromosome numbers in different species were examined, no obvious patterns emerged. Closely allied species might show either similar or widely different chromosome numbers. Looking at the evolutionary tree more broadly, Bateson remarked: "Low forms may have many; highly complex types may have few" (Bateson, 1907, p. 165).

Bateson's view was that chromosomal phenomena were an effect, just as many other differences in organisms were, of deeper causes of hereditary differences. The one concession he made was that of Wilson's work with sex chromosomes. It was the one instance in which a "chromosome difference has been proved to be associated with a somatic difference" (Bateson, 1907, p. 174); however, he interpreted it within a Mendelian account of sex:

> Whether the accessory body is in these types the "cause" of femaleness or only associated with that cause, we have at last the long expected proof that sex is determined in the germ cells, so far as these specific cases are concerned. In those cases we may even go farther and declare that the female is homozygous in femaleness [*XX*], while the male is heterozygous [*XY*] in sex. Such a result accords well, I think, with the general conclusions to which breeding experiments, on the whole, point. (Bateson, 1907, p. 175)

Bateson's objections to the chromosome theory extended throughout much of his career. I will return to Bateson's views in Chapters 9 and 10 with discussion of his (and Punnett's) discovery of coupled characters. Bateson and Punnett explained them without appealing to chromosomes.

Morgan's objections to Mendelism, the chromosome theory, and the chromosomal basis of sex determination between 1900 and 1910 have been enumerated by Allen (1978, pp. 125-144), and also discussed by Gilbert (1978, pp. 336-349). Baxter (1974) and Gilbert (1978) have tried to account for the different attitudes of Wilson and Morgan, who were colleagues with very similar training and were usually familiar with the same evidence. We will not pause to consider these interesting discussions here, except to note that reaching a full understanding of why *an individual scientist* assessed a given theory at a given time as well-supported and worthy of acceptance is difficult. No easier is it to fully understand why *several contemporary scientists*, such as Bateson, Wilson, and Morgan, differed in their assessments of the evidence for a theory. The biographer is in a better position to suggest plausible reasons for a given person's attitude than the historian attempting to track the "public knowledge" at intervals of time, unless the person explicitly states reasons in a publication. Chapter 15 will discuss the role of an individual's methodological and metaphysical commitments in theory assessment, commitments that biographers sometimes succeed in uncovering.

At any rate, in a paper in 1910, Morgan assessed current evidence and presented several reasons for his skepticism about the chromosome theory of Mendelian heredity and his views about the relation of chromosomes to sex determination. Morgan echoed some of Bateson's criticisms about the numbers of chromosomes in different species and the seeming lack of differences among chromosomes in different tissues. He questioned Boveri's evidence for the individuality of the chromosomes and suggested instead that the observational evidence was equally compatible with the view that they might fuse and then different materials might separate before germ cell formation. In assessing the relation of the chromosomes to Mendelism, he argued that only a few of the predicted correlated characters had been found. And the few that were known provided "no evidence that they Mendelize in groups commensurate with the number of chromosomes" (Morgan, 1910a, p. 468).

Morgan was much more favorably disposed to the claim that sex determination was associated with specific chromosomes; he also considered the possibility that sex was a Mendelian factor. In support of the chromosome view of sex, he cited the work of Wilson and Stevens on X chromosomes. Morgan himself had worked on sex determination in parthenogenetic forms that could switch to sexual forms, given certain external conditions. He examined the chromosomes and showed that males had fewer chromosomes. The process by which they were produced involved unusual cell divisions in which two chromosomes were lost. Those eggs with fewer chromosomes produced males (for more detail, see Morgan, 1909a; 1909d; 1910d). Thus, external conditions somehow triggered an internal mechanism for sex determination, a mechanism associated with the chromosomes. Morgan concluded: "The facts make out a strong case in favor of the view that we have probably found the mechanism by means of which sex is determined" (Morgan, 1910a, p. 493).

This evidence was not sufficient to convince him of the general claim that chromosomes provided the entire physical basis of heredity or that they carried preformed units that produced all the characters. He wished, he said, for a more "epigenetic" view, not such a "preformation" theory. He wanted a theory that explained embryological development by reference to "differences in reaction, and not to separation of mixed materials" (Morgan, 1910a, p. 479). He, too, at this point, was impressed by the role of the egg cytoplasm in development and was unwilling to attribute the control of embryological development to the chromosomes (Morgan, 1910a, p. 453). Morgan changed, from being a critic of both Mendelism and the chromosome theory to being a major developer of both, because of new evidence. Morgan found correlations between sex and characters showing Mendelian segregation, as will be seen in Chapter 9.

Allen, in discussing Morgan's opposition to the chromosome theory of Mendelian heredity between 1903 and 1910, said that the "close correlation between cytological and Mendelian results was not so obvious as it might appear in retrospect.... [I]n 1909 it involved an act of faith to make the transition between one theory and the other" (Allen, 1978, p. 142). In a similar vein, Cock defended Bateson's opposition to the chromosome theory: "Throughout most of the first decade of the century the chromosome theory of heredity had no more support than a simple, if impressive, analogy" (Cock, 1983, p. 40). Correlation, we are often warned, does not necessarily imply causal connection. The strikingly similar properties and behaviors of the chromosomes and Mendelian factors were seen by some biologists as evidence that the chromosomes physically contained the factors as their parts. Alternatively, it was viewed by other biologists as evidence that they might both be caused by some third, as yet unknown, thing or process. From an evidential point of view, the lack of predicted correlated characters was strong evidence against Sutton's view, and no cytological evidence existed in favor of de Vries's view of exchange of pangens between paired chromosomes. Nonetheless, the exploration of connections between cytology and genetics looked fruitful for further work. It had, as Morgan put it, "aroused expectation to a high pitch of interest in the application of the observations of cytology to the conclusions in regard to Mendelian segregation" (Morgan, 1910a, p. 465).

7.6 Strategy of Using Interrelations

Making the interfield connection between genetics and cytology that resulted in the formation of the chromosome theory of Mendelian heredity was discussed in previous work on interfield theories (Darden and Maull, 1977; Darden, 1980b). The concept of a scientific field fits the genetics and cytology examples well—they were clearly differentiated areas of research in science. Thus, designating the chromosome theory an "inter*field* theory" is appropriate.

If the two bodies of knowledge that are connected can be viewed as being at different structural or functional levels in a hierarchy, then such a bridging theory can be called an "inter *level* theory." (See Staats, 1983, and Bechtel, 1986, for further discussion of the problems of differentiating levels and the analysis of interlevel theories.) The levels of organization studied by genetics and cytology are related in complicated ways. It is not obvious which field should be considered at the "lower" level. Mendelian genetics studied visible characters and postulated causal factors for those characters. The location and "physical basis" of the factors were unknown, except that factors had to be carried by germ cells. Mendelian genetics thus involved three "levels" of organization (if they may be properly called "levels"): observable characters, unobservable differences among germ cells, and unobserved causal factors of unknown size. Cytology, on the other hand, studied cells and their parts using the microscope (with the occasional speculation by cytologists about the functions of the structures they observed). De Vries (1903) and others termed the genetics work "physiological," and the cytological "morphological." The relation between cytology and genetics is thus like the relation between physiology and anatomy, between the study of functional or causal processes and visible structures that participate in those processes. Thus, the label interfield theory applies well to the chromosome theory, but interlevel is not so clearly applicable.

When theories can be ordered as to levels, philosophers have often discussed the relation between them as being a reduction. The higher level theory is reduced to the lower level theory. A theory about "wholes" would be reduced to a theory about its "parts." Theoretical reduction is a method for unifying two theories, in the strong sense of eliminating one or deriving one from the other. The result of reduction is one theory rather than two. (For a sampling of the extensive philosophical discussion of reduction, see, Oppenheim and Putnam, 1958; Schaffner, 1967; 1969; 1974b; Maull, 1977; Kitcher, 1984.) Cytology was not reduced to genetics, however, nor was genetics reduced to cytology in any of several senses of reduction. Cell theory was not derived from Mendelian theory, nor vice versa. Cell theory was not replaced by Mendelian theory, nor vice versa. The relation between genetics and cytology was not that of reduction in the sense that either was eliminated in favor of the other. Instead, the two fields were related to each other by the postulation of an interfield theory bridging the two fields and allowing each to provide new ideas for solving problems in the other.

Aspects of the chromosome theory case suggest a general strategy. This will be called the strategy of **postulating interrelations between two bodies of knowledge**, in short, the strategy of **using interrelations**. The term "interrelation" is used here for a relation that is proposed between two different bodies of knowledge developed using different techniques, and perhaps developed independently of each other. If the two bodies of knowledge can be said to be part of two different scientific fields, then the interrelation is an "interfield" relation. If they can be ranked as referring to different hierarchical levels, then the interrelation is an "interlevel" relation. If the relation can be said to be part of a theory (as opposed to a looser idea about some sort of relation between the two), then postulating the interrelation can be said to be an instance of the formation of an interfield or an interlevel "theory."

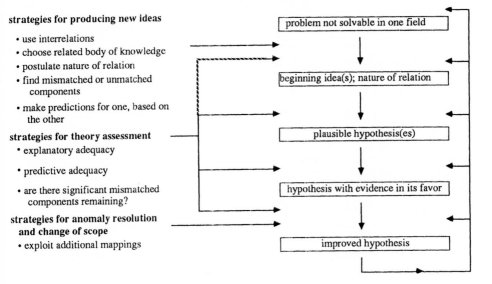

strategies for producing new ideas

- use interrelations
- choose related body of knowledge
- postulate nature of relation
- find mismatched or unmatched components
- make predictions for one, based on the other

strategies for theory assessment

- explanatory adequacy
- predictive adequacy
- are there significant mismatched components remaining?

strategies for anomaly resolution and change of scope

- exploit additional mappings

problem not solvable in one field

beginning idea(s); nature of relation

plausible hypothesis(es)

hypothesis with evidence in its favor

improved hypothesis

Fig. 7-1. Refined stages and strategies, IV

Dividing a broad area of science into mutually exclusive fields (with all the concomitant sociological aspects) or into clearly different hierarchical levels is sometimes easy and at other times difficult, depending on the case. In addition, the connection established between two different bodies of knowledge may be obviously termed a "theory," as the chromosome theory was. In other cases, however, the connection between two different bodies of knowledge may not have been developed into a complete *theory* and historically may not have been called a "theory." Thus, I am generalizing earlier discussions of interfield theories (Darden and Maull, 1977; Darden, 1980b) to a discussion of interrelations between two different bodies of knowledge. That terminology focuses on the nature of the interrelation and use of the strategy to form such a relation, rather than on the details of distinguishing the two things which are related. It is, I believe, a helpful refinement in order to make the strategy of using interrelations a more general one.

The strategy of using interrelations has several steps: (1) finding a problem that requires an interrelation to another body of knowledge, such as asking about the physical location of the Mendelian factors; (2) locating an appropriate other field to aid in solving the problem, such as determining that cytology, but not embryology, held a key to solving the problem about the location of Mendelian factors; (3) postulating a *kind* of interrelation, such as *identity* of factors and chromosomes or a *part-whole* relation between them; (4) using the interrelation to guide hypothesis formation, such as the prediction that groups of Mendelian characters would be correlated in inheritance because their factors were parts of the same chromosomes; (5) testing the new hypotheses and assessing the adequacy of the postulated interrelation; for instance, the lack of sufficient numbers of correlated characters provided evidence against the chromosome theory, prior to 1905; and finally, (6) assessing the result of forming a successful interrelation, such as concluding that the specific relation that genes are parts of chromosomes (once there was good evidence in its favor) provided a bridge between cytology and genetics. These steps in using the strategy of interrelations are illustrated in Figure 7-1.

Each step in the strategy of using interrelations will be discussed in more detail in subsequent chapters (see especially Chapter 13, and also Chapters 9, 10, and 15). The discussion here will focus on Step 3, postulating a specific interrelation, and Step 4, using the interrelation to guide hypothesis formation in one or both fields.

A number of different kinds of specific interrelations might be postulated. The following questions illustrate possibilities: Can entities in one field be *identified* with those in another? Can entities in one field be claimed to be *parts* of entities in another? Can a *functional* or *causal* process in one field be associated with structures in another? Can phenomena in one field be seen as the cause of those in another? Thus, interrelations between two different bodies of knowledge might be of the following types: identity, part-whole, structure-function, or causal. Other cases are likely to give examples of additional types. As Chapter 2 discussed, however, some sort of analogical relation is not called an "interrelation," in the technical sense in which I use the term.

Postulation of such interrelations can then guide the formation of hypotheses; knowledge from one field can be used to formulate hypotheses in the other, and vice versa. Table 7-1 illustrates relations among properties of chromosomes and allelomorphs. Some similar properties were known before the formation of the interrelation and served to make postulating an interrelation plausible; these include Property 1, individuality, Property 2, pairing, and Property 3, one-half in germ cells (gametes). Such similar properties made plausible the postulation of an identity relation: perhaps allelomorphs were to be identified with chromosomes. Property 8, however, indicates that the number of chromosomes was small compared to the number of allelomorphs (or the number of characters on the basis of which Mendelians inferred allelomorph numbers). Thus, a more plausible interrelation was that allelomorphs were parts of chromosomes. That interrelation guided the formation of new hypotheses about both chromosomes and allelomorphs. Property 4, the mixing of maternal and paternal characters in offspring, led to the prediction that maternal and paternal chromosome sets would be mixed in the formation of germ cells. Conversely, Property 8 about the small chromosome number led to the prediction that groups of Mendelian characters would be linked in inheritance.

A fruitful area for additional research is how postulating a specific kind of interrelation can, in general, guide hypothesis formation. For example, if a functional entity is identified with a structural one, is there anything in general that would be expected to be the case? All the functions attributed to the functional entity might be expected to have corresponding properties in the structural one. Failure to find functional differences (e.g., dominance-recessiveness) mirrored in structural differences (e.g., homologous, autosomal chromosomes in hybrids appeared identical) may count as evidence against the postulated identity relation. As another example, consider aspects of part-whole relations generally. Can anything be said to be expected of wholes, based on properties of their parts, and vice versa? If a part has a certain chemical composition, then one knows that the whole also has that as a component of its chemical makeup. This generalization of the interrelations strategy is an active area of current research (Darden and Rada, 1988a; 1988b).

The strategy of interrelations is a strategy for producing new ideas, using knowledge from two different bodies of knowledge. It also serves as a strategy for resolving anomalies, when an anomaly in one field can be solved by considering aspects of the related phenomena in the other. Unmatched properties, not previously used to form new hypotheses, may be used at a later stage of theory development. In Chapters 9 and 10, additional mappings between properties of chromosomes and properties of Mendelian factors will be discussed. We will see that the strategy of using interrelations proved fruitful in theory development in genetics.

7.7 Conclusion

This chapter has discussed the proposal in 1903-1904 of the chromosome theory of Mendelian heredity, the claim that genes are parts of chromosomes. Its proposal was based on the numerous similar properties between Mendelian factors and chromosomes, but it was a controversial theory. It did not win acceptance from all Mendelians, some of whom, although committed to the idea of differences between the germ cells of hybrids, did not attribute those differences to different factors in different chromosomes. The development of the chromosome theory, an interfield theory, and the theory of the gene, an intrafield theory, had closely related, but somewhat separate histories. The theory of the gene, as stated by Morgan in 1926 (quoted in Chapter 4; Morgan, 1926, p. 25), did not refer to chromosomes; all its components were supported by evidence from Mendelian breeding experiments. Morgan and his students used the chromosome theory in developing the theory of the gene. Yet other Mendelians, who opposed the chromosome theory, formed competing hypotheses that did not appeal to chromosomes. The development of the chromosome theory, as opposed to the theoretical components of Mendelism generated to explain the results of breeding experiments, will thus be discussed separately in the following chapters. The ability to exploit interrelations between genes and chromosomes proved to be the key to the success of the Morgan group, in contrast to their critics (discussed in Chapters 9 and 10). The attempts to form interrelations between genetics and other fields will also be discussed in Chapters 9, 10, and 13. The strategy of forming interrelations between different, but closely related, bodies of knowledge, is a very powerful strategy for theory change.

CHAPTER 8

Tests of Segregation

8.1 Introduction

This chapter discusses challenges to the Mendelian theoretical components of segregation, also called "purity of the gametes." That hybrids form germ cells of two types, each pure with respect to one or the other of paired, alternative parental characters, was the central new claim of Mendelism in 1900. Testing the generality of segregation and forming hypotheses to account for exceptions were major activities of Mendelians, especially between 1900 and 1910, but continuing until about 1920. This chapter discusses two challenges to the purity hypothesis: Cuénot's 2:1 mice ratios and Castle's cases of modified characters that resulted from hybrid crosses in mammals, especially rats. By 1920, these exceptions were explained without modifying the central claim that pure Mendelian factors segregate. Confidence in the general applicability of segregation was increased by its having successfully faced these challenges.

Alternative hypotheses for anomalous 2:1 ratios, found by Cuénot, will be considered in Section 8.2. Three alternatives were proposed historically by Cuénot, Morgan, and Castle. By using the strategy that I devised to reproduce their three hypotheses, I found another hypothesis that could have accounted for the anomaly, but was not proposed by anyone at the time (so far as I have been able to find). Section 8.3 discusses that strategy of **delineate and alter**: delineating (possibly implicit) separable theoretical components as candidates for modification in the face of an anomaly and systematically altering them, either by **specialization** or by **proposing their opposites**.

Section 8.4 considers another challenge to segregation by Castle, who proposed that factors contaminated each other while they were together in the hybrid, resulting in modified characters. Such modified characters showed the effects of having had a hybrid ancestry. An alternative explanation of modified characters in terms of additional modifying factors is discussed, along with assessments of the relative merits of the two hypotheses. Finally, Section 8.5 incorporates the strategy discussed in Section 8.3 into a set of general steps for resolving anomalies. These include localization of plausible components to modify, strategies for producing new hypotheses at these various sites, and finally, strategies for theory assessment to choose among the competing alternatives. Figure 2-1 of the stages and strategies for theory change is once again refined in Figure 8-2, adding strategies for resolving anomalies.

8.2 Cuénot's 2:1 Ratios

The theoretical components at issue in this section are the ones for explaining 3:1 ratios, namely the segregation components. My characterization of them in Table 5-2 is repeated here, with the information about lack of dominance added as a qualification.

Component 5. Segregation

C5.1 In the formation of germ cells in a hybrid produced by crossing parents that differed in a single pair of characters, the parental characters segregate or separate, so that the germ cells are of one or other of the pure parental types. (Each germ cell has either *A* or *a* but not both; called "purity of the gametes.")

C5.2 The two different types of germ cells are formed in approximately equal numbers. (Equal numbers of germ cells with *A* and *a*.)

C5.3 When two hybrids are fertilized (or self-fertilization occurs), the differing types of germ cells combine randomly. Symbolically, $(A + a)(A + a) = AA + 2Aa + aa$; in appearance, 3:1 ratios, given complete dominance, or 1:2:1 ratios, if the hybrid differs in appearance from the parents.

Tests of segregation, Component 5, produced numerous cases that showed 3:1 or 1:2:1 ratios. This work thus confirmed the wide applicability of the segregation components of Mendelism in many hybrid crosses of many varieties of plants and animals (e.g., Bateson, 1909, ch. 2). Some crosses, however, resulted in exceptions to the 3:1 ratios. Such anomalous data posed a challenge to the segregation components.

Cuénot (1905) found an exception to the expected 3:1 ratios while doing Mendelian experiments with mice. On breeding yellow mice with those having other colors, yellow was found to be dominant; yet, when the hybrids were bred, the percentage of yellow in the F_2 generation was smaller than expected. The ratios were between 2:1 and 3:1, about 2.55 yellow. When Cuénot bred the yellows so produced, he was unable to obtain any homozygous yellows, that is, no pure dominants (no *AA*) were found. Castle and Little (1910) repeated Cuénot's experiment with a larger sample, confirmed the anomalous results of no pure dominants, and showed that the ratio was closer to 2:1.

It was agreed that the segregation components of Mendelism, listed here as Components 5.1, 5.2, and 5.3, were the site of the problem within the theory, but Morgan, Cuénot, and Castle proposed different hypotheses to account for this exception. Their alternative hypotheses can be seen as occupying a scale from a radical challenge of the general claim of purity to no change in the theoretical components (but an explaining away of the anomaly). Morgan's (1905; 1909b) challenge entailed the most fundamental modification, a denial of the purity of the germ cells in general (i.e., a denial of Component 5.1), with this exceptional case from mice providing the evidence for their impurity. Cuénot left the component (C5.1) of the purity of the germ cells intact, but proposed a modification to the claim that germ cells combined randomly in fertilization (Component 5.3). Instead, he suggested that selective fertilization (to be explained later) occurred in this case. In yet another alternative, Castle and Little explained away the exceptional data as an unusual case of inviability of some gametic combinations. Their hypothesis left the generality of Components 5.1 to 5.3 intact, but uncovered the implicit assumption that all zygotes (fertilized eggs) are equally viable. This section examines these three alternative hypotheses in more detail.

In order to explain his exceptional results, Cuénot proposed that not all germ cells combined randomly (Component 5.3); instead, he claimed that, sometimes, selective fertilization occurred. In his yellow mice case, the germ cells bearing factors for yellow selectively combined with germ cells bearing different factors, but not with each other. Symbolically, if germ cells with the factor for yellow are represented by A and non-yellow by a, $(A + a)(A + a) =$ no AA, only $2Aa:1aa$. Had it been accepted, the hypothesis of selective fertilization would have entailed a limiting of the generality of Component 5.3. Sometimes the germ cells combine randomly and produce 3:1 ratios and sometimes they do not and produce anomalous ratios. Thus, Cuénot's alternative could have been encompassed within Mendelism by specializing the universal claim of randomness and adding another possibility.

Lock presented Cuénot's explanation of the 2:1 ratios in terms of selective fertilization, in both the first and second editions of his textbook (Lock, 1906, p. 196; 1909, p. 209). He recognized that Cuénot's hypothesis would be an added component of Mendelism, not a change that threatened the entire theory:

> It must be recorded as a distinctly exceptional case, though not, be it noted, as an exception to Mendel's law. The gametes obey the law [i.e., segregate purely], as was shown by crossing yellow with non-yellow, and it is only in their manner of combination that a complication has been introduced. (Lock, 1906, p. 196)

Lock was somewhat reserved in his endorsement of selective fertilization, but he said it gained some support because selective fertilization was a hypothesis used by others in attempts to provide a Mendelian explanation of sex determination (Lock, 1906, pp. 256-257; 1909, p. 274). However, no explanation was proposed by the defenders of selective fertilization to account for the incompatibility of certain types of germ cells for each other, and, therefore, it was a controversial hypothesis.

Morgan (1909b) criticized Cuénot's hypothesis of selective fertilization as "a conception entirely foreign to the whole Mendelian scheme. There is no evidence of selective fertilization *in this sense* known elsewhere and it seems a very questionable advantage to introduce the factor into the Mendelian process" (Morgan, 1909b, p. 503). Morgan seems to be criticizing the hypothesis for being ad hoc and lacking in generality. Bateson and Punnett also criticized selective fertilization by countering that it would not even explain the exceptional ratios. Because more sperm are produced than eggs, there would be sufficient numbers of "non-yellow" sperm to fertilize all the "yellow" eggs. So, one would expect a 3:1 ratio of yellow to other color (i.e., $3Aa:1aa$), even though no pure yellows (AA) were among the proportion of yellows (Bateson, 1909, p. 119).

Morgan (1905; 1909b) formulated an alternative hypothesis. At the time, he was a critic of Mendelism and he interpreted Cuénot's exceptional results as evidence against the "assumed Mendelian purity of the germ cells" (Morgan, 1905, p. 877). Morgan, as others, found it puzzling that, for example, a hybrid yellow pea could give rise to a pure breeding green strain that would never show the effects of having been produced by a hybrid yellow. The idea was an old one that mongrels would show effects of their hybrid ancestry in later generations. Morgan was not alone in questioning the generality of the central claim of purity of the gametes; purity was a surprising finding and certainly not immune from various challenges in the early days of Mendelism (see, e.g., Spillman, 1902).

Morgan's alternative hypothesis can be localized as a *universal* denial of the purity of the gametes in Component 5.1, in contrast to Cuénot's selective fertilization, which was a *specialization* of the randomness in Component 5.3. Morgan proposed that Mendelian factors never segregated into pure parental forms (Component 5.1), but that hybrids always produced germ cells with both dominant and recessive factors present. In order to account for the usual instances of 3:1 ratios (which had been explained by assuming segregation of pure dominant and pure recessive factors), Morgan proposed that in half the germ cells the recessive was latent; in the other half, the dominant was latent. Symbolically, instead of A and a, Morgan proposed that germ cells of hybrids should be represented by $A(a)$ and $a(A)$, with latent factors indicated in parentheses. In future generations, Morgan predicted, the effect of the hybridization would be evident. No pure breeding dominant or recessive forms would emerge after hybridizations, he predicted. What appeared to be pure breeding forms contained the factor for the other grandparental character latently; for example, the supposedly pure breeding yellow peas (AA) and green peas (aa) in the F_2 generation in Mendel's experiments actually, on Morgan's view, were not pure breeding at all. At some later point, with continued self-fertilization, they would show the effect of their mongrel ancestry. The supposedly pure dominant yellow peas were to be symbolized by $A(a)A(a)$ and the green character would reappear in subsequent generations as the latent (a) somehow became active.

To explain Cuénot's exceptional yellow mice, Morgan suggested that Cuénot had found no pure breeding dominants because the latent recessives had appeared sooner than is usually the case. (Morgan did not suggest why the latents appeared sooner.) Cuénot found no pure breeding dominant forms because no pure breeding dominants (i.e., AA) were ever produced after hybrid crosses, according to Morgan; seemingly homozygous dominants in other cases really contained the recessives latently [i.e., $A(a)A(a)$]. Cuénot's anomaly, thus, from Morgan's perspective, brought into question the correctness of the central claim of the purity of the gametes.

Had Morgan's "latent" hypothesis become accepted, it is questionable whether anything called "Mendelism" would have survived. Segregation was labeled by Bateson the "essential discovery which Mendel made" (Bateson, 1909, p. 13). If the purity of the gametes, as Bateson also called segregation, had been shown to be incorrect, then it is unlikely that Mendel would have been credited with a discovery. Thus, Morgan's alternative hypothesis to account for Cuénot's 2:1 ratios was a much more radical change to the theory than Cuénot's hypothesis of selective fertilization. Morgan's hypothesis entailed the denial of a central claim of Mendelism, while Cuénot's was an added component that merely involved limiting the generality of a less central claim of Mendelism.

Morgan's alternative to purity of the germ cells gained some adherents (e.g., Davenport, 1907). When Morgan tested his prediction with other strains of mice, however, he did not find the predicted reappearance of dominants in the recessive strains. His conclusion: "It is evident that the hypothesis failed when tested and must therefore be abandoned" (Morgan, 1911d, p. 95).

Castle and Little (1910) proposed yet another hypothesis that did not entail the modification of any of the theoretical claims. They suggested that embryos formed by the mating of two germ cells with yellow factors were inviable. Given this hypothesis one would expect Cuénot's 2:1 ratio (AA died, $2Aa:1aa$). As evidence for their hypothesis, they appealed to the smaller numbers of young produced in these crosses and also pointed out that decreased viability had been found in other cases. They cited work by Baur on snapdragons which also showed that one class of zygotes did not develop. Thus, they argued that their hypothesis explained the evidence, as well as being a kind of explanation confirmed elsewhere, so it had some general applicability beyond this one case. By 1914 (e.g., Morgan, 1914), Castle's explanation of Cuénot's anomalous results had been accepted. Later, dissection of dead embryos confirmed the Castle and Little hypothesis beyond any doubt (Kirkham, 1919; Robertson, 1942).

Castle and Little's inviability hypothesis entailed the least change to Mendelism. No change was needed in the theoretical components. Consequently, the 2:1 ratio turned out to be an anomaly that did not require a modification in the theory in order to account for it. Instead, it was "explained away" as a malfunction, as an understood exception to the normal Mendelian process. It did serve, however, to make explicit an implicit, unstated assumption: all zygotes (fertilized eggs that develop into embryos) are equally viable. Symbolically, AA and Aa and aa forms all live. With unequal viability, 3:1 ratios do not result; however, even after the discovery of lethal gene combinations, the assumption of equal viability did not become an explicitly stated component of the theory. It remained as an implicit assumption that in normal segregation cases the zygotes resulting from random fertilization were equally viable, unless otherwise stated.

Other instances of homozygous lethal combinations of genes were discovered. Thus Cuénot's anomaly provides an example of what I will call a "monster anomaly," namely, an anomaly that is not resolved by modifying the theory; instead, the case is shown to be a monstrous instance in which the usual process specified by the theory fails in a known way. (I have taken the term "monster" from Lakatos's, 1976, delightfully named strategy of "monster-barring," as I will discuss in Chapter 15. "Monster" is a wonderful term for use for genetic anomalies, since monster-like mutants may result.) When monster anomalies are not unique, but are part of a class of such failures of the usual process (as are homozygous lethals), I will call the class a "malfunction class." In other words, malfunction classes are known kinds of anomalies that can be explained away without altering the components of a theory, a sort of expected monster. Morgan and Cuénot considered the anomaly more serious; their alternative hypotheses, in contrast to Castle and Little's, would have required modifying the theoretical components. Distinguishing monsters from more serious anomalies that require theoretical modification is important in resolving the anomaly. These and other kinds of anomalies will be discussed more fully in subsequent chapters.

As we have seen, in response to the 2:1 anomalous ratios, Morgan proposed changes in Component 5.1, Cuénot to Component 5.3, and Castle and Little uncovered an implicit assumption about viability. What about Component 5.2, the claim that the two different types of germ cells are formed in approximately equal numbers? As far as I have been able to find in the historical record, no one proposed an alternative hypothesis to this claim to account for Cuénot's 2:1 ratios. Still, an alternative could have been generated by specializing this general component to the claim that, in this case (and perhaps others), some gametes were not formed or were impotent in fertilization.

Figure 8-1 provides my depictions of the alternative hypotheses. Diagram (d) shows a Punnett square with unequal numbers of gametes of the two types forming in the two sexes. In one sex (either male or female), no gametes with A form; only germ cells with a participate in fertilization. In the other, twice as many with A are produced than with a. Such a pattern would result in $2Aa:1aa$ in the zygotes, as the diagram shows. Of course, questions would arise as to why A gametes were not viable in one sex but were produced in a double dose in the other.

My hypothesis of unequal numbers raises additional questions, but so did the historical ones. Morgan's hypothesis of impurity of gametes raised the question of why in the mouse case the recessives became active earlier than in other 3:1 cases. Morgan offered no explanation. Similarly, proponents of selective fertilization had no answer to why some gametes did not combine with others. In addition, Castle and Little's hypothesis of inviability raised the issue of why the yellow-yellow combination was lethal. Nonetheless, empirical evidence eventually favored the inviability hypothesis and resulted in its acceptance, even with its promissory note

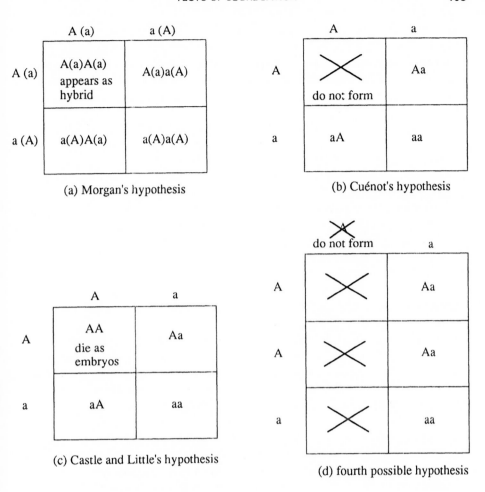

Fig. 8-1. Alternative hypotheses to explain 2:1 ratios.

for somehow explaining the cause of death. I suppose I have to concede that my hypothesis involves two different changes: something happens in one sex such that no *A* gametes are formed and something else happens in the other to give a double dose. If two independent (would they necessarily be independent?) variations would be less likely to occur than one, then my hypothesis of unequal numbers is less plausible than the three historical ones.

8.3 Strategy of **Delineate and Alter**

Each of the alternative hypotheses to explain the 2:1 ratios can be analyzed as providing an alternative to one of the theoretical components that I included as part of the segregation components. As discussed in Chapter 5, some of the theoretical components could be seen as needed in the theory in order to account for specific domain items. In Chapter 6, we saw how exceptions to the domain item of the appearance of dominance in the F_1 generation resulted in changes to the theoretical component claiming that dominance was universal. Similarly, the segregation components explained the 3:1 ratios. They were thus the components most "directly" connected to the domain item of 3:1 ratios (or 1:2:1, without dominance) in the F_2 generation. When Cuénot found exceptional 2:1 ratios, the most plausible localization of the difficulty with the theory was in the segregation components.

My analysis of the segregation components for the period 1900-1910 states them as three separable claims: (5.1) separation of factors into different types of pure germ cells, i.e., segregation; (5.2) equal numbers of the types of germ cells; and (5.3) random combination of the types of germ cells at fertilization. These segregation components functioned importantly in the explanation of the 3:1 ratios. When an exception was found to the expected 3:1 ratios, then the most plausible localization of the difficulty was in these segregation components. Plausibility, of course, is not necessity: other explicit or implicit assumptions might be implicated because they played a role in explaining the 3:1 ratios. But a first step is to consider the explicitly stated components that "directly" accounted for the ratios.

The different attempts to explain the anomaly show that the scientists can be seen, within the framework of my analysis, to have focused on different, separable, theoretical components associated with segregation: Morgan with the purity component (5.1) and Cuénot with the random component (5.3). Castle, on the other hand, uncovered an implicit assumption (equal viability) that, so far as I have been able to find, had not been explicitly recognized as an additional theoretical component necessary for the other components to explain 3:1 ratios. These alternative modifications show the usefulness of delineating the separable components and uncovering implicit ones, so that alternative modifications can be considered in the different components if an anomaly arises.

Historically, I suspect that the scientists came to understand the separateness of the assumptions as a result of the attempts to account for anomalies. As I warned in Chapter 5, I presented the components as of 1900-1903 a bit more clearly than I believe any one biologist stated them at the time. Purity, randomness, and equal numbers were among my explicitly stated components because one or another of the early Mendelians discussed them. Equal viability, however, was not included because I found no mention of it until this anomaly arose. Anomalies, I believe, play an important role in forcing a careful examination of previously implicit assumptions about segregation during the consideration of what could be going wrong.

By using the strategy of carefully delineating separable assumptions and systematically considering alternatives to each, I was able to generate a fourth alternative hypothesis. I systematically considered whether any of the separable assumptions I had delineated had not been used in hypothesis formation and if any could be. Equal numbers of different types of gametes (Component 5.2) was not a component explicitly stated and altered to account for Cuénot's 2:1 ratios, so far as I have found in the historical record.

From this case, we can extract the lesson that more careful explicit delineation of the separable components of a theory is a useful first step in trying to localize difficulties. I use the term "separable," rather than "independent," because the assumptions are not all logically independent. Segregation is a temporal process and the separable components refer to different steps or features of that process. If an earlier step is altered, for example, unequal numbers of types of gametes are formed, then later steps are affected. The systematic interconnections among the theoretical components versus their modularity become important issues during anomaly resolution, a point discussed in Chapter 5 and to be returned to in Chapter 12. Changes in one segregation component may or may not affect the others, depending on the nature of the connection of that component with others and the nature of the change made to it.

The 2:1 anomaly not only guided the localization of the problem in the segregation components, it also helped to guide the nature of the changes. The type of anomaly to be accounted for provided a constraint—not just any change would produce ratios of no homozygous dominants to 2 parts heterozygotes to 1 part homozygous recessives. If the ratios had not been 2:1, but had been, for example, 1 pure dominant to 1 pure recessive with no heterozygotes, then the hypotheses to explain 1:1 would have been different from those for 2:1. Trying to determine what went wrong helped to suggest certain kinds of modifications to the separable components. In constructing my hypothesis about unequal numbers, I had to juggle the proportions so that more germ cells of A type were formed than a type in the mating partner, in order to produce $2Aa:1aa$. So, the nature of the anomaly constrained (or more positively, helped direct the search for) a hypothesis that could account for it. Nevertheless, the anomalous 2:1 ratio alone was not sufficient for generating the alternative changes.

The various ways the changes were made can be seen as exemplifying the following method: take one of the separable, theoretical components about segregation, consider the key idea in that component, and propose its opposite. The first step in producing a new hypothesis is finding a seminal idea that can be built into a plausible hypothesis. When a theoretical component is localized as a site for change because of an anomaly, then one way of getting a seminal idea with which to start hypothesis formation is to **propose the opposite** of the key idea in that component. The four alternative hypotheses can be analyzed as built on the following oppositions: purity-impurity (Morgan's), random combination-selective fertilization (Cuénot's), equal viability-unequal viability (Castle and Little's), and equal numbers-unequal numbers (mine). A seminal idea is just the beginning point. The strategy of propose the opposite thus is just a strategy for getting a vague type of hypothesis that will have to be filled in with more detail in order to generate a plausible, testable hypothesis.

A strategy that can be of use in moving from a seminal idea to a plausible hypothesis is the strategy of **specialization**, discussed in Chapter 6. When a hypothesis to account for an anomaly is constructed, the decision must be made whether it is a universal, or a more limited, change to the component to be altered. Specialization is a way of making a more limited change. The strategy of **specialization** can be seen as exemplified in the generation of Cuénot's and Castle and Little's hypotheses. Cuénot specialized the opposite of the random fertilization component: "all combine randomly" became "some do not combine randomly." The random fertilization component was specialized to limit its generality, with selectivity added as another theoretical component that applied to some cases; in other, nonanomalous cases, the random component still applied. Cuénot's change can thus be seen as another example of the **specialize and add** strategy discussed in Chapter 6: the generalization that "*all* combine randomly in fertilization" was specialized to "only *some* combine randomly in fertilization." Then, another component was added: "*the rest* show selective fertilization" (selective is an opposite of randomness). Castle also specialized a previously implicit assumption and added an additional possibility: "all equally viable" to "most equally viable and a few not equally viable."

Morgan's hypothesis was not a *specialization* of one claim; it was a *universal* denial of purity. The range of the responses shows differences in how seriously each scientist took the anomaly. Morgan saw it as evidence for a universal denial of the purity of the gametes in all purported Mendelian crosses. He was a critic of Mendelism in general at the time, and that may help explain why he proposed such a radical change. Cuénot and Castle, on the other hand, probably wanted to preserve most of the Mendelian claims and thus they generated less radical alternatives. The strategy of **specialization** is a strategy to employ if a goal is to leave much intact; the strategy of **proposing the universality of the opposite** of a component is a strategy to use if the goal is to make a more major change in the theory.

In constructing my alternative hypothesis of unequal numbers of types of gametes, I found it very useful to use the symbolic notation of A and a. I manipulated it to see what gametic types and proportions were necessary to produce 2:1 ratios. A diagram of that symbolic representation is presented in Figure 8-1(d), along with my depictions of the other hypotheses using those symbols. Although I have no evidence that symbolic representations and their manipulations figured in the reasoning to hypotheses historically, I can imagine how they might have. Certainly the symbolic representations were used in books and papers to illustrate the theoretical ideas. Explaining Mendelian genetics without using symbolic representations for types of germ cells is difficult, although it is possible to do so. I will discuss geneticists' use of symbolic notations in Chapter 11 as clues for understanding their claims about genes and for ferreting out implicit assumptions. Chapter 12 also introduces various uses of a structural, diagrammatic representation of the theory of the gene.

From this analysis of my own method of hypothesis formation, another strategy for producing new ideas can be suggested—**introduce and manipulate a symbolic representation** of the theoretical entities and processes. This strategy for producing new ideas was introduced in Chapter 3 in considering hypothetical steps that could have produced Mendel's law. It was suggested that 3:1 ratios might call to mind their representation in the mathematical formula $(A + a)(A + a)$. Using the same symbolic notation introduced in original theory construction and manipulating it in various ways provided a means of constructing my alternative hypothesis of unequal numbers of A and a gametes. Thus, **manipulating a symbolic notation** associated with the theory is another strategy of potential use in doing anomaly resolution.

Historically, *different scientists* proposed the different alternative hypotheses to account for the anomaly. I, as yet another investigator, managed to construct another one. I used the strategies of delineating separable claims, systematically considering alternatives to each by asking what its opposite would be, and deciding if the alternative should be a universal denial of the component or only a specialization. However, nothing about the nature of the reasoning in using the strategy (**delineate and systematically alter**) requires that *different people* construct the various alternative hypotheses. *One person* could delineate all the separable assumptions and systematically consider alternatives to each. Hull (1988) suggested a model for scientific change that made different research groups the sources for variant hypotheses; that is certainly one possibility and fits the facts in this historical case. There is no necessity, however, of assuming that a single scientist can generate only one alternative hypothesis. Sometimes, in fact, papers propose several alternative hypotheses to account for an anomaly, and, then, discuss experiments that aided in choosing among them (e.g., Bateson, 1903). The strategy of **delineate and alter** is a method that a single scientist (or a single research group) could use to be more systematic in generating alternatives.

In each of the alternative hypotheses, it was predicted that other cases would be found that showed the same anomalous behavior. Morgan predicted general impurity; selective fertilization was thought to occur in other cases. Neither of these predictions was confirmed during this period. Castle and Little's hypothesis of homozygous lethals, on the other hand, did turn out to be a class of anomalies that distorted 3:1 ratios. Each biologist thus claimed some applicability of his hypothesis beyond the one anomaly that it was created to explain.

Philosophers sometimes advocate a conservative strategy be taken when an anomaly or a new belief challenges currently accepted views. When faced with an anomaly, they suggest making a change that has the least effect on current views. Proceed to question more central claims only if the slight modifications fail (e.g., Lakatos, 1970; Shapere, 1980, p. 79; Harman, 1986). Kuhn (1970), on the other hand, might be interpreted as claiming that the response to a number of anomalies, as opposed to mere puzzles, is to change one's entire belief structure. In so far as this one example of an anomaly in Mendelism can be interpreted as being relevant to the issue of choosing a general strategy, this particular case of the 2:1 ratios does not suggest proposing only one change, whether most conservative or most radical. Instead, it suggests devising a set of plausible candidates. In this case a range of hypotheses was proposed, from a radical denial of a central claim (Morgan), to a specialization of a less central component (Cuénot), to no modification in the theoretical components as stated, but the discovery of a quirky class of instances (Castle). All three hypotheses were proposed and tested almost simultaneously. My fourth alternative could have been as well. The strategy of **delineate and alter,** thus, suggests generating a set of plausible modifications, regardless of how peripheral or fundamental the change would be. Decisions about which to test depend on other considerations; strategies for devising experiments are outside the scope of this analysis.

In summary, from this case we get the general strategy for anomaly resolution that I call **delineate and alter.** (A less refined version of the **delineate and alter** strategy was discussed in Darden, 1982b.) It may be characterized as having the following steps. When faced with an anomaly, verify that the anomalous data is correct. Localize the components of the theory involved in explaining the domain item for which the anomaly is an exception. Delineate the separate components of the potentially faulty module and consider successively altering each one to produce alternative hypotheses. Strategies for making the changes will vary depending on the nature of the anomaly and the nature of the components themselves, but the delineation step will focus the task. Three strategies for producing new hypotheses by altering old ones can be seen as exemplified in this case. First, **propose the opposite.** Second, **specialize** one component and **add** a closely related component to account for the anomaly (the strategy of **specialize and add** from Chapter 6). Third, **uncover an implicit assumption** and show how it is violated in the anomalous case, and, perhaps, explain away the anomaly as an instance of a malfunction class without actually altering any general component of the theory. These three strategies could be coupled with the strategy of **manipulating a symbolic representation** of the normal process to see what step(s) could be failing and how, as we will discuss in more detail in Chapter 12.

After the alternative hypotheses were generated, they were assessed using various strategies for theory assessment. Eventually Castle and Little's hypothesis of inviability was confirmed as the correct explanation of the 2:1 ratios. The final step in resolving an anomaly is to assess the various alternative hypotheses and choose the best, or else suspend judgment, if no single best one emerges at a given point in time. (Strategies for theory assessment will be discussed in the final section of this chapter.)

Mendelism's survival of the challenge to segregation posed by Cuénot's 2:1 ratios served to strengthen the evidence in its favor. No claim had been immune from testing, but the components of the separation of pure factors and random fertilization had survived the tests. Using Popper's (1965) language, the theory had been corroborated as a result of surviving severe tests and not being falsified.

8.4 Castle and Contamination

The alternative hypotheses to be considered in this section were formulated primarily as a debate about the purity of the gametes, namely, Component 5.1:

Component 5. Segregation
C5.1 In the formation of germ cells in a hybrid produced by crossing parents that differed in a single pair of characters, the parental characters segregate or separate, so that the germ cells are of one or other of the pure parental types. (Each germ cell has either *A* or *a* but not both; called "purity of the gametes.")

Also at issue in the hypotheses we are about to discuss was the claim about the nature of unit-characters and the multiple factor hypothesis, whose development was discussed in Chapter 6. In order to have them before us in this discussion, the versions from Table 5-2 for 1900-1903 and the newly refined version as of 1910 (which I have labeled C1.2a' and C1.2b') are restated here:

> As of 1900-1903:
> Component 1. Unit-characters
> C1.1 An organism is to be viewed as composed of separable unit-characters.
>
> As of about 1910:
> Component 1'. Factors and characters
> C1.1' Characters are produced by factors.
> C1.2a' One factor may produce one character or
> C1.2b' Multiple factors may interact in the production of one character.

As we saw in Section 8.2, Morgan proposed and retreated from an early challenge to the claim of purity of the gametes. The most serious challenge to that claim (Component 5.1) came from Castle and the results of some of his breeding experiments with mammals. Castle found characters in mammals that showed dominance, in the F_1 hybrids; for example, long coat length dominated over short coat length in guinea pigs. Castle argued that such characters "mendelized." Yet, after breeding the hybrids he found a deviation from the expected 3:1 ratios. More recessives were produced than expected. Even more importantly, the recessive characters were not pure; that is, the long haired coat was not as long as the grandparental coat had been. The character thus showed effects of having been produced by a hybrid cross; a greater range of variability in coat length was present after the hybrid cross than before. As discussed earlier, it is indeed puzzling that a recessive, grandparental character reappears in the F_2 in pure form (i.e., *aa*) and never again shows any evidence of having been through a hybrid cross. Castle and his associates believed that they had evidence that such purity did not exist for a number of characters in a number of species (e.g., Castle and Forbes, 1906; Castle and Phillips, 1914).

In order to explain the anomalous results, Castle proposed that some sort of modification or contamination occurred during the hybrid generation. Consequently, when the gametes formed in the hybrid, they were not pure. He referred to this as "impurity of the gametes" (Castle and Forbes, 1906, p. 5). Furthermore, Castle believed that, by selecting modified characters and continuing to interbreed the animals having the characters with the most modification, the character would be further modified because of continued contamination. He and his associates carried out an extensive series of experiments with pigmented rats that at first seemed to confirm his expectations (Castle and Phillips, 1914). Castle's interest in practical breeding and in evolution made the question of the role of selection in producing new races a particular interest of his (Dunn, 1965b; Taylor, 1983).

Putting this work into the framework of the analysis here, Castle can be seen as having empirical evidence that provided exceptions to the original empirical generalization of 3:1 ratios: most importantly, he believed he had found exceptions to the claim that pure breeding forms reemerged after hybridization. Castle viewed this evidence as providing a challenge to a part of Component 5.1, namely, the purity of the gametes. Again we see the localization of an anomaly to a domain item in the theoretical component that directly accounts for that item: exceptions to 3:1 were localized in one of the segregation components.

Nevertheless, although the evidence pointed to lack of *pure* unit-characters reemerging from the hybrid, the evidence did not, in Castle's view, challenge the claim that segregation occurred, in the sense of *separation* of factors during germ cell formation. From Castle's viewpoint, what segregated were *contaminated* factors, not *pure* parental factors. He called them "determiners" of "mendelizing characters" and objected to calling them "pure genes" (Castle and Phillips, 1914, pp. 5-6). Castle's response to this anomaly shows that Component 5.1 actually contained two separable assumptions: one asserts that parental factors are not modified by being together in the hybrid (to be labeled C5.1.1'); another makes the separable claim that in the formation of germ cells in the hybrid, allelomorphic factors segregate so that the germ cells have one or the other but not both (to be labeled C5.1.2'). Castle denied the first claim about lack of modification, but still asserted the claim of paired determiners segregating. Again, we see an anomaly forcing the refinement of a single theoretical claim into two, previously implicit, separable ones: one could be asserted while denying (or limiting the generality of) the second.

Bateson, for example, stressed that purity of the gametes was among the most important of the new conceptions of Mendelism. Castle, on the other hand, argued that it was not "a necessary part of mendelism, not even an original part; but it is very important for us to know whether it is true or not." His experiments, he continued, were designed to test the "validity of the theory of pure gametes" (Castle and Phillips, 1914, p. 5).

Castle did not address the question of whether all Mendelian crosses resulted in contaminated factors, even the ones that seemed to show 3:1 ratios with pure recessives emerging from the cross. He was focused on his own data. His hypothesis of modified factors was designed to explain his own cases. Thus, it is unclear whether Castle proposed a *universal* impurity of the gametes or whether he was merely proposing a *specialization* of the general claim of purity to allow some cases in which modification occurred. Taylor (1983) argued that the lack of concern for explaining data other than his own was a failing in Castle's methodology. (Although, it is worth noting, that methodology succeeded with the inviable yellow embryo case.)

Another hypothesis, also proposed to account for Castle's anomalous data, did show concern for other Mendelian cases. As discussed in Chapter 6 (Section 6.2), in 1910, Castle's colleague at Harvard, East, accounted for some characters in corn by proposing that multiple factors interacted to produce a character. Castle discussed the possibility that multiple, modifying factors could account for the modifications observed in the coat pattern of rats, but, he claimed, the multiple factor hypothesis was insufficient to account for all the variability (Castle and Phillips, 1914). Members of the Morgan group, Muller (1914) and Sturtevant (1918), disagreed with Castle's assessment of the multiple factor hypothesis. They provided arguments in favor of the hypothesis of relatively stable, noncontaminating factors with additional modifying factors to account for the type of variability that Castle found. Castle's crosses, in their view, did not result in contaminated factors; instead, the "impure" characters were caused by the interactions of several pure factors.

From the perspective of my analysis of the components of Mendelism, the modifying factor hypothesis can be characterized as localized in Component 1. A modification of that component, namely, the multiple factor hypothesis, had already been proposed to account for another anomaly (as discussed in Chapter 6; the modifications to Component 1 were stated at the beginning of this section). Invoking the multiple factor hypothesis to account for Castle's modified character data thus used an existing hypothesis to account for yet another anomaly. The Morgan group's localization of the explanation for the anomaly in a modified version of Component 1, namely, in the (C1.2b') multiple factor component, provides an exception to the localization strategy discussed thus far. The strategy discussed in previous cases suggested focusing on the theoretical components that "directly" accounted for the anomalous domain item. That strategy worked for the locus of Castle's change: lack of pure recessives focused on the assumption of purity of the gametes. In contrast, the Morgan group's hypothesis of multiple, modifying factors was not an alternative to the segregation components; it was an alternative to Component 1, the unit-character concept.

Why was this an alternative locus for modification? The answer is that the unit-character concept was what might be called a more general or basic (I'm not sure of the best terminology) theoretical component. The unit-character component was assumed in the statement of the segregation components. The one unit-one character concept can be seen as playing a role in Castle's alternative hypothesis that a modified character implies a modified factor. From Castle's perspective, *one* modified character indicated the presence of *one* modified unit.

A method for localizing other components of the theory that can be questioned when an anomaly arises is to consider if more "general" or "basic" theoretical components figure in the explanation. They, too, become potential sites in the theory to be modified. Again, the modularity of the components is limited by their interconnections. On the other hand, we again see that not all the components were sites for modification; only two different hypotheses were proposed.

For the changes to the one unit-one character concept, no new alternative had to be constructed. An already existing hypothesis, used to account for another anomaly, was invoked. It thus already had some evidence in its favor. Nevertheless, debate raged between Castle and the Morgan group on the issue of whether the modifying factor hypothesis had sufficient evidence in its favor to be an adequate explanation for the modified characters that Castle had found.

Castle explicitly discussed the problem of drawing conclusions about invisible factors on the basis of data about visible characters: "Since the supposed 'factors' of inheritance are invisible, we can not hope to deal with them directly by experiment, but only indirectly" (Castle, 1916b, p. 95). He explicitly assessed the evidence in favor of multiple factors and found it wanting:

> . . . modifying genes. In some cases they are known to have other functions also. Thus the gene proper of one character may function also as a modifying gene for another character. But in the majority of cases the only ground for hypothecating the existence of modifying genes is the fact that characters are visibly modified.

> As an alternative to the theory of modifying genes, the theory has been considered that genes may themselves be variable and if so, genes purely modifying in function might be dispensed with. (Castle, 1919d, p. 127)

Thus Castle assumed a tight connection between the nature of the character and the unit producing it. (See Carlson, 1966, ch. 4, for further discussion of what Carlson called the "unit-character fallacy.") It was a reasonable methodology: infer the nature of the factors on the basis of the properties of the visible characters. As Taylor (1983) pointed out in discussing Castle's reasoning, the data did show modified characters. Two explanations existed: the genes themselves were modified or additional modifying genes could be postulated. Taking into account only Castle's data, Castle's contamination hypothesis was arguably the simpler hypothesis of the two in the sense of the number of genes postulated. (See Carlson, 1966, ch. 4, on Castle's use of simplicity.) Additionally, Castle may be seen as arguing against the ad hocness of postulating additional modifying genes if no independent evidence for their existence was found, other than the evidence of the existence a modified character.

Muller, however, was concerned about the generality of the conclusion about gene contamination for cases other than Castle's rats. Muller argued that numerous cases from numerous crosses in numerous varieties (especially *Drosophila*) were known in which the characters showed no evidence of modification by hybridization. More specifically, Muller argued that the multiple factor hypothesis was to be preferred because it was consistent with previous work and did not involve the "radical" denial of the conclusion to which "all our evidence points," namely, "the conclusion that the vast majority of genes are extremely constant," and change only by occasional mutation (Muller, 1914, pp. 61-62). Muller seems to have been considering a wider domain than Castle. Thus, he too could invoke the charge of ad hocness: Castle's claim of contaminating genes was an explanation only for Castle's cases, but was not supported by much of the other evidence in the domain of Mendelism. The modifying factor hypothesis, Muller claimed, was to be preferred as being more generally applicable to the entire domain of Mendelism.

Morgan and his students did much to develop the chromosome theory of Mendelian heredity, as Chapters 9 and 10 discuss. Their chromosome theory located the genes along chromosomes. Castle paid little attention to the cytological aspects of his breeding experiments and questioned the assumption of the linear arrangement of genes along chromosomes (see Chapter 10). In the light of the Morgan group's assumptions, Muller thus criticized Castle's assumption that allelomorphic genes for the same character would be likely to affect each other:

> To thus suppose that independent genes *fuse* or induce changes in one another, merely because they happen to produce similar *end effects* upon the organism, and in spite of the fact that they usually lie in different chromosomes and are apt to differ from each other as much as do other genes, is utterly teleological. (Muller, 1914, pp. 67-68)

Castle (1914) responded to Muller's critiques. He reasserted that the characters he had studied were found to be variable. Furthermore, he had a student investigate various characters in stocks of *Drosophila* that Morgan had sent him. In those, as well, variable characters were found. Castle reiterated his claim that variable characters were produced by variable genes. He denied that the evidence was adequate to support a "fundamental principle" that "Mendelian factors are constant!" (Castle, 1914, p. 40).

Sturtevant (1918) undertook extensive tests in *Drosophila* of Castle's view that characters could be shown to vary because contaminated genes could be selected to increase variability through successive crosses. Thus, Sturtevant tested Castle's claim of contamination with *Drosophila* crosses that showed increased variability in the F_2 of the same kind as in Castle's mammal cases, but the results did not support Castle's conclusions. Instead, Sturtevant was able to provide evidence for the presence of multiple, modifying genes. He used methods developed by the Morgan group to localize the relative positions of genes in linear linkage maps (which will be discussed in the next two chapters). Sturtevant criticized Castle's contamination hypothesis: "The theoretical conclusions reached by Castle are not in agreement with those arrived at by various other investigators, including the author, although for the most part the data obtained are very similar" (Sturtevant, 1918, p. 47). He summarized his conclusions: "The conclusions are drawn that selection is usually effective only in isolating genetic differences already present; and that genes are relatively stable, not being contaminated in heterozygotes, and mutating only very rarely" (Sturtevant, 1918, p. 3).

In 1919 Castle, acting on the suggestion of his student Wright, designed a crucial experiment to choose between the hypothesis of contamination (and increased modification by selection) and the hypothesis of multiple, modifying genes. Because his organisms did not lend themselves to use of the Morgan group's methods, Castle did not map the genes for coat color in rats in linear linkage groups associated with specific chromosomes. Instead, Castle and Wright designed a breeding experiment to test the contamination hypothesis, which Wright viewed less favorably than Castle (Provine, 1986, pp. 70-72). They crossed a wild strain of rats with rats having the purportedly contaminated factors. If the multiple factor hypothesis was correct, they predicted, then some offspring should show the original, unmodified character. The outcrossing would separate the original factor from the ones modifying it in some cases, and the original factor would be again functioning in an unmodified way. If the contamination hypothesis was correct, however, then no organisms with the original, unmodified character should appear because no pure factors remained. All would have been contaminated in the previous hybrid crosses.

The original pigmented coat color reappeared. The experiment confirmed the hypothesis of multiple, modifying genes. Unmodified genes were extracted by outcrossing. Castle claimed that a "crucial experiment" had been performed and that his earlier view of contamination (and its increase by selection) was wrong. In 1919, he published a paper entitled "Piebald Rats and Selection, A Correction," in which he admitted that his previous hypothesis was wrong and that the multiple, modifying factor hypothesis was confirmed for his rat case (see Carlson, 1966, ch. 5, and Provine, 1986, pp. 48-50, 69-73, for more discussion of Castle's hooded rat experiments).

(This experiment was decisive in showing that unmodified characters, and thus unmodified genes, could be extracted from the purportedly modified strains. It is not clear to me, however, that this would have been a crucial experiment if the outcome had been the opposite. Crucial experiments decisively rule in favor of one hypothesis and against the other. Would a failure to find unmodified characters have confirmed Castle's hypothesis of contamination? It seems to me the proponents of multiple factors could have said that the modifying factors just hadn't been sorted out by the outcrossing.)

Provine said of Castle's style and his student Wright's reaction to it:

> Castle's basic attitude was that science progressed most rapidly by the publication of bold hypotheses that all could examine with the greatest critical acumen. A bold hypothesis refuted by later research was not a failure but an advance in knowledge. Wright, with his quiet self-assurance, did not fit well with Castle's style—he preferred instead to get the hypothesis right the first time and to dispense with the retractions. The spectacle of Castle publicly retracting his hypotheses was not a model Wright would follow. Given his own personal style and his reaction to Castle, it will be no surprise to learn that Wright sometimes did not make clear where or when he had changed his mind. (Provine, 1986, p. 38).

1. Reproduce anomalous data
    a. Repeat experiment
    b. Find relevantly similar situation

2. Localize potentially problematic components
    a. "Direct" strategy: find components that directly account for anomalous domain item (or its normal, nonanomalous counterpart)
    b. "Explanation" strategy: trace all the components involved in the explanation of the anomalous item (or its normal, nonanomalous counterpart)
    c. "Implicit" strategy: uncover implicit assumptions in the explanation of the anomalous item (or its normal, nonanomalous counterpart)

3. Generate alternative hypotheses to account for anomaly
    a. Use strategies for producing new ideas, with these modifications:
        i. Add the constraint of preserving unproblematic theoretical components
        ii. Consider whether any other modified component can be used in explaining the anomaly
        iii. Systematically alter all the localized, potentially problematic components
        iv. Consider the nature of each old component when generating its replacement and consider how to alter it
    b. Examples of strategies for generating alternative components
        i. Propose opposite
        ii. Specialize
        iii. Introduce and manipulate symbolism

4. Assess among the alternative hypotheses
    a. Use the usual strategies for theory assessment
    b. Devise a crucial experiment to choose among two or more alternative hypotheses

5. Evaluate the nature of the hypothesis with evidence in its favor; several possible outcomes:
    a. The anomaly was "explained away" as a quirky case or a malfunction class (i.e., it was barred as a monster)
    b. The anomaly was explained using an existing theoretical component
    c. The anomaly resulted in a change to one or more theoretical components
    d. The anomaly was quite severe and the entire theory had to be discarded

Table 8-1. Steps in anomaly resolution.

Empirical evidence produced by a crucial experiment settled the dispute about contaminating factors. Thus, the "purity of the gametes" once again withstood challenges. This debate showed that two separable assumptions were contained in what had been lumped together as segregation, namely purity (lack of contamination) and separation of factors in germ cell formation. In addition, the debate served to strengthen the evidence in favor of the hypothesis of multiple factors and the evidence in favor of the constancy of Mendelian factors. Finally, as Carlson (1966, ch. 5) stressed in his analysis, the resolution of this debate provided additional support for severing the one unit-one character assumption. Distinctions between what Johannsen (1909) called the "phenotype," the visible characters of an organism, and the "genotype," the underlying genes, were becoming clearer. Means for getting independent evidence about the nature, number, and locations of genes, other than direct inference from the character that the gene produced, were being developed, as we will discuss more fully in subsequent chapters.

8.5 Strategies for Resolving Anomalies

The resolution of the Cuénot and Castle anomalies illustrates a series of steps in anomaly resolution. This section will discuss five steps in anomaly resolution; they are summarized in

Table 8-1. First, the anomalous data must be verified to be anomalous. That verification was easy in the Cuénot mouse case, but proved much more difficult in the Castle contamination case. After verifying the correctness of the data, the second step is localization. The problems for the theory posed by the anomalies were localized in one or more theoretical components of the theory. In both cases, several components became the sites for generating alternative hypotheses. Third, alternative hypotheses are generated to account for the anomaly. Fourth, the various alternatives are assessed using various strategies of theory assessment, including, but not limited to, empirical testing. Finally, one hypothesis is confirmed as the correct means for resolving the anomaly. In these two cases, no change in the theoretical components of Mendelism resulted, but several components were refined, with the uncovering of implicit separable assumptions.

8.5.1 Reproducing Anomalous Data

The first steps in both the Cuénot yellow mice case and the Castle contamination case were attempts either to replicate the exact experiment or to find relevantly similar cases in other species. It was important to be certain that the anomalous results were accurate. Cuénot's experiments and data were easier to replicate than Castle's cases of modified characters. Cuénot's case only involved one character in one species, yellow versus non-yellow coat color in mice. Castle and Little easily repeated Cuénot's experiments, confirmed that no pure dominants were produced, and refined the data to show that the ratios were approximately 2:1 (rather than Cuénot's average of 2.55:1).

In Castle's contamination cases, testing by others was much more difficult. Numerous characters through many generations in hundreds of animals in numerous species were investigated by Castle and his colleagues. Finding relevantly similar modified characters in other species, for instance, in *Drosophila*, left open the possibility that they were not adequate replications of Castle's mammal experiments. Thus, verifying the correctness of the contamination data was more difficult. Eventually, however, Castle showed that characters only *appeared* to be fundamentally modified; in fact, unmodified characters, namely, pure grandparental characters, could be extracted after hybrid crosses. Other responses to his anomaly were debated for several years before its final resolution.

8.5.2 Localization of Anomaly

After the anomalies were accepted as sufficiently serious to require new hypotheses to account for them, specific theoretical components of Mendelism became sites for modification. Mendelism, as a whole, was not considered to be threatened by the anomalies, with the possible exception of the hypothesis Morgan proposed to account for Cuénot's results. Problems were *localized* in specific theoretical components and various alternatives to these specific components were formulated as candidate hypotheses for becoming modified components.

The dominance case in Chapter 6 and the hypotheses discussed for the Cuénot case in this chapter both exemplify a simple strategy for localization of the components in the face of an anomaly. Consider which theoretical component(s) can "directly" account for the domain item for which an anomaly has now been found. A means of detecting direct connections is to remove the normal domain item and then consider which theoretical components would be unnecessary. For example, if 3:1 ratios were not part of the domain, which components would be unnecessary? Hypotheses to account for anomalies to dominance were localized in the dominance component, which directly accounted for dominance phenomena (or at least promised to do so). Similarly, Cuénot's 2:1 ratio anomaly focused the search for alternatives in the segregation components because they directly accounted for 3:1 ratios. Castle's modified character anomaly can also be interpreted as exemplifying the direct localization strategy. Castle's inability to extract *pure* recessives can be interpreted as an anomaly for the domain item of 3 dominants (both *AA* and *Aa*) and 1 *pure* recessive (*aa*). Thus, he localized the difficulty posed by modified characters in the segregation component (5.1) that asserted *purity* of the gametes as a means of explaining 3:1 ratios.

However, the alternative hypothesis of multiple, modifying factors to account for Castle's seemingly contaminated characters was not an alternative to the segregation components. The location of the problem in the unit-character component cannot be analyzed as exemplifying the simple direct localization strategy found in the other cases. Not all components of the theory as they were stated in Chapter 5 were directly connected to domain items. Some components, such as Component 1, the unit-character concept, were general or basic or prior assumptions. They were used in formulating the other components, which were more directly associated with the domain items. To the extent that those components were used in a given explanation, they also became candidates for modification if an anomaly arose. Such general assumptions might be called into question in the light of an anomaly.

Yet more problems arise for the direct localization strategy. Not only does the modifying factor hypothesis indicate that the direct localization strategy fails, Castle and Little's hypothesis of lethals shows that laying out all the possible sites for modification may entail uncovering previously implicit assumptions. Their hypothesis was not even the denial of a more general theoretical component that I stated for the theory in 1900-1903, such as the unit-character concept. Instead, as a first step in generating their alternative hypothesis, they had to uncover the *implicit*, unstated component of equal viability.

It may be a difficult task to uncover all the theoretical components used in explaining data, especially if they were implicitly assumed. It may also take effort to come to understand all the systematic connections between theoretical components; however, some modularity did exist in this case. Not all theoretical components were candidates for change in the face of each anomaly. In this historical case, geneticists managed to localize problems at a few plausible locations. Although localization may not have been easy, neither was it impossible. Scientists did not throw up their hands and retreat to saying that the entire Mendelian theory was falsified by either one of these anomalies.

Glymour pointed out similar cases in physics that he examined: "Scientists often claim that an experiment or observation tests certain hypotheses within a complex theory but not others" (Glymour, 1980, p. 133). He suggested drawing diagrams that illustrate the skeleton of the argument for explaining a particular domain item. Then, the theoretical components used in the explanation can be seen and localized (e.g., Glymour, 1980, p. 197). In AI, localizing a problem within knowledge explicitly represented in a system is called "credit assignment." The components used in explaining an item are "assigned the credit" for accounting for it; if an anomaly arises for that item, then those are the components that are localized for potential modification. Actually, "blame assignment" might be a better term, because credit assignment is generally done to try to localize failures (see, e.g., Charniak and McDermott, 1985, p. 634, on "credit assignment" and their Section 11.3, pp. 635-638 on failure-driven learning).

My diagrams in Chapter 5 showing which theoretical components directly accounted for which domain items were a first step in this direction. Additional components could have been added when they were shown to be relevant in a particular case, such as the unit-character component's role in Castle's claim that one modified character implied one modified unit. In addition, a careful consideration of how a domain item is explained by the theory might bring to light implicit, hidden assumptions, as Castle and Little managed to do in uncovering the equal viability assumption. (AI systems usually have explicit representations of knowledge; they, thus, may be unable to localize an anomaly in tacit assumptions that knowledge engineers failed to uncover and to explicitly build into the knowledge base. A promising line of research is to explore computer representations of diagrams that may contain implicit assumptions; see Larkin and Simon, 1987; Waltzman, 1989.)

Thus, one method for localization can be devised. The first, direct strategy for localization is to focus on the components directly connected with the anomalous item. The next "explanation" strategy is to explore what other, more general, explicitly stated, theoretical components play a role in the explanation of the anomaly. Finally, an attempt can be made to state explicitly any implicit assumptions that figure in the prediction for which an anomaly has emerged. The philosophical holist could argue that by indicating that implicit assumptions may be involved, I have opened the door to the view that anything can be modified in the face of any anomaly. Whether or not that is the case in principle, in fact, adopting such a holistic view is not a useful way to generate strategies for resolving specific anomalies. Historically, for numerous specific anomalies in genetics, the problems were localized and were solved.

8.5.3 Generating Alternative Hypotheses

Figure 8-2 is a refined version of the diagram of the stages and strategies of theory change introduced in Chapter 2. Testing a hypothesis can be seen as resulting in an anomaly, which then poses a new problem, requiring strategies for anomaly resolution. After the empirical results are confirmed to be problematic, then the problem is localized in one or more components of the theory. The strategy of **delineate and alter** may produce several candidate components to change.

The step after localization is the generation of alternatives to the potentially problematic components (if the simple strategy of **deletion** of the problematic component cannot be used, as it was for resolving the anomalies to dominance). Generating new hypotheses uses strategies for producing new ideas, with added constraints of preserving components of the theory that are not considered to be problematic. When strategies for producing new ideas are used to create a totally new theory, the possibilities are much more open. With anomaly resolution, however, potentially problematic theoretical components exist. Altering an old component may be easier than constructing a hypothesis de novo. It is usually easier to alter something old rather than generate something entirely new. The process of anomaly resolution also differs from de novo construction in that it has the constraint of preserving the unproblematic components for which evidence exists. Both de novo construction and anomaly resolution must result in a hypothesis satisfying the constraint of explanatory adequacy. The nature of the domain item to be explained (the anomaly in this case) constrains the kinds of hypotheses that will adequately explain it. Whatever hypothesis is generated must have explanatory adequacy, in the sense of explaining the anomaly; however, neither the localization of components to change, nor the nature of the anomaly, is sufficient for generating the new hypotheses. Other information is needed to generate alternative hypotheses, after focusing on the delineated locations and constraining the search by the nature of the anomaly. Strategies for finding and using such information are discussed below.

In Section 8.3, we discussed the strategy for anomaly resolution that I call **delineate and alter**. This strategy involves systematically considering alterations to each separately delineated component at the site of the problem. Exactly how a separable component is modified differs from case to case. One strategy for producing a new hypothesis by changing an old component is to **propose its opposite**: purity to nonpurity, random to selective, equal numbers to unequal

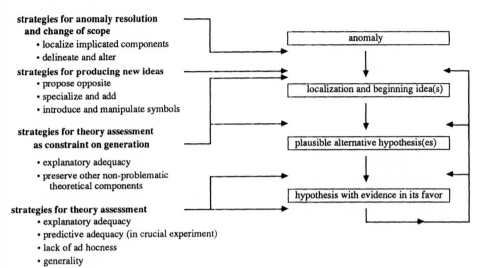

strategies for anomaly resolution
and change of scope
  • localize implicated components
  • delineate and alter

strategies for producing new ideas
  • propose opposite
  • specialize and add
  • introduce and manipulate symbols

strategies for theory assessment
as constraint on generation
  • explanatory adequacy
  • preserve other non-problematic
    theoretical components

strategies for theory assessment
  • explanatory adequacy
  • predictive adequacy (in crucial experiment)
  • lack of ad hocness
  • generality
  • simplicity
  • number of additional problems raised

anomaly

localization and beginning idea(s)

plausible alternative hypothesis(es)

hypothesis with evidence in its favor

Fig. 8-2. Refined stages and strategies, V.

numbers, equally viable to not equally viable. The seminal idea produced by the idea of an opposite is only the first step in getting a viable hypothesis. As Figure 8-2 shows, seminal, beginning ideas occupy a stage between the unsolved problem and a plausible hypothesis.

"Nonpurity," for example, can be interpreted as a seminal idea exemplified in two different specific hypotheses. Morgan's impurity hypothesis was that latent recessives occurred in the seemingly pure dominants, as did latent dominants in the seemingly pure recessives. Castle, on the other hand, proposed an impurity hypothesis by claiming that factors contaminated each other. Morgan and Castle proposed the opposite of purity, but their hypotheses differed in the actual details of the postulated impurity. Considering the opposite of a delineated assumption thus only generates a possible *type* of hypothesis in an *abstract* way; other ideas supplied by the case will be needed to develop the type into a concrete hypothesis capable of providing an explanation for the anomaly.

In addition to proposing its opposite, another way of using the old component to generate the new one is to **specialize** the component and **add** a related one. This strategy of specializing an overgeneralization and adding a new component to account for the anomaly was discussed and illustrated in Chapter 6. If Castle proposed a universal denial of the claim that gametes are ever pure, then his hypothesis of contamination was a *universal, general* denial of purity, exemplifying the strategy of **propose universal opposite**. In other words, he would have changed "all pure" to "none pure." On the other hand, he may have believed that sometimes the gametes resulting from a hybrid cross were pure, but in his cases showing modified characters, the gametes contained contaminated factors. If the latter interpretation is correct, then his hypothesis can be seen as an example of the strategy of **specialize and add** (discussed in Chapter 6).

In contrast to Castle's contamination hypothesis, the multiple, modifying factor hypothesis required no new hypothesis generation. A hypothesis proposed in an earlier case of anomaly resolution was invoked to account for the new anomaly. Extending the scope of a new hypothesis to include the anomaly is thus another way of resolving the anomaly.

In sum, these examples of anomaly resolution exemplify the strategy of **delineate and alter**. They also exemplify additional general strategies for how to do the alterations, namely **propose the opposite, specialize and add**, and **extend the scope** of an existing hypothesis.

8.5.4 Testing Alternative Hypotheses

Once several alternative hypotheses had been proposed for modifying the theory in different ways to account for the anomaly, the alternatives were compared and evaluated. I am labeling criteria used in evaluating hypotheses as "strategies for theory (or hypothesis) assessment." Their place in the stages of theory change is illustrated in Table 8-1 and Figure 8-2. I call these "strategies" too because a given scientist must decide which means of evaluation to employ. And, as we have seen, different scientists sometimes assess the adequacy of hypotheses differently. Furthermore, the strategies for theory assessment can be used in two different ways. First, they can serve as a constraint in the generation of alternative hypotheses. Alternatively, the generation process may be relatively unconstrained, giving rise to numerous, perhaps not very plausible, alternatives; then, the criteria of theory assessment are employed to choose among that array. Thus, strategic decisions in theory construction have to be made about which criteria of assessment to apply and when to apply them.

The strategies for theory assessment most often discussed by philosophers of science are those of **empirical adequacy**, either explanation of known domain items or prediction of new ones. Certainly the appeal to empirical adequacy can be seen to have figured in the debates discussed in this chapter. In order to be a successful resolution of an anomaly, the proposed hypotheses had to explain the anomaly adequately. As discussed in Section 8.2, in assessing Cuénot's selective fertilization hypothesis, Bateson and Punnett alleged that the hypothesis was an inadequate explanation for 2 heterozygotes to 1 pure recessive, because it would have predicted 3 heterozygotes to 1 pure recessive. **Explanatory adequacy** was certainly a criterion functioning in these assessments. Additionally, the cases show numerous instances of **predictive adequacy** used as the means of testing the alternatives. Failed predictions forced both Morgan and Castle to abandon their impurity hypotheses.

Other strategies for theory assessment, besides empirical adequacy, also figured in the debates. One was the issue of the ad hocness of proposed alternatives, or conversely, the **generality** of proposed hypotheses. Numerous cases in these debates show arguments in favor of a hypothesis based on the claim that it explained other, known instances, in addition to the anomaly it was introduced to explain. An obvious concern was to avoid postulating a hypothesis for which no evidence existed other than the anomaly it was proposed to explain.

Castle argued in favor of his contamination hypothesis on the grounds that it was less ad hoc and also that it was simpler. He argued that it was ad hoc to introduce modifying factors to account for modified characters if no independent evidence for such factors existed. Fewer genes, he claimed, had to be postulated, given his hypothesis, than the number required by the multiple, modifying factor hypothesis. Considering just his data showing modified characters as the domain to be explained, Castle plausibly argued that his view was simpler and less ad hoc.

Muller, however, was concerned about the wider domain of results from all Mendelian crosses. From a broader perspective of the domain of Mendelism as a whole, there is a question as to whether Castle's contamination hypothesis was simpler and less ad hoc than the modifying factor hypothesis. If contamination occurred with some genes, but not with others, then an additional theoretical component would have been added to Mendelism. For any given gene, testing would have to be done to see if it behaved in a contaminating or pure way. From Muller's perspective, Castle's contamination hypothesis appeared to be ad hoc for Castle's cases of modified characters and not applicable to Mendelism as a whole. Moreover, other instances of multiple factors had evidence in their favor. Thus, Muller's charge of ad hocness against Castle's hypothesis was a plausible assessment from the perspective of the wider domain. Furthermore, as we will see in the next chapter, the Morgan group developed techniques for providing independent evidence for the location of modifying factors. Gene mapping enabled them to avoid the ad hocness that rightly concerned Castle.

Simplicity issues are clearest when something can be counted and when one alternative can be claimed to contain fewer than the other. Castle's contamination hypothesis was simpler in the sense that, in order to explain a particular cross producing modified characters, fewer genes had to be postulated. The modifying factor hypothesis, however, was simpler if what was counted were the required number of theoretical components of Mendelism about the types of possible relations among genes in the hybrid. To have introduced a hypothesis about allelomorphic factors sometimes contaminating each other in a hybrid, but other times not producing such a contamination, would have added another major theoretical component. Castle seemed to have been too focused just on his own particular anomalous data. Taylor, in assessing Castle's reasoning, plausibly suggested that Castle was not sufficiently concerned with reconciling his (apparently) contradictory data with the orthodox explanation before proposing a radical alternative (Taylor, 1983, p. 211).

Thus, various strategies used in assessing alternative hypotheses included **explanatory adequacy, predictive adequacy, lack of ad hocness**, the **generality of the scope** of the domain items covered by the hypothesis, and the issue of **simplicity**.

In summary, this chapter discussed two anomalies that were successfully resolved without changing explicitly stated theoretical components of Mendelism. Their resolution added knowledge about homozygous lethals (what I call a "malfunction class" of an explainable kind of anomaly) to the body of knowledge in genetics and strengthened the evidence for the multiple factor hypothesis. Consideration of the alternative responses to the anomalies served to illustrate various steps in, and strategies for, anomaly resolution. These steps and strategies are enumerated in Table 8-1 and illustrated in Figure 8-2. The next chapter discusses additional means of generating alternative hypotheses in the face of a much more severe anomaly for Mendelism, namely, the anomaly of correlated or "linked" characters.

CHAPTER 9

Reduplication, Linkage, and Mendel's Second Law

9.1 Introduction

As we discussed in Chapter 5, in the early days of Mendelism, what is now referred to as the "independent assortment" of two different traits, or "Mendel's second law," was not separated from the law of segregation. In 1900, segregation appeared to hold for dihybrid crosses involving two different traits, as well as in the usual monohybrid cases. Mendel's crosses of yellow tall peas with green short ones, for example, produced ratios of 9 yellow tall:3 yellow short:3 green tall:1 green short in the F_2. All possible combinations occurred independently. (See Chapter 3 and Figure 3.4 for a review of independent assortment.) Exceptions to 9:3:3:1 ratios for two traits began to be discovered which did not also involve exceptions to 3:1 ratios for the same traits considered separately. These exceptions suggested that some traits appeared to be associated with each other in inheritance, not assorting independently. These anomalies thus showed the need to separate cases of segregation from independent assortment.

Between 1906 and 1911, various hypotheses were proposed to account for the exceptional "coupled" or "associated" or "linked" characters. Section 9.2 discusses Bateson and Punnett's discovery of anomalous ratios and the two different hypotheses they proposed: coupling of factors and reduplication of germ cells. Section 9.3 discusses the strategies exemplified in Bateson's hypotheses, including the strategies of **delineate and alter** and **propose the opposite** that were discussed in Chapter 8. It also introduces the new strategy of **move to a new level of organization** and briefly mentions the strategy of **using analogies**. Section 9.4 takes up Morgan's alternative to Bateson's hypothesis, in which Morgan extended the Boveri-Sutton chromosome theory. Morgan explained the anomalous ratios by claiming that groups of Mendelian factors were parts of the same chromosomes. Furthermore, Morgan claimed that segments of homologous chromosomes were exchanged (crossed-over) to produce "partial coupling." Section 9.5 discusses the strategies of **using interrelations** and **move to a new level of organization** exemplified in Morgan's association hypothesis. Finally, 9.6 discusses **strategies of theory assessment** that were used to assess the reduplication hypothesis, and to show its inadequacies compared to the "theory of association," which became known as "linkage and crossing-over."

The successful resolution of the anomalies to 9:3:3:1 ratios produced significant additions to the theoretical components of Mendelism. These additions included a clear delineation of

Mendel's second law of independent assortment as separate from segregation, and the addition of an entirely new set of theoretical components to explain linked traits. These theoretical changes will be discussed in this and in subsequent chapters.

9.2 The Reduplication Hypothesis

The focus in the early days of Mendelism was on segregation. Two allelomorphic characters, such as yellow and green peas, produce 3:1 ratios. Segregation explains the 3:1 ratios in terms of the segregation of factors in the formation of different types of pure germ cells. Segregation was the new discovery of Mendelism that attracted the attention of the early Mendelians and their critics. As discussed in Chapter 5, only some of the early statements of Mendelism mentioned 9:3:3:1 ratios produced by dihybrid crosses with two different traits (two pairs of four allelomorphic characters). The 9:3:3:1 ratios were not a focus of attention.

(A note on terminology is helpful here. The ambiguity of the term "character" is a problem in distinguishing segregation from independent assortment. "Character" is used both for a "character type," such as pea color, and a "character state" of that character type, such as green colored peas. In discussions of segregation of yellow peas and green peas, yellow and green are referred to as two characters, or more precisely, two alternative "allelomorphic" characters. However, when dihybrid crosses involving two different, nonallelomorphic, character types are to be discussed, such as pea color and height of pea plants, sometimes the same phrase "two characters" is used in a different way to refer to the two different character types that, all together, have four different states. Geneticists did not have clear and unambiguous terminology to distinguish between these two uses of the phrase "two characters." Context and examples usually make clear a particular usage. In this chapter I will be clearer than the historical sources. "Two allelomorphic characters" will refer to two states of a given character type and the term "trait" will refer to a type of character.)

As discussed in Section 5.7, Castle (1903) and Correns (1900) mentioned instances of "coupled characters" or "linked traits" in discussing exceptions to Mendelism. Castle mentioned that in mammals white hair and pink eyes commonly occurred together (Castle, 1903, p. 402). Correns investigated linkage in further work after 1900 (Saha, 1984). In retrospect, Sturtevant (1965, p. 39) said that the genetics of the strains of *Matthiola* (stocks) plants that Correns investigated turned out to be very complicated, but the linkage that Correns proposed at the time was *complete linkage* of two traits. Correns struggled to sort out the explanation of the anomalous phenomena of coupled characters. Because of these and other anomalies, he came to question the general applicability of Mendelism (Saha, 1984).

In 1905, Bateson, Saunders, and Punnett reported cases of "coupling" of traits in what is usually credited with being the first report of data that later was interpreted by Morgan to be *incomplete linkage* of two traits (Morgan, 1926, p. 10). Bateson and his associates proposed two different theoretical explanations for this coupling. This section will discuss these two hypotheses. The first, in terms of factors that "coupled" or "repulsed," was never fully developed into a theoretical explanation. It was rapidly abandoned, along with the terminology of "coupling" for describing the data. Bateson's second hypothesis, the reduplication hypothesis, was developed and expanded from 1911 until about 1916.

In 1905, Bateson and his associates reported results from crosses with sweet peas *(Lathyrus odoratus*, not to be confused with Mendel's *Pisum sativum*). Flower color and pollen shape were traits that showed normal Mendelian segregation. Purple flowers crossed with red showed purple to be dominant in the F_1; self-fertilization produced approximately 3 purple to 1 red in the F_2. Similarly, long pollen grains proved dominant to round and produced approximately 3 long to 1 round. Anomalous data, however, resulted from dihybrid crosses in which the two traits were considered together. When plants with purple flowers and long pollen were crossed with plants with red flowers and round pollen, the expected 9:3:3:1 ratios were not found. Instead the results were 1,528 purple long, 106 purple round, 117 red long, and 381 red round (Bateson, Saunders, and Punnett, 1905).

Bateson and his associates characterized these results as showing a "partial coupling" between purple and long and also between round and red. In order to explain this result, they proposed that the gametes containing the various combinations of factors were found in unequal numbers. They proposed that the "gametic output" of the F_1 was approximately $7AB + 1Ab + 1aB + 7ab$, instead of the equal numbers ($1AB + 1Ab + 1aB + 1ab$) that produced 9:3:3:1 ratios. Calculating the expected numbers, given such a gametic output and random matings, gave fairly close agreement to the data collected (Bateson, Saunders, and Punnett, 1905, p. 141). They suggested that the anomaly resulted from a coupling of the dominant factors for purple and long, although they did not develop this hypothesis in detail.

Bateson and Punnett continued to investigate such phenomena and found other cases that they designated "repulsion." They characterized the two kinds of cases by 1911:

> Expressed in a general form, the conclusion to which we have been led is that if A, a, and B, b, are two allelomorphic pairs subject to coupling and repulsion, the factors A and B will repel each other in the gametogenesis of the double heterozygote resulting from the union Ab x aB, but will be coupled in the gametogenesis of the double heterozygote resulting from the union AB x ab. The F_1 heterozygote is ostensibly identical in the two cases, but its offspring reveals the distinction. We have as yet no probable surmise to offer as to the essential nature of this distinction, and all that can yet be said is that in these special cases the distribution of the characters in the heterozygote is affected by the distribution in the original pure parents. (Bateson and Punnett, 1911b, p. 216, my emphasis)

The terminology used to describe the data reflects one and perhaps two of Bateson's theoretical assumptions. As discussed in Chapter 6, Bateson proposed that dominance was to be explained by the presence and absence theory. According to that hypothesis, A and B represented dominant factors that were present, but a and b represented the absence of dominant factors. Consequently, if the A and B characters were found together more frequently than expected, then the factors A and B were said to couple. On the other hand, if, in different crosses, the A and b characters (as well as the a and B combination) occurred more frequently than expected, then A and B were said to repulse. Whether coupling or repulsion occurred depended on which combinations were together in the original pure parents. In the sweet pea case, for example, if the original plants crossed were purple long with red round, then the dominant factors for purple and long were said to couple. The coupling of factors caused the character combination of purple and long to be more frequent in the F_2.

Although no explanation was given for this abnormal coupling or repulsing behavior of the factors, the terminology of coupling and repulsion reflected Bateson's predilection for using analogies from physics when speculating about explanations for biological phenomena. He often referred to "waves," "vortices," and "charges" (e.g., Bateson, 1913, chs. 2 and 3; 1916, p. 462; for discussion of these views, see Coleman, 1970; Darden, 1977; Cock, 1983). Unfortunately, Bateson never developed or published a clearly articulated hypothesis about the coupling and repulsing of factors. Nonetheless, the terms with which he characterized the anomalous data—coupling and repulsing characters—hinted at his underlying theoretical ideas. Nevertheless, actual values of the anomalous numerical results could be, and in fact were, extracted from the "theory-laden" terminology of coupling and repulsing. In other words, the data showing the anomalous ratios of coupled characters (e.g., 1,528 purple long, 106 purple round, 117 red long, 381 red round) could be labeled with other "theory-laden" terminology, such as calling them associated or linked characters rather than coupled ones, as will be discussed more fully in Section 9.6.

Others found additional data that they attempted to explain in terms of coupling and repulsion. Not only were traits such as flower color and pollen shape coupled, some traits seemed to be completely coupled to sex. Bateson and Punnett (1908) took up the attempt to explain results in the moth *Abraxas*. Doncaster had found a color variety that showed unusual inheritance: when a paler colored *laticolor* female was crossed with a darker colored *grossulariate* male, all the F_1 hybrids were *grossulariate*. Anomalies appeared in the F_2, however: no *laticolor* males were formed. Doncaster attempted to provide an explanation in terms of coupling between femaleness (interpreting sex as a Mendelian factor) and the factor for *laticolor*. Several additional assumptions about dominance and selective fertilization were needed to explain the results; we won't pause to discuss the complex details here (see Bateson and Punnett, 1908, p. 179).

After discussing the complexities of Doncaster's explanation, Bateson and Punnett argued for an alternative with fewer assumptions in which they proposed that a repulsion occurred between femaleness and the *grossulariate* factor (Bateson and Punnett, 1908, p. 179). Again, we will not pause here to follow all the details of this explanation, other than to note that attempts were being made to explain the coupling (or repulsion) of Mendelian traits with sex. The *Abraxas* case was the first of many such instances of relations between Mendelian traits and sex. Several attempts were made to explain it (see discussion of four different explanations, including Doncaster's and Bateson and Punnett's, in Morgan, 1910, "Chromosomes," pp. 486-488).

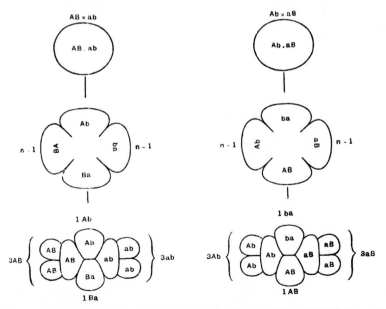

Fig. 9-1. Bateson and Punnett's (1911) reduplication diagrams (From Punnett, R.C. (ed.) (1928), *Scientific Papers of William Bateson*. Cambridge, England: Cambridge University Press, v.2, p. 213)

In 1911, Bateson and Punnett developed the hypothesis of "reduplication" to explain the production of unequal numbers of types of gametes. They proposed explaining exceptions to 9:3:3:1 ratios, not as caused by factors coupling and repulsing, but as a result of differential duplication of some germ cells. This differential reduplication, they claimed, produced the unequal numbers of types of gametes. Figure 9-1, taken from their 1911 paper, graphically shows how abnormal numbers were produced by reduplications of some cells. For larger numbers of gametes than the 3:1:1:3 depicted in the diagram, they realized many cell divisions would have to occur after segregation had occurred. No evidence for such divisions had been found by cytologists studying gametogenesis. Sutton and Boveri had proposed, and numerous others accepted, that segregation occurred during the reducing division, in which the chromosomes were halved in the formation of germ cells (see Chapter 7). In contrast, Bateson and Punnett proposed that segregation occurred "at some early stage in embryonic development" prior to the formation of germ cells (Bateson and Punnett, 1911a, p. 212).

They generalized their results by proposing that the numbers of gametes could be represented either as

$(n-1):1:1:(n-1)$ or as $1:(n-1):(n-1):1$

with $2n$ the number of gametes in the series and $4n2$ the number of zygotes formed by fertilization with two gametes. They implicitly restricted n to a power of 2, so their formula predicted odd numbers in the gametic series, such as 3:1:1:3 or 7:1:1:7 (this point was made by Cock, 1983).

The new hypothesis of reduplication led them to suggest a change in the language for reporting the phenomena:

> In view of what we now know, it is obvious that the terms "coupling" and "repulsion" are misnomers. "Coupling" was first introduced to denote the association of special factors, while "repulsion" was used to describe dissociation of special factors. Now that both phenomena are seen to be caused not by any association or dissociation, but by the development of certain cells in excess, those expressions must lapse. It is likely that terms indicative of differential multiplication or proliferation will be most appropriate. At the present stage of the inquiry we hesitate to suggest such terms, but the various systems may conveniently be referred to as examples of *reduplication*, by whatever means the numerical composition of the gametic series may be produced. (Bateson and Punnett, 1911a, p. 214)

In 1913, Trow discussed additional cases of apparent reduplication that did not fit Bateson and Punnett's prediction of odd numbers in the gametic series. He proposed a more general formula and extended the idea of reduplication to cases with three, four, or more factors:

> Given three factors A, B, and C and the occurrence of reduplication between A and B in the form
> $n:1:1:n$
> and between A and C in the form
> $m:1:1:m$,
> where n may be equal to, greater, or less than m, is there necessarily a form of reduplication between B and C, and if so, of what type must it be? We need only state now that the answer is that there is reduplication between B and C of the type
> $nm + 1 : n + m : n + m : nm + 1$. (Trow, 1913, pp. 313-314, my emphasis and spacing)

In order to accommodate all the cell divisions that would have to occur, Trow proposed an elaborate series of primary and secondary reduplications. He did not speculate as to when during germ cell formation the cell divisions occurred.

Punnett discussed Trow's extensions and provided a possible cell division method to explain the more general ratio of n:1:1:n. Figure 9-2 shows Punnett's reduplication diagrams. However, Punnett was puzzled as to how to accommodate the case of three factors (Punnett, 1913, p. 94).

Anomalies for 9:3:3:1 ratios showed the need to separate segregation (which produced 3:1 ratios) from independent assortment of different traits, although Bateson did not explicitly stress that distinction in his discussions. The problem posed by the anomalies was localized in the details of the processes producing the germ cells in hybrids. The next section discusses this localization of the anomaly and the generation of alternative hypotheses to account for it.

9.3 Strategies, including Delineate and Alter

Anomalies for 9:3:3:1 ratios showed the need to delineate the theoretical components explaining 3:1 ratios from those explaining 9:3:3:1 ratios. In Chapter 5, the theoretical component for independent assortment was stated as Component 6:

C6. Explanation of dihybrid crosses [later called "independent assortment"]

C6.1 In crosses involving parents that differ in two pairs of differentiating characters [dihybrid crosses], the pairs behave independently. Symbolically, (AB + Ab + aB + ab) (AB + Ab + aB + ab) = complicated array that in appearance [given complete dominance] reduces to 9AB:3Ab:3aB:1ab.

As discussed in Chapter 5, the usual way this theoretical component was stated in the early literature was merely to say that segregation held when two different characters (traits) were considered together. The domain of the law of segregation, thus, included not only crosses with one allelomorphic pair but also crosses in which two different traits were followed, namely, in

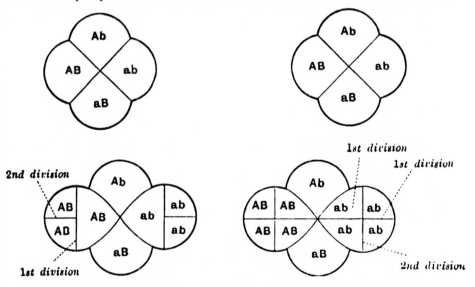

Fig. 9-2. Punnett's (1913) reduplication diagrams. (From Punnett, R. C. (1913), "Reduplication Series in Sweet Peas," *Journal of Genetics* 3, p. 94)

dihybrid crosses. Although no historical source (that I have found) did so, it is possible to state explicit theoretical components for what later came to be called "the law of independent assortment." In developing this set of components, I make use of the factor hypothesis that was developed between 1905 and 1910, and I explicitly apply each of the separable segregation components to dihybrid crosses.

C6 (1910) Explanation of dihybrid crosses:

C6.1 In the formation of germ cells in a hybrid produced by crossing parents that differed in two or more traits, the parental factors segregate or separate, so that the germ cells are of all possible pure parental combinations. Symbolically, $AABB$ x aabb = F_1 hybrid $AaBb$, which produces germ cells symbolized by AB, Ab, aB, and ab.

C6.2 The different types of germ cells are formed in equal numbers. Symbolically, equal numbers of germ cells of types AB, Ab, aB, and ab are formed, i.e., gametic proportions are 1:1:1:1.

C6.3 When two hybrids are fertilized (or self-fertilization occurs), the differing types of germ cells combine randomly. Symbolically, $(AB + Ab + aB + ab)$ $(AB + Ab + aB + ab)$ = complicated array that in appearance [given complete dominance] produces 9:3:3:1 character ratios in the F_2 generation.

I doubt that geneticists in the period between 1906 and about 1920 saw these components for explaining 9:3:3:1 ratios as so explicitly separated from the components for segregation (that explained the 3:1 ratios) as stated here. Nonetheless, this explicit statement of the independent assortment components will serve the purpose of showing in which theoretical components Bateson localized the coupling anomaly. This delineation of components will also aid in analyzing exactly what alterations Bateson made when he proposed his two different hypotheses, coupling and reduplication. Note that in normal cases of independent assortment, Component 6.2 claims that the ratio for the F_1 *gametic* proportions is 1:1:1:1, and Component 6.3 claims that the expected, normal F_2 *character* ratio is 9:3:3:1.

In order to analyze the nature of Bateson's hypotheses, it is useful to distinguish (1) independent assortment (as it was later called), (2) complete linkage (as it was later called), and (3) the anomaly Bateson called partial "coupling" (later called "incomplete linkage"). Independent assortment, as the components state, results in the gametes of hybrids forming equal numbers of all possible pure parental combinations. Complete linkage is the opposite of independent assortment. Component 6.1 is violated in complete linkage: not all possible parental combinations occur. For example, assume that white hair and pink eyes were always inherited together, and further that they were characters caused by different factors. If A symbolizes white hair and B pink eyes, then only gametes symbolized by AB and ab form. (A complication is introduced if one factor can cause two characters; then, distinguishing one factor-two characters from complete linkage of two factors for two different characters is difficult. Actually, albinism, the lack of pigment causing white hair and pink eyes, is now known to be due to a single gene difference. For the purpose of this example, however, assume that complete linkage of two factors produces the correlation of the characters. I would prefer a real example of complete linkage, but it is now known that, in general, complete linkage between separate genes does not occur, unless there is some inhibition of crossing-over.)

As seen in other cases, Mendelians inferred the nature of the germ cells in the hybrid from proportions of characters in the F_2. In all the previous hypotheses discussed, equal numbers of germ cells of each type were postulated (except for my own hypothesis for explaining Cuénot's 2:1 ratios in Chapter 8). In the case of the independence of two traits in a dihybrid cross, equal numbers of the four classes of germ cells explained the 9:3:3:1 ratios. Symbolically, assuming complete independence, the hybrid was expected to form germ cells in the proportions of 1AB to 1Ab to 1aB to 1ab. At the opposite extreme, if complete linkage occurred between two dominant characters of two different traits, then the explanation would be that the proportions of germ cells in the hybrid would be 1AB to 0Ab to 0aB to 1ab. If complete linkage occurred between a dominant character in one trait and a recessive character for another trait, then the expected gametic proportions would be 0AB to 1Ab to 1aB to 0ab.

Bateson's anomalous data for sweet peas, however, did not show either independence or complete linkage. More purple flowers with long pollen shape (AB) and red flowers with round pollen shape (ab) appeared than expected, given complete independence; however, some purple round and red long combinations did occur. Instead of the 9:3:3:1 character ratios expected given complete independence, Bateson and his colleagues found the anomalous ratio of 1,528:106:117:381. Bateson characterized the data as showing "partial coupling."

Bateson and his colleagues argued that the anomaly was to be explained by postulating unequal numbers of types of gametes. In other words, in order to account for the coupling anomaly, Bateson introduced the *new* idea that the ratios of the types of gametes in the F_1 hybrids could be inferred to be *unequal*. Based on the numerical anomalies in the data about observable, coupled characters in the F_2, he inferred the proportions of types of gametes in the F_1 hybrids. In sum, the 3:1 ratios of segregation and the 9:3:3:1 ratios in normal dihybrid crosses allowed Mendelians to infer that different types of pure gametes formed in equal numbers in hybrids. The same kind of reasoning can be used to infer the abnormal proportions of gametic types for the abnormal character ratios. The predicted 1:1:1:1 proportion for the four types of gametes was violated in this anomalous case.

Bateson postulated gametic ratios of approximately 7AB to 1Ab to 1aB to 7ab. When such proportions of types of gametes combined randomly, character ratios of 177:15:15:49 (approximately 11.8 to 1 to 1 to 3.2) would be expected. In other words, a cross represented by (7AB + 1Ab + 1aB + 7ab) (7AB + 1Ab + 1aB + 7ab) produces (with complete dominance) predicted character ratios (in the sweet pea case) of 177 purple long to 15 purple round to 15 red long to 49 red round. For the hundreds of characters observed, Bateson claimed close agreement between this prediction and the data from the sweet pea cross (Bateson, Saunders, and Punnett, 1905, p. 141). My calculation of the proportions found in his data of 1,528:106:117:381 equals approximately 13.7 to 1 to 1 to 3.4. Bateson believed that the hypothesis of unequal numbers of types of germ cells was adequate for accounting for the anomaly. (Bateson did not analyze his data into ratios with decimals; I calculated the character ratios this way to emphasize how the anomaly differed from the expected 9:3:3:1 Mendelian ratios.)

Bateson was able to use the anomalous ratios of characters to estimate the numerical values of proportions of types of gametes. The numerical data of the anomaly thus aided in what may be called "parameter fixing" in the hypothesis, that is, in estimating the values of the numerical, unequal proportions of types of gametes. "Reasoning backward" from the numerical ratios of characters to be explained to the explanation in terms of the unequal numbers was a method for refining the unequal numbers hypothesis. In other words, not just any proportion of unequal numbers of types of gametes would have explained the specific anomalous data; the specific, numerical proportions of gametes in the anomalous case could be calculated from the anomalous character ratios. For other character data, Bateson and his colleagues predicted and found other proportions, such as 3:1:1:3. The nature of the anomaly itself is once again seen to play a specific role in reasoning to the hypothesis provided to account for it, as did Cuénot's anomalous 2:1 ratios discussed in Chapter 8.

Bateson's hypothesis of unequal numbers of types of gametes can be seen as localizing the anomaly in what I have labeled Component 6.2, the assumption of equal numbers of types of germ cells in the hybrid to explain 9:3:3:1 ratios. Thus, the components that "directly" accounted for 9:3:3:1 ratios were the site of modification. The strategy of **delineate and alter**, introduced in Section 8.3, is also exemplified in my analysis of this response to the coupling anomalies. By delineating the separable assumptions made to explain 9:3:3:1 ratios, one component is seen to have been the site for alteration. Further, the strategy of **propose the opposite**, discussed in Chapter 8, can be seen as a means of producing the seminal idea of unequal numbers of types of gametes. By explicitly realizing that the normal cases required equal gametic proportions, the idea of the opposite of equal proportions could be considered to account for the abnormal, anomalous case. Making explicit the assumption of equal numbers could give the idea of searching for hypotheses for producing unequal numbers. Both of Bateson's hypotheses, coupling and reduplication, were hypotheses to account for the production of unequal gametic proportions (as was Morgan's linkage hypothesis which will be discussed in the next section). Thus, the type of hypothesis that could be generated by the strategies of **delineate and alter** and **propose the opposite** was a hypothesis of a type to produce unequal numbers of gametes in cases of coupled characters.

Because numerous crosses showed the nonanomalous 9:3:3:1 ratios, Bateson was not proposing a universal denial of equal gametic proportions. His hypothesis can be interpreted as specializing Component 6.2 to apply to fewer domain items. In other words, the 9:3:3:1 ratios in the domain were still to be explained by the component of equal gametic proportions, but the equality component (C6.2) did not apply to all dihybrid crosses any more. The previously anomalous ratios in cases of coupled characters became a new domain item. In order to account for such ratios, a new theoretical component postulating unequal gametic ratios was added to the theory. Thus, again, the **specialize and add strategy** is exemplified: "all gametic types occur in equal numbers" was specialized to "in some dihybrid crosses, all gametic types occur in equal numbers" and "in other dihybrid crosses, some gametic types occur in unequal numbers."

Although Bateson was less clear about this than Morgan, the coupling anomalies showed the need to specialize what had been called "Mendel's law." Theoretical components for dihybrid (and multihybrid) crosses were delineated from those of segregation. The anomalies showed that the behavior of the factors in the formation of germ cells was not the same in dihybrid crosses as in crosses involving only one trait. Ratios of characters for two or more traits required a different explanation than the segregation of two allelomorphic characters for one trait. Mendel's law was thus specialized to "Mendel's law of segregation" for the segregation of two allelomorphic factors, and another "law" was needed for dihybrid cases. Bateson documented the first case of partial coupling in 1905. In 1919 Morgan explicitly claimed that independent assortment was a separate "second law," as will be discussed in more detail in Section 9.5. The delineation of independent assortment from segregation, thus, took some time to articulate fully. Its development was driven by the discovery of anomalies to independent assortment.

Bateson did not stop with merely calculating the unequal numbers of types of gametes formed. The hypothesis of unequal numbers raised the further question—what caused the unequal numbers to form? Bateson's first hypothesis to account for the sweet pea case vaguely postulated some sort of coupling between the factors A and B to produce larger numbers of AB gametes. For new data about other characters, which showed larger than expected combinations between the dominant character for one trait and the recessive for the other, he proposed a repulsion between the dominant factors. Symbolically, A and C repulsed each other to produce larger numbers of Ac and aC gametes. (A and c could not couple because, according to Bateson, c represented the absence of the factor C; similarly for C and a. A present factor could not couple with the absence of another one; thus, the presence and absence theory necessitated the second process of repulsion.) Although Bateson never developed this seminal idea of the coupling and repulsing of factors into a fully articulated hypothesis, the way he discussed it showed that he might have been thinking of a factor as having a forcelike nature that was capable of attraction or repulsion. His language suggests analogies to physical phenomena. These analogies to physics did not serve to provide a fully articulated hypothesis. (For further discussion of Bateson's use of vague analogies to physics in hypothesis formation see Coleman, 1970; Darden, 1977; Darden, 1980b; Cock, 1983.) Sometimes the strategy of forming hypotheses by **using analogies** is a fruitful one (see e.g., Hesse, 1966), but Bateson never managed to use his analogies to physics to develop his seminal ideas about forces into plausible hypotheses.

Bateson's hypothesis of coupling factors might be seen as localizing the cause of unequal numbers in some process that subverts the independent separation described in Component 6.1. Actually, Component 6.1 is not stated in a way that easily allows it to be altered to account for partial coupling of factors, because it claims that the factors separate so as to form all possible combinations of pure parental factors. Even in partial coupling, all combinations form, but they somehow form in unequal numbers. Thus, the coupling hypothesis seems to require a new theoretical component between C6.1 and C6.2 to describe some process by which some factors partially couple to produce unequal numbers in the next step. This problem of finding the appropriate localization in explicitly stated components for the cause of unequal numbers shows the difficulties that anomalies can pose. Theoretical components carefully crafted to account for one set of domain items may need significant alterations and additions in the light of anomalies or in the face of new kinds of cases not even considered as possibilities at the time of construction of the original components.

The change from the hypothesis of coupling and repulsing factors to the reduplication hypothesis involved a change in the *level of organization*. Bateson and Punnett were still attempting to provide a cause for unequal gametic proportions, a violation of Component 6.2; however, they proposed a different answer to the question of how the unequal numbers formed. Instead of considering a process at the level of the *factors* themselves (such as coupling and repulsion was), they moved to the level of the entire germ *cell*. They proposed that equal numbers of the types did form, but then some types of germ cells selectively reduplicated, that is, underwent additional cell divisions, to produce unequal numbers of types. Coupled *characters* were no longer explained by the *factors* coupling; the anomalies were instead explained by differential reduplications of types of *germ cells*. (Coleman, 1970, p. 262, stressed the importance of the cell as a level of organization in the "Cambridge school of morphology," in which he placed Bateson.)

The differences between coupling and reduplication show that alternative hypotheses can be generated by considering an alternative level of organization. The strategy of **move to another level of organization** is thus another method to use in generating alternative hypotheses, if the case is such that different hierarchical levels can be identified.

As will be seen in the next section, yet another level of organization existed, between the individual factors and the germ cells, namely, groups of factors on pieces of chromosomes. Finding the appropriate level of organization to use in constructing an explanatory hypothesis is an important and sometimes difficult process. Bateson used levels—factors and cells—that had already been postulated to exist prior to his use of them in this case of anomaly resolution. Sometimes new levels of organization are discovered in the process of solving the problem, such as the level of groups of genes on chromosomes for solving the problem of partial coupling. Alternatively, new levels may be discovered in another field and then used to solve a pending problem, such as the level of macromolecules for solving the problem of gene reproduction. (Chapter 15 will provide further discussion of the idea that certain problems get resolved by finding appropriate units at appropriate levels of organization.)

The change in theory proposed by the reduplication hypothesis can be localized in between Component 6.1, the formation of all possible types of germ cells, and C6.2, the final result of equal numbers of all types. Reduplication was a process that was proposed to occur after the formation of all types; it subverted the equality and produced the final result of unequal numbers of types of germ cells in the hybrids.

As is typically the case with new hypotheses, the reduplication hypothesis itself raised new questions—especially about the time and location of the numerous cell divisions. Bateson and Punnett did not accept Sutton and Boveri's chromosome theory that claimed that the factors were part of the chromosomes, nor did they accept the claim that Mendelian segregation occurred during the reducing division of germ cell formation in which the chromosome number was halved. Instead, they speculated that segregation occurred earlier in embryological development than the time of germ cell formation; the reduplications also occurred in an early embryological stage. Thus, although they appealed to a known level of organization, namely the cell and cell divisions, they did not postulate an interrelation to specific behavior at that level already discovered by cytologists. Instead, they predicted that instances of differential cell division would be discovered by embryologists investigating early stages. That prediction was never confirmed.

Explicitly stating three theoretical components for independent assortment allowed us to localize Bateson's two hypotheses. My statements of the theoretical components as of about 1910 are, as I indicated, more explicit than was present in the historical record. This explicit separation of different claims allows possible sites for anomaly resolution to be found. Bateson localized the anomaly in what I labeled Component 6.2, equal numbers of gametes. He proposed an alternative to C6.1 with the idea of coupling. His reduplication hypothesis added a component between C6.1 (all combinations) and C6.2 (equal numbers) with the claim that some cells with some combinations reduplicated.

Could employing the strategy of **delineate and alter** produce another hypothesis? In other words, could some other alternative hypothesis, localized in another component, also have explained the anomalies? Component 6.1 could be altered to claim that factors that are part of the same chromosome would not separate independently, but complete linkage of factors on the same chromosome would not account for the data of characters only partially coupling. It would have been possible to refine the idea (which de Vries may have proposed) that pangens jumped between chromosome threads. The claim that all pangens jumped with equal frequency to produce all possible combinations in equal numbers (which explained independent assortment) would have to have been limited to account for partial coupling. Some pangens would have been more tightly bound to the chromosome threads than others.

Somewhat later, Goldschmidt (1917) proposed another alternative (which will not be discussed here in detail) that resembled a limited jumping hypothesis. He suggested that genes were loosely carried in a linear arrangement by the chromosomes but the genes left the chromosomes to move to the cytoplasm (to explain gene action). When the genes returned to reconstitute the chromosomes before cell division, sometimes a particular gene returned to the wrong location, hence partial coupling. (For more discussion of Goldschmidt's hypothesis, see Carlson, 1966, pp. 77-79; Wimsatt, 1987, pp. 47-50.) In the next section we will discuss Morgan's resolution of the problem with his hypotheses of linkage of genes along the chromosomes and crossing-over between homologous chromosomes.

What about Component 6.3, random combinations at fertilization? Could the gametes be formed in equal numbers but not combine randomly? Could selective fertilization again have been invoked as an explanation in this case, as it was in the Cuénot 2:1 case? Could some combinations of the *AB*, *Ab*, *aB*, and *ab* gametes unite more often at fertilization than others? Could some hypothesis about the selective combinations of these types explain the anomaly? Figure 9-3 shows the usual 9:3:3:1 case and indicates which combination classes would have to be increased or decreased for partial coupling. The figure shows how complicated a hypothesis of selective fertilization (i.e., a denial of C6.3) would have to be to explain the anomaly of partial coupling. Cuénot only had to assume one kind of selectivity (*A* and *A* gametes did not combine) when he invoked selective fertilization to explain the 2:1 anomaly (discussed in Chapter 8). In order to explain Bateson's anomalous data, numerous instances of selectivity between numerous types of gametes would be required. Probably a good many ad hoc assumptions would be needed

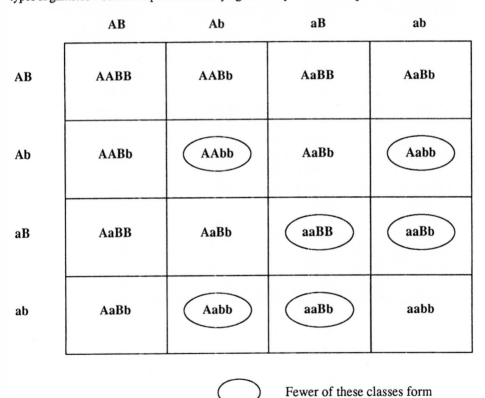

Fewer of these classes form

Fig. 9-3. Selective fertilization? Can selective fertilization be used to explain the partially coupled characters A and B? To do so, combinations of factors AB and ab would have to be more numerous than those with aB and Ab. Numerous ad hoc assumptions would be needed to produce fewer of these specific classes of factor combinations.

to make it work. At least, I cannot see a simple hypothesis of selective fertilization. I am not surprised that no one (so far as I know) proposed one historically.

Bateson's two different hypotheses to explain unequal numbers of gametes exemplify two different strategies for hypothesis formation, **appeal to analogies** and **use an interrelation** to predict phenomena in another field. The latter served him better than the former in this case for generating a plausible, testable hypothesis.

Bateson's appeal to coupling and repulsing may be seen as exemplifying the strategy of **appealing to analogies** from another field, in this case electrical phenomena. As I have argued elsewhere (Darden, 1980b), the strategy of **appealing to vague analogies** is a weak one. Although Bateson may have been making analogies to phenomena in physics, he was not appealing to specific physical forces known to occur in cells. No physical measurements of coupling forces were known or were suggested. Thus, I do not label Bateson's appeal to attracting and repulsing forces as an example of an *interrelation* between two established bodies of knowledge. He was trying to explain genetic data using analogies to physical forces, but he was not connecting that data to any actual or predicted data from physics about forces. Thus, his coupling hypothesis does not exemplify the strategy of **using interrelations** between two bodies of knowledge to generate an interfield hypothesis.

In contrast, the reduplication hypothesis can be analyzed as an attempt to use interrelations between genetics and cytology, because it postulated cell divisions. The theoretical unit in the reduplication hypothesis was the cell, a known entity in cytology. Bateson's appeal to cytology, however, did not make use of any known phenomena of numerous cell divisions; it made a prediction that cytologists would find numerous selective reduplications of cells sometime during embryological development. Such an interrelation to cytology was thus a weaker use of the **interrelations strategy** than was the chromosome theory, which postulated connections between properties of factors and observed chromosomes. Independent evidence for both factors and chromosomes already existed. (For further discussion of differences between appeal to vague analogies in another field and an actual interfield theory, see Darden, 1980b; Staats, 1983.)

In summary, the strategies exemplified in Bateson's hypotheses to resolve the anomaly of partial coupling followed the steps in anomaly resolution introduced in Chapter 8. The anomaly was localized in components directly accounting for the anomalous data, and an alternative to one of the problematic components was proposed. The formation of the hypothesis of unequal numbers of types of gametes exemplified the strategies of **delineate and alter**, **propose the opposite**, and **specialize and add**, discussed in Chapter 8. The formation of causal hypotheses to explain the production of unequal numbers exemplified the strategies of **appeal to analogies** and **use an interrelation** to phenomena in another field. Finally, the strategy of **move to another level of organization** can be seen as exemplified in the change from coupling to the reduplication hypothesis.

We now turn to the successful use of the strategy of **using interrelations** in Morgan's hypothesis about interrelations between sex chromosomes and sex-heredity. Further development of the chromosome theory ultimately resolved the anomalies of the complete coupling of traits with sex and the partial coupling of some traits (other than sex) with each other.

9.4 Morgan and Sex Linkage

As discussed at the end of Chapter 7, from 1900 until early 1910, Morgan was a critic of Mendelism and the chromosome theory, although he had a more favorable view of the connection between chromosomes and sex than did Bateson (e.g., Morgan, 1910a).

Morgan began breeding insects, including fruit flies (now called *Drosophila melanogaster*), to look for mutations of the sort that de Vries had discovered in *Oenothera*, the evening primrose (Allen, 1978, pp. 144-148). He discovered a male fruit fly that did not have the normal red eyes but was a mutant white (Morgan, 1910c). Although a change in eye color was a smaller scale change than the new species-producing mutants that de Vries found in *Oenothera*, Morgan used de Vries's term "mutant" (Morgan, 1910b). Morgan often used existing terms and gradually changed their meaning rather than developing new terms for the new phenomena and the new theoretical explanations that he and his students discovered and proposed.

Morgan's now famous discovery of this white-eyed mutant marked the beginning of the work that led to his change of view about the nature of Mendelian factors and their relations to chromosomes (Carlson, 1966, Ch. 6; Allen, 1978, pp. 148-153). Even so, his 1910 paper discussing his results reads as if it were just another in the series by Bateson and others trying to provide a Mendelian explanation of yet another instance of "sex-heredity," in Bateson's terminology, or a "sex-limited" character as Morgan called it (Morgan, 1910c). (The terminology soon shifted to "sex-linked" with Morgan's development of a new explanation. Today, "sex-limited" refers to different types of genes on autosomes, namely non-sex chromosomes, that are only expressed in one sex, such as genes affecting milk production in cows. See, e.g., Strickberger, 1985, pp. 166-167.)

Morgan (1910c) discussed the Mendelian breeding experiments with the white-eyed male. He bred it with normal red-eyed females. All the F_1 offspring were red-eyed. When the F_1 hybrids were interbred, the F_2 generation showed the expected Mendelian proportion of 3:1 red to white, but, unexpectedly, all the white-eyed flies were males. The results were, as Morgan said, the converse of the *Abraxas* moth case (discussed in Section 9.2), in which only females showed a particular character.

Morgan explained the results by proposing that sex was a Mendelian factor that was homozygous in females (XX) and heterozygous in males (X with no pairing factor) in *Drosophila*. In 1910, Morgan used the symbol X for a *Mendelian factor*, not for the X chromosome that cytologists associated with sex determination. He proposed, further, that the X factor and the R factor for red were completely coupled. The mutant fly was assumed to have lost the red factor (assuming the presence and absence theory). Figure 9-4 shows the crosses and results. Bateson might well have proposed the explanation himself: it used his terminology and theoretical assumptions and was the kind of explanation of *Abraxas* that Bateson and Punnett (1908) had given, except that in that species the female was the heterozygote. In this first paper on the white-eyed mutant, Morgan, like Bateson, did not mention chromosomes (Morgan, 1910c). Morgan used his scheme to make predictions that were confirmed in various test crosses.

By the next year, Morgan (1911b) had found additional sex-limited mutations that behaved similarly to white eyes. Furthermore, they showed partial coupling to each other. Morgan's *Drosophila* research thus brought together the Mendelian work on sex-heredity and partial coupling. By proposing an alternative to Bateson's explanation of partial coupling, Morgan also tied this work to the cytological findings about sex chromosomes. His skepticism gone in this 1911 paper, he located the Mendelian factors in the chromosomes. He clearly recognized the data as showing exceptions to what he called "Mendel's law of inheritance," which, he claimed,

> . . . rests on the assumption of random segregation of the factors for unit characters. The typical proportions for two or more characters, such as 9:3:3:1, etc., that characterize Mendelian inheritance, depend on an assumption of this kind. In recent years a number of cases have come to light in which when two or more characters are involved the

proportions do not accord with *Mendel's assumption of random segregation*. The most notable cases of this sort are found in sex-limited inheritance in *Abraxas* and *Drosophila*. (Morgan, 1911b, p. 384, my emphasis)

Morgan, along with other biologists in this period, did not separate segregation from independent assortment or distinguish two Mendelian laws, as his usage of "random segregation" in the quotation showed. Using our modern distinctions, ones that Morgan proposed by 1919, his cases obeyed Mendel's first law of segregation (3:1 ratios), but violated Mendel's second law of independent assortment (9:3:3:1 ratios).

After discussing Bateson's explanations in terms of coupling and repulsion, Morgan proposed his alternative:

In place of attractions, repulsions and orders of precedence, and the elaborate systems of coupling, I venture to suggest a comparatively simple explanation based on results of inheritance of eye color, body color, wing mutations and the sex factor for femaleness in *Drosophila* . (Morgan, 1911b, p. 384)

Morgan continued by postulating that the factors for those characters were located linearly along the chromosome. Citing the cytological work of Janssens (1909) which showed visible intertwinings of chromosomes during the formation of germ cells, Morgan proposed that sometimes the homologous chromosomes exchanged parts. He thus provided an explanation for what Bateson called "partial coupling." He argued that the anomalous cases were thus explained by

P_1 WX-W   x   RRXX
                   white-eyed male   red-eyed female

gametes   WX - W   (male)

                      RX - RX   (female)
                      ─────────────────────

F_1 zygotes   RWXX (50%) - RWX (50%)
                   Red female   Red male

F_1 flies mated

gametes   RX - WX   (F_1 female)

                             RX - W   (F_1 male)*   *note: in male, complete
                             ─────────────────────   coupling of R with X; no
                                                                          WX gametes formed
F_2 zygotes   RRXX - RWXX - RWX - WWX
                   (25%)   (25%)   (25%)   (25%)

                   Red   Red   Red   White
                   female   female   male   male

Results   3 red:1 white   (no white-eyed females)

Fig. 9-4. Crosses with the white-eyed male *Drosophila*.

... a simple mechanical result of the location of the materials in the chromosomes, and of the method of union of homologous chromosomes, and the proportions that result are not so much the expression of a numerical system [i.e., Bateson's gametic formulas] as of the relative location of the factors in the chromosomes. *Instead of random segregation in Mendel's sense we find "association of factors" that are located near together in the chromosomes. Cytology furnishes the mechanism that the experimental evidence demands.* (Morgan, 1911b, p. 384)

In another paper in 1911, Morgan explained how his discovery of sex-limited characters and his explanation of them caused him to change his negative view of the chromosome theory:

For several years it has seemed to me that the chromosome hypothesis, so called, could not be utilized to explain the Mendelian results in the form presented by Sutton, because, if it were true, there could be no more Mendelian pairs in a given species than the number of chromosomes present in that species. Even if this objection could be avoided ... the more serious objection still remained, namely, that with a small number of chromosomes present many characters should Mendelize together, but very few cases of this sort are known. De Vries was the first, I believe, to point out that this objection could be met if the genes are contained in smaller bodies that can pass between homologous pairs of chromosomes; and Boveri has admitted this idea is compatible with his conception of the individuality of the chromosomes. In the case of the inheritance of two sex-limited characters in the same animal we have an experimental verification of this hypothesis. (Morgan, 1911c, p. 77-78)

Morgan's hypothesis was more original than his credit to de Vries and Boveri suggested. The partial coupling anomalies for the 9:3:3:1 ratios were unknown in 1903 and 1904 when de Vries and Boveri proposed their hypotheses, which was discussed in Chapter 7. De Vries was accounting for the independent assortment of characters, and may have proposed that pangens could jump between nuclear threads, in contrast to Janssens's and Morgan's idea that actual pieces of homologous chromosomes crossed-over. Morgan's hypothesis of the association of grouped factors with pieces of chromosomes was a new hypothesis, proposed to resolve the anomalies of partial coupling and sex-limited characters. He extended the interrelations between Mendelian factors and chromosomes that had been previously proposed to provide a new, unified explanation for sex-heredity and partial coupling. These were new interrelations, extending previous ideas to account for new anomalies for Mendelism. For the first time particular Mendelian factors, for example, factors for red and white eye color in fruit flies, had been associated with a specific chromosome, the X chromosome.

Morgan's own independent lines of research on mutations and on sex determination came together with the discovery of the white-eyed mutant. Such convergence of results, allowing new interrelations between mutants showing Mendelian segregation and sex-linked heredity has been interpreted as another reason, in addition to that mentioned earlier by Morgan, for Morgan's change in his opinion of the adequacy of both Mendelism and the chromosome theory (Manier, 1969; Allen, 1978, p. 145; for further discussion of Morgan's "conversion," see Lederman, 1989).

Morgan's work progressed rapidly with the discovery of additional sex-limited mutants in *Drosophila* and breeding experiments producing data about their associations. Morgan developed his "theory of association" (Morgan, 1911a), as he originally called what was soon labeled "linkage and crossing-over." He proposed that factors (he used the term "gene" consistently only after 1917) lying "lineally along the chromosome" would be associated in inheritance; but "the associations are not absolute for occasionally the twisting of the chromosomes will be such that even regions lying lineally near together will come to lie on opposite sides of the united chromosomes" (Morgan, 1911a, p. 404). He indicated that the experimental results "accord completely" with the predictions of his theory (Morgan, 1911a, p. 404). The further development of the linkage hypothesis and the idea of crossing-over between homologous chromosomes will be discussed in the next chapter.

9.5 Strategies: Interrelations and Levels of Organization

The anomalies for 9:3:3:1 ratios created the need to separate Mendel's law into two parts, the law of segregation and the law of independent assortment. Morgan gradually developed a clear understanding of their differences, developed the terminology of "independent assortment," and explicitly stated them as two separate laws, as will be discussed in the next section.

The strategies of **delineate and alter** and **specialize and add** are exemplified in this development. In order to explain the anomalies for 9:3:3:1 ratios that were not also anomalies for 3:1 ratios, the theoretical components explaining them needed to be separated. Mendel's law was not one, but two separable claims. They had not been explicitly separated during the first decade of Mendelism. No need had arisen to define "random segregation" as applying only to crosses involving one trait but not to those in which two or more traits were followed. The new coupling anomalies, however, created such a need for delineating two different "laws." The universality of the law of segregation was preserved by delineating a second law to which exceptions had been found. Then, the anomaly was localized in the newly delineated theoretical components that directly accounted for the 9:3:3:1 ratios. Simultaneously with its statement, that second law was specialized to account for exceptions to it. Not all dihybrid crosses showed independent assortment, only those involving traits from different linkage groups.

Interestingly, Mendel's second law was not even seen to be a separate "law" until it was also seen not to be a *universal* law. (Philosophers sometimes restrict the usage of "law" to universal generalizations for which no exceptions are known.) The anomalies to 9:3:3:1 showed that dihybrid (and multihybrid) crosses were grouped into two different kinds of cases—those with traits that assorted independently and those that showed incomplete (or very occasionally complete) linkage. Thus, in its very formulation, the law of independent assortment was specialized to "only some Mendelian factors for different traits show independent assortment" and another condition was added, "Mendelian factors belonging to the same linkage groups do not show complete independence."

Delineating segregation from independent assortment and explaining the anomalies for the latter were important changes in Mendelism. These changes produced the following results: preservation of the universality of the law of segregation, formulation of the more limited second law of independent assortment, and the addition of theoretical components about linkage and crossing-over to account for exceptions to the second law.

As did Bateson, Morgan "proposed the opposite" of equal numbers of types of gametes; that is, he localized the anomaly in C6.2, the component claiming equal numbers of gametes. He also developed a new hypothesis as to the *cause* of the unequal numbers—association of factors on pieces of chromosomes. With occasional splitting of chromosomes into different pieces and their rejoining, a few germ cells of what came to be called the "cross-over" classes were produced. This new hypothesis about the cause of unequal numbers of types of gametes exemplifies at least two strategies that have already been discussed. The strategies are **using interrelations** to other bodies of knowledge and **moving to a new level of organization**.

Morgan extended the interrelations between cytology and Mendelism begun by the Boveri-Sutton chromosome theory proposed in 1903 and 1904. His development of that theory resolved the anomaly that the numbers of correlated characters were fewer than expected, given Sutton's prediction of complete linkage. Morgan's new hypothesis of sex linkage connected at least three different lines of research: first, the cytological work on sex chromosomes by Stevens, Wilson, and others; second, the work on "sex-heredity" (to use Bateson's term) for some Mendelian characters, which Mendelians had been investigating by doing breeding experiments; and third, the discovery of partial coupling and repulsion that Bateson and Punnett had tried to explain with the coupling and reduplication hypotheses. In addition, he drew on new cytological observations of intertwining chromosomes by Janssens. His hypothesis can thus be seen as extending the use of the strategy of **using interrelations** between Mendelism and cytology.

The visible cytological work on sex chromosomes was thus related to the data from breeding experiments with *Drosophila*. Table 9-1 shows Morgan's extensions to the mappings between chromosomes and Mendelian factors begun by Sutton, Boveri, and de Vries (and listed in Table 7-1). Sex chromosomes and Mendelian characters associated with sex were related. Also, the pairing and intertwining of homologous chromosomes were correlated with linkage groups of Mendelian factors and coupling due to exchanges between paired chromosomes. Morgan's hypothesis predicted that non-sex linkage groups (i.e., autosomal linkage groups) would be found. That prediction was soon confirmed, as will be shown in the next chapter. The strategy of **using interrelations** between different bodies of knowledge, between genetic and cytological results, was a powerful one in this case.

In postulating that factors were associated in what came to be called "linkage groups" along chromosomes, Morgan hypothesized the existence of a *new level of organization* of the factors.

Chromosomes	Factors
1. Sex chromosomes sex chromosomes differ from autosomes	1. Sex-limited factors several sex-limited characters found in breeding experiments; Prediction: corresponding factors associated with sex chromosomes
2. *Drosophila:* homozygous females, heterozygous males *Abraxas:* heterozygous females, homozygous males	2. Only one allelomorphic factor for a sex-limited trait in heterozygous forms, e.g., male *Drosophila*
3. Intertwining of paired homologous chromosomes Prediction: pieces of chromosomes cross-over during intertwining	3. Partial coupling or linkage of factors Prediction: occasional interchange of allelomorphic factors between homologous link age groups (explanation of anomalies for 9:3:3:1 ratios)
4. Number of homologous chromosome pairs is related: 1 pair of sex chromosomes, other pairs of autosomes	4. Prediction: number of linkage groups species corresponds to the number of pairs of chromosomes in a species; one group sex-limited

Table 9-1. Relationships between chromosomes and Mendelian factors, 1910-1911.

This new hypothesis postulating a new level of organization simultaneously resolved the anomaly for Sutton's theory of too few completely linked characters and removed the anomaly for Mendelism of partially coupled characters. Morgan's hypothesis of association thus exemplifies the strategy of **postulating a new level of organization** in the formation of a new hypothesis. The idea for that level was furnished by the field of cytology, but intertwining chromosomes only suggested that pieces might break and rejoin. Janssens's diagrams of intertwining threads provided no way of observing actual switching of pieces of chromosomes. Morgan predicted new cytological phenomena to account for the genetic anomaly.

The new level of organization investigated with techniques from two different fields constituted a very successful line of research after 1911. The method of using two different experimental methodologies from two different fields and correlating the results in a combined research program was carried out by Morgan and especially by his students. The *Drosophila* group was composed of students who were trained in both Mendelian and cytological techniques by Morgan and the cytologist Wilson at Columbia. **Using interrelations** can thus be seen to be a strategy not only for forming hypotheses but also for developing a research methodology for correlating results generated by two different methods and for predicting additional correlations. Successfully carrying out such a research program may necessitate new training of scientists in both areas.

The next chapter discusses the numerous empirical results, analytical methods, new hypotheses, and the new model of "beads on a string" that resulted from the work that further developed the chromosome theory of Mendelian heredity and investigated the new level of organization, that is, linkage groups.

In summary, the strategies exemplified in Morgan's "theory of association" included **delineate and alter, specialize and add, propose the opposite, use interrelations to another body of knowledge**, and **move to a new level of organization**.

9.6 Assessments: Reduplication versus Linkage

The hypotheses of reduplication and linkage (with crossing-over) were competitors during the period between 1910 and about 1915. (For additional historical discussion of this dispute, see Carlson, 1966, ch. 7; Allen, 1978, pp. 156-164; Cock, 1983.) As discussed in Section 9.2, Trow and Punnett extended reduplication to account for cases in which three characters were coupled. In doing so, however, the formula for the numbers of gametes and the claims about numbers of cell divisions became quite complicated. Three-factor crosses were easily explained in terms of associations of three factors along a chromosome, with occasional crossing-over between them. (This difference in ease of extendability of the two hypotheses was pointed out by Cock, 1983.)

Between 1910 and 1915, Morgan and his students found the predicted linkage groups of characters not associated with sex, and proposed that they were part of non-sex chromosomes. The *Drosophila* group accounted for both coupling and repulsion as a result of factors linked on the same chromosome. In 1915, Morgan and his students, Bridges, Sturtevant, and Muller, published their now classic *Mechanism of Mendelian Heredity*. Their further development of the chromosome theory will be discussed in detail in the next chapter. Here the relevance of that book is that they included a section that evaluated the reduplication hypothesis as an alternative to their hypothesis, called "linkage" and "crossing-over" (Morgan et al., 1915, pp. 74-77).

The reduplication hypothesis, they said, had continuously encountered anomalies to its predictions that necessitated adding new, ever more complicated assumptions. The original prediction of the theory was that gametic ratios would be in the series 1:1:1:1, 3:1:1:3, 7:1:1:7, 15:1:1:15, 31:1:1:31. However, character ratios had been found that could not be explained using these ratios, so more elaborate formulas had to be proposed. Moreover, the denial that segregation occurred at the reducing division necessitated assumptions about cell divisions early in the development of the embryo that, they claimed, had not been supported by evidence from experimental embryology. They detailed yet other anomalies that necessitated yet other changes to the theory. Some of the new evidence that they had discovered in *Drosophila* would, they suggested, necessitate yet further extensions (Morgan et al., 1915, pp. 74-77). This evaluation showed that the Morgan group used several theory assessment strategies, including **complexity**, **ad hocness**, and lack of easy **extendability** in the light of new coupling data.

Cock, as a historian evaluating Bateson's reduplication hypothesis with hindsight, also assessed it as inadequate:

> The theory, with its concentration on cell-division, reflects Bateson's embryological mode of thought, and the diagram with which he first expounded it [see Figure 9-1], showing parental types at opposite ends of one axis and recombinants at the ends of the perpendicular axis, betrays his concern with pattern. It was not a very good theory; quite apart from proving inadequate to fit the facts, it interpreted one unexplained phenomenon (the excess of non-recombinant offspring) in terms of another (the propensity of non-recombinant cells to undergo extra divisions) invoked **ad hoc** and equally unexplained. (Cock, 1983, p. 37)

Thus, in addition to these critiques, Cock added the criticism that reduplication predicted the existence of a cytological phenomenon which itself would be in need of explanation. Why would some cells that were precursors to some germ cells reduplicate while others did not? Bateson and his colleagues had suggested no answer to this additional question raised by the reduplication hypothesis.

When Bateson (1916) reviewed *Mechanism*, he expressed reservations about the chromosomal interpretation of linkage (which will be discussed in Chapter 10). He conceded, however, that the "hypothesis of reduplication" was "admittedly a very crude conjecture" which, nonetheless, had the advantage of being "non-committal" (Bateson, 1916, p. 460). Presumably he meant it was noncommittal about the nature of the factors, in that it did not claim they were material units arranged along chromosomes. The 1911 edition of Punnett's *Mendelism* contained a chapter on repulsion and coupling. In the 1919 edition, Punnett added a chapter on the chromosome theory. By the 1927 edition, the chapter on coupling and repulsion had been replaced by chapters on sex-linkage and chromosomal linkage. Coupling, repulsion, and reduplication were replaced by linkage and the chromosome theory.

In *Mechanism*, only one "Mendel's law" was stated (Morgan et al., 1915, p. 1). Reduplication was explained as producing exceptions to "segregation" (Morgan et al., 1915, p. 74). Although linkage was proposed to account for anomalies to the 9:3:3:1 ratios, it was not discussed in the context of a second Mendelian law or as an exception to a general process called "independent assortment."

According to Monaghan and Corcos (1984), Morgan first used the term "independent assortment" in 1913. He did not explicitly formulate a second Mendelian law until 1919 in his book, The *Physical Basis of Heredity*. In that book, Morgan detailed recent progress in genetics and used as chapter titles: "Mendel's First Law—Segregation of The Genes" and "Mendel's

Second Law—The Independent Assortment of The Genes." The two laws were discussed separately before the anomalies explained by linkage were introduced. In crediting the second law to Mendel, Morgan once again obscured his role in extending Mendelism, just as he did with the use of de Vries's term "mutant" and his claim that linkage confirmed predictions of de Vries's 1903 hypothesis (which explained independent assortment and did not predict linkage). Explicit separation of the two laws was yet another case of the role of anomalies in forcing implicit assumptions to be articulated and refined as the theory changed. Morgan explicitly separated and named the previously "implicit" second law.

This dispute between reduplication and linkage provides a possible test of a claim in philosophy of science. Philosophers of late have been given to talking about the "theory-ladenness" of observations, of the ways theories affect what is observed. Sometimes the claim goes so far as to imply that all observations are so thoroughly imbued with theory that data explained by one theory could not be explained by another. (See Suppe, 1977, for discussion and critiques of the "theory-ladenness" of observation.) This case did show "theory-laden" terminology for characterizing the anomalous data, and the terminology did change with the changing hypotheses. The numerical ratios, however, could be "unpacked" from the theory-laden terminology. The ratios counted as data for both hypotheses.

Bateson first used the terms "coupling" and "repulsing" to refer to the anomalous ratios. The hypothesis of coupling and repulsing factors (with whose terminology the observation language was laden) divided the phenomena into two separate domain items—the coupling of two dominant factors (when AB and ab classes were more common) and the repulsing of the dominant factors (when Ab and aB classes were more common). The nature of the proposed factor interactions (and thus the phenomenal, observation language dependent on them) was also influenced by the presence and absence theory, as was discussed earlier. Recessive factors could not couple, because, Bateson proposed, recessive characters were caused by the absence of the corresponding dominant allelomorphic factor. The shift to the reduplication hypothesis provided a unified explanation of coupling and repulsion ratios in terms of selective reduplications of cells. With that change, Bateson explicitly said the phenomenal language of "coupling" and "repulsing" characters should be abandoned.

Morgan used the phenomenal language of "coupled characters" when he criticized the hypothesis of "attractions and repulsions" (Morgan, 1911b, p. 384); however, for his own hypothesis he discussed "association" of factors. Perhaps he avoided "coupling" of factors so as not to evoke Bateson's physical analogies. With the development of the linkage hypothesis, the *Drosophila* group dropped the phenomenal language of "coupling" and "repulsion" of characters; they referred to both cases as the "linkage" of characters, which reflected their alternative theoretical view. No problem was posed in extracting the anomalous data from Bateson's work, which was reported in terms of numbers of organisms with certain combinations of characters. Bateson's numerical data could be explained by the hypothesis that factors were incompletely linked on the chromosomes. With the acceptance of the linkage hypothesis, the phenomena of coupled characters took on the new "theory-laden" terminology, which is today still in use, namely "linked characters." Similarly, with the development of the linkage hypothesis, Morgan himself changed from the terminology of "sex-limited" characters, that is, characters found in one sex but not the other, to "sex-linked." These cases showed agreement as to the actual numerical data, although the phenomenal language changed with changing theory.

Several strategies for theory (or hypothesis) assessment are reflected in the judgment that association was better than reduplication. Both hypotheses could explain the anomalous data for two trait linkage; thus, they did not differ significantly as to **explanatory adequacy** at the outset. When reduplication was extended to crosses with three or more traits, however, its formulas became quite complex and it necessitated elaborate differential cell divisions for which no cytological evidence existed (Cock, 1983, p. 48). Thus association was **simpler** and more **extendable**. Furthermore, Morgan's association hypothesis had in its favor that it appealed to visible structures, the chromosomes, and their observed intertwining. Counting against the hypothesis that chromosomes split and rejoined, on the other hand, was the problem that switching of pieces of chromosomes (later called "crossing-over") was only *inferred* from observed intertwinings of chromosomes. (Crossing-over was not observed cytologically until the 1930s; see Creighton and McClintock, 1931.) Thus, some empirical evidence existed that favored linkage over reduplication, but the empirical grounds for choosing between the two hypotheses were not decisive in 1915. No crucial Mendelian breeding experiment was proposed to adjudicate between the two hypotheses. This case contrasts with Castle's crucial experiment that ruled out the existence of modified genes (discussed in Chapter 8). Thus, the most important strategies for theory assessment in this case were not explanatory or empirical adequacy.

Instead, reduplication failed because it required numerous ad hoc assumptions to account for new evidence that made the hypothesis ever more complex. It failed to be easily extendable to multitrait crosses. Furthermore, as we will shortly see, the linkage hypothesis proved to be extremely fruitful in the development of a new line of research. Thus, the reduplication hypothesis failed when assessed using the strategies of **predictive adequacy, extendability, simplicity**, and **fruitfulness**.

Does the hindsight that linkage was a better hypothesis than either coupling or reduplication provide any lessons for evaluating strategies for producing new hypotheses? As I have argued elsewhere (Darden, 1980b), **using interrelations** to items in another field for which evidence exists is a good strategy. Bateson first appealed to vague analogies from physics with coupling and repulsion of factors. Bateson improved his method when he changed from coupling to reduplication—reduplication appealed to cytological processes of cell division. The cell divisions were unknown, however. Bateson did not interrelate the anomalous Mendelian ratios to any known cytological phenomena. Had his prediction of numerous differential cell divisions been confirmed by cytologists, then the reduplication hypothesis would have been an interfield theory for which evidence was subsequently found. Thus, Bateson's appeal to cytology was not as good a use of the interrelations strategy as was Morgan's because Bateson predicted cytological phenomena for which no evidence existed. Morgan's use of interrelations to cytology was better because he postulated (or extended an earlier) interrelation to known cytological entities, the chromosomes. Furthermore, he appealed to a visible phenomenon, chromosome intertwining. However, he too had to appeal to a *predicted* process, the breaking and rejoining of chromosomes. As will be discussed in the next chapter, the lack of cytological evidence for this behavior of the chromosomes served as a point of critique against Morgan's hypothesis of crossing-over until cytologists confirmed the prediction. The strategy of **using interrelations** can thus be refined to suggest that connections to known phenomena be explored before predicting as yet unknown items in the other field.

Figure 9-5 summarizes the strategies discussed in this chapter. Testing Mendel's law had produced the coupling anomalies. These anomalies resulted in the delineation of segregation from independent assortment. Then, the problem to be solved was how to explain coupled characters. Various strategies for producing new ideas and building them into plausible hypotheses are listed. Some theory assessment strategies can be seen to play a role in limiting the

testing of segregation produced anomalous ratios

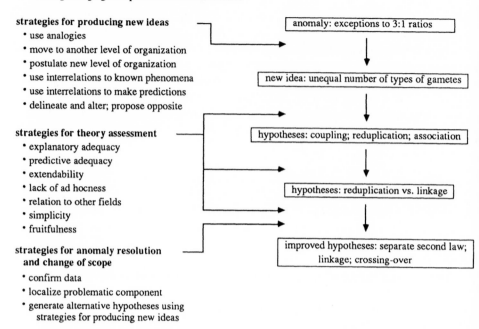

strategies for producing new ideas
* use analogies
* move to another level of organization
* postulate new level of organization
* use interrelations to known phenomena
* use interrelations to make predictions
* delineate and alter; propose opposite

strategies for theory assessment
* explanatory adequacy
* predictive adequacy
* extendability
* lack of ad hocness
* relation to other fields
* simplicity
* fruitfulness

strategies for anomaly resolution and change of scope
* confirm data
* localize problematic component
* generate alternative hypotheses using strategies for producing new ideas

anomaly: exceptions to 3:1 ratios

new idea: unequal number of types of gametes

hypotheses: coupling; reduplication; association

hypotheses: reduplication vs. linkage

improved hypotheses: separate second law; linkage; crossing-over

Fig. 9-5. Refined stages and strategies, VI

plausible hypotheses to those that explain the anomaly. The coupling, reduplication, and association hypotheses were evaluated. Coupling and reduplication failed; linkage and crossing-over became the new subjects of tests.

The empirical testing of the hypotheses of linkage and crossing-over will be discussed in the next chapter. Then, new anomalies arose and became the problems to be solved; the problem stage (as depicted in the diagram in Figure 9-5) is reached again. We now turn to the discussion of the development of the linkage hypothesis.

CHAPTER 10

The Chromosome Theory
and Mutation

10.1 Introduction

In 1912, Spillman, an early proponent of Mendelism, published a review article surveying the "present status of the genetics problem," as he put it. He reviewed the history of research on the problem of heredity, including Mendelism. In the first decade of the century, Mendelian investigations had focused on the issue of the universality of the law of segregation. That question, he claimed, was now settled. He went on to discuss puzzles facing geneticists, including the new findings about numerous instances of "pairing of unrelated characters" being found in animals and plants. Then he expressed hope that "Mendelists" would find new "working theories" and "stumble on to facts of a new kind that will give meaning to those we already have" (Spillman, 1912, p. 766). The field was in need of a fresh approach. Morgan and his students supplied it.

The work of the *Drosophila* group progressed rapidly between 1911 and 1915. In 1915, they published their (now famous) *Mechanism of Mendelian Heredity*. In that book they laid out a strong defense of the chromosome theory of Mendelian heredity. Their work leading up to that book, according to Morgan's biographer Allen, proceeded in "three directions": first, the construction of chromosome maps; second, "modifying and refining the basic Mendelian rules," such as work on multiple allelomorphs and multiple factors; and third, work initiated by the discovery of non-disjunction, the occasional abnormal distribution of chromosomes in the formation of germ cells (Allen, 1978, p. 172). Some of the *Drosophila* group's work on the second line of research has already been discussed in Chapters 6 and 9. In addition to arguing that the one unit-one character concept needed to be altered to allow multiple factors to affect one character, they also discussed instances in which one factor affected several characters. Thus, the one-to-one relationship was expanded to include not only many-to-one, but also one-to-many relations between factors and characters.

Section 10.2 will discuss mapping and non-disjunction, which were major developments associating Mendelism and chromosomes. Section 10.3 discusses Bateson's review of *Mechanism*, showing the opposition of a staunch Mendelian to the *Drosophila* group's chromosome theory. Section 10.4 discusses Castle's different assessment of the chromosome theory. Castle came to accept the view that factors were parts of chromosomes, but he objected to the assumption that factors were *linearly* arranged along the chromosome threads. Section 10.5 surveys the range of modular hypotheses that were explored as alternatives to the hypotheses proposed by the *Drosophila* group. Section 10.6 introduces the new problem of mutation that arose for Mendelism and indicates the new research problems that Muller developed. These sections complete the historical development of the components of the theory of the gene between 1900 and 1926, except for the next chapter's discussion of the development of claims about genes. Finally, Section 10.7 discusses the strategies exemplified in these changes, especially the strategies of **using interrelations** and **introducing an analog model**. Interrelations between genes and chromosomes were represented by the analog model of "beads on a string." The interrelations and analog model functioned in providing new ideas, in developing new terminology, and in resolving anomalies. Various theory assessment strategies exemplified in Bateson's and Castle's assessments of, and challenges to, the chromosome theory are summarized. The section closes with discussion of the strategy exemplified in the mutation case—**take an old idea and change it** to produce a new theoretical component.

These developments added significant, new theoretical components to Mendelism. Linkage turned out to be an anomaly that could not be explained away as a monster; instead, it was a "model" anomaly (to be defined in the next section), whose resolution required extensive theory modification. New components to account for linkage ratios became important parts of the theory of the gene.

10.2 Mapping and Non-disjunction

Sturtevant developed a technique for mapping the linear relations among factors in linkage groups, using Mendelian breeding data. Although genetic maps are sometimes referred to as "chromosome" maps, it is important to realize that they are constructed using data from Mendelian breeding experiments, not by using cytological techniques to observe chromosomes.

A suggestion of Morgan's provided the basis for the mapping techniques. He suggested that the degree of linkage between two characters could be explained in terms of the linear distance of their factors along the chromosome. Factors closer together would produce characters more often linked in inheritance than factors further apart (Morgan, 1911a). Sturtevant developed Morgan's idea into a quantitative technique (Sturtevant, 1965, p. 47). He used data from breeding experiments showing linked characters (with crossing-over) to calculate the relative positions of the factors to each other. This method yielded quantitative predictions. Symbolically, assume the factors A, B, and C are linked in a line in that order; if A and B are linked with a (relative) value of 2, and B and C are linked with a value of 3, then the prediction would be that A and C are linked with a value of 5. Sturtevant calculated linkage values for six sex-linked factors in *Drosophila*. He summarized his results in a linear diagram showing the relative positions of the factors and quantitative values for their "strength of association," that is, relative linkage values (Sturtevant, 1913b). We will not pause here to go into the details of how the number of cross-overs were used to develop relative linkage values. These methods were further developed by the *Drosophila* group and the refined methods are familiar to students of Mendelian genetics today, as the methods became a standard tool in genetic analysis (for further discussion, see Carlson, 1966; Strickberger, 1985).

The *Drosophila* group found, as was expected, four groups of linked factors, corresponding to the four pairs of chromosomes in fruit flies. They thus were able, by 1915, to produce four linear

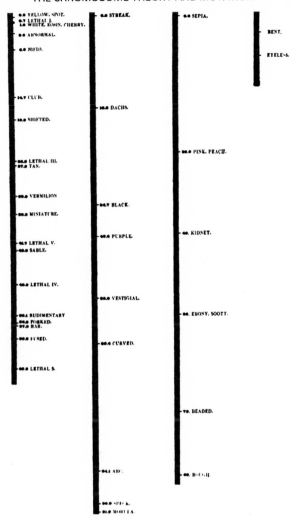

Fig. 10-1. Linkage maps for *Drosophila* in 1915 (From Morgan, Thomas H.; A. H. Sturtevant; H. J. Muller; and C. B. Bridges (1915), *The Mechanism of Mendelian Heredity*. New York: Henry Holt and Company, frontispiece)

linkage maps showing, they claimed, the relative positions of the factors along the four chromosomes. Their 1915 maps are shown in Figure 10-1. The visual model that they developed for the factors located along the chromosomes was that of beads along a thread. Beside the diagram of the bead model, they showed crossing-over between homologous chromosomes (see Figure 10-2). The fine structure of the chromosome that the hypothetical model depicts could not be *observed* to have beadlike parts or to break and rejoin during intertwining. The diagrams depict *hypotheses* about chromosomes that explain data from genetic breeding experiments.

Some of the predicted linkage values were smaller than expected for factors distant from each other along the maps. To explain such anomalies in their data, the *Drosophila* group suggested two cross-overs occurred in the anomalous cases. They encountered a different kind of anomaly when their predictions failed for factors very close together. This anomaly was explained by an additional hypothesis that they called "interference," in which one cross-over was postulated to

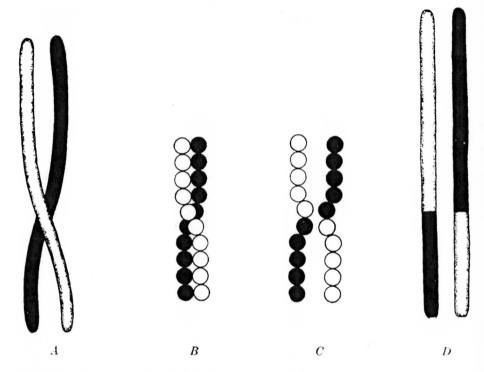

Fig. 10-2. Diagram to represent crossing-over (From Morgan, Thomas H.; A. H. Sturtevant; H. J. Muller; and C. B. Bridges (1915), *The Mechanism of Mendelian Heredity*. New York: Henry Holt and Company, p. 60)

interfere with the formation of the other. They explicitly used their analogy to intertwining threads in presenting (and perhaps in constructing) this hypothesis:

> . . . Suppose that crossing over results from a breaking of the threads at some point due to the strain of a very tight twisting, and that a break at one point relieves the strain in the vicinity, thus tending to prevent another nearby. (Morgan et al., 1915, p. 64, footnote 1)

In sum, additional hypotheses of double cross-overs and interference explained various anomalies in the linkage values. The chromosome theory proved fruitful in allowing the development of the new technique of mapping, using cross-over data from breeding experiments to make inferences about the relative positions of factors along the chromosomes. The model of beads on a string associated with this further development of the chromosome theory proved fruitful in explicating crossing-over and useful in developing the new hypotheses of double cross-overs and interference.

In addition to anomalies that resulted in these new hypotheses, Bridges found yet other anomalies for predicted ratios of inheritance of sex-linked characters in *Drosophila*. As Morgan (1910c) had shown, when a female with white eyes was mated to a male with red eyes, the normal result was that the daughters had red and the sons white eyes. As we discussed in Chapter 9, Morgan explained this result by claiming that eye color was completely linked with the *X* chromosome. Since the F_1 male received its one *X* chromosome from the white-eyed mother, then all males were white, whereas the F_1 female received one *X* chromosome from its father, which carried the dominant red factor. Bridges repeated Morgan's experiments and occasionally found anomalies: about 5 percent of the daughters showed, not the expected red, but white; also about 5 percent of the males showed red eyes (Bridges, 1913, p. 588).

Bridges predicted that the abnormalities in the breeding ratios were the result of what he called "non-disjunction." In non-disjunction, a small percentage of the eggs contained two sex chromosomes instead of the normal one, while corresponding eggs contained no sex chromosomes. Thus, the anomalous flies resulted from the abnormal eggs being fertilized: females with two X chromosomes from the mother and none from the father would be white eyed, while the eggs with no X chromosomes that were fertilized by sperm with the X carrying the factor for red would produce the 5 percent red-eyed males.

Bridges (1913) conceded that no direct cytological evidence yet existed for non-disjunction in *Drosophila*. In support of his hypothesis, however, he cited evidence in species other than *Drosophila* of abnormal numbers of sex and autosomal chromosomes; such abnormalities in chromosomes had been observed by Wilson, Stevens, and others (Bridges, 1913). He then turned from the breeding experiments that had produced the anomalies to careful cytological work on the abnormal fruit flies to test his prediction. Thus, he used abnormal ratios produced in *Mendelian* breeding experiments to make a prediction about the nature of the *cytological* phenomena. Taking up the tools of the cytologist, he tested his prediction by examining the chromosomes of the fruit flies showing the abnormalities in breeding. His work showed the facility that Morgan's and Wilson's students developed in using the techniques of both fields.

Bridges confirmed his prediction. Instead of the expected sex chromosome pairs, he found flies with two X chromosomes along with the Y, namely, XXY. The failure of the sex chromosomes to separate normally was called "non-disjunction." The anomalous Mendelian ratios thus were explained by monstrous chromosome abnormalities. The correspondence of abnormalities between the inheritance of sex-linked characters in Mendelian breeding experiments and abnormalities in the visible sex chromosomes was a striking correlation (Bridges, 1914). Bridges indicated his view of the importance of this correlation for the chromosome theory in a paper entitled "Non-disjunction as Proof of Chromosome Theory of Heredity" (Bridges, 1916).

Chromosomes	Factors
1. Number of homologous chromosome pairs is species-specific	1. Number of linkage groups is species-specific and corresponds to the number of homologous chromosome pairs
2. Abnormalities in chromosome number, e.g., non-disjunction XXY	2. Abnormalities in character inheritance caused by abnormal numbers of factors, e.g., abnormal sex-heredity in non-disjunction
3. Linear, threadlike appearance of chromosomes	3. Linear arrangement of factors in linkage groups
4. Intertwining of chromosomes during germ cell formation; Prediction: breaking and joining at cross-over points	4. Incomplete linkage of characters in linkage groups caused by crossing-over of factors in linkage groups
5. Prediction: double cross-overs; interference of one cross-over with another	5. Mapping yields predicted values for distances, but some anomalies

Table 10-1. Relations between chromosomes and factors, 1915.

The new hypotheses about double cross-overs, interference, and non-disjunction expanded the list of interrelations between chromosomes and factors. Table 10-1 shows the new additions to the list that expanded the claims of Sutton and Boveri in 1903-1904 (Table 7-1) and Morgan in 1911 (Table 9-1). Cytology, as Morgan put it, provided the "mechanism" of Mendelian heredity. The factors were part of the chromosomes. Chromosome movement and crossing-over provided physical, mechanical processes during germ cell formation that resulted in the subsequent characters in offspring. Ratios of those characters were the results produced in genetic breeding experiments.

The chromosome theory of Mendelian heredity had, the *Drosophila* group claimed, much evidence in its favor, based on the Mendelian and cytological results in *Drosophila*. Correlating the breeding data with the cytological observations, playing off results from one field against those in the other, using techniques from one field to get independent evidence about the findings in the other, developing a visual analog model for the chromosome theory—these were the fruits of the interrelations between genetics and cytology. Even so, not everyone was convinced that the *Drosophila* group had provided adequate proof for their version of the chromosome theory, as the next two sections will discuss.

In Chapter 5, I stated a component of Mendelism for the explanation of dihybrid crosses as of 1900-1903 in the following way:

C6. [later called "independent assortment"]

C6.1 In crosses involving parents that differ in two pairs of differentiating characters [dihybrid crosses], the pairs behave independently. Symbolically, (AB + Ab + aB + ab) (AB + Ab + aB + ab) = complicated array that in appearance [given complete dominance] reduces to 9AB:3Ab:3aB:1ab.

In Chapter 9, I developed a set of components that were more explicit than any that I found in the historical record for the assumptions about dihybrid crosses using the factor hypothesis, as of about 1910:

C6. Explanation of dihybrid crosses:

C6.1 In the formation of germ cells in a hybrid produced by crossing parents that differed in two or more traits, the parental factors segregate or separate, so that the germ cells are of all possible pure parental combinations. Symbolically, $AABB$ x $aabb$ = F_1 hybrid $AaBb$, which produces germ cells symbolized by AB, Ab, aB, and ab.

C6.2 The different types of germ cells are formed in equal numbers. Symbolically, equal numbers of germ cells of types AB, Ab, aB, and ab are formed, i.e., gametic proportions are 1:1:1:1.

C6.3 When two hybrids are fertilized (or self-fertilization occurs), the differing types of germ cells combine randomly. Symbolically, (AB + Ab + aB + ab) (AB + Ab + aB + ab) = complicated array that in appearance [given complete dominance] produces 9:3:3:1 character ratios in the F_2 generation.

By 1915, the claims in these components of 1910 were known to be too general. Anomalous data for the 9:3:3:1 ratios, as Chapter 9 discussed, resulted in the separation of the law of segregation and what came to be called the "law of independent assortment." Components 6.1 to 6.3 state the law of independent assortment. The linkage anomalies can be analyzed as causing those components to be specialized: not all traits assort independently because those in different linkage groups (on different chromosomes) show independent assortment. In contrast to the anomalies for segregation discussed in Chapter 8, the anomalous ratios for independent assortment showing linked characters could not be explained away. They were not due to an incorrect interpretation of the data or to monstrous, quirky cases. Anomalies of linked characters appeared frequently in many crosses in many species; the linkage anomalies resulted in a modification of the Mendelian theory. By analogy with the term "model organisms," the linkage anomalies in *Drosophila* may be called "model anomalies." They served as models for the normal cases of linkage. They also showed how to modify the theoretical components of independent assortment. Monster anomalies (introduced in Chapter 8) could be explained away as one quirky case (e.g., some of de Vries's *Oenothera* mutants) or as a malfunction class, not requiring modification of the theoretical components (e.g., homozygous lethals and non-disjunction). Model anomalies could not be explained away without theory modification. Responses to the model linkage anomalies exemplify the **specialize and add** strategy: the newly delineated law of independent assortment was specialized, followed by the addition of new components to explain the linkage anomalies.

Morgan's statement of the theory of the gene in 1926 included the newly delineated second law and claims about linkage and crossing-over:

> The theory . . . states that the members belonging to different linkage groups assort independently in accordance with Mendel's second law; it states that an orderly inter-change—crossing-over—also takes place, at times, between the elements in corresponding linkage groups; and it states that the frequency of crossing-over furnishes evidence of the linear order of the elements in each linkage group and of the relative position of the elements with respect to each other. (Morgan, 1926, p. 25)

Although these components of the theory were formulated by the *Drosophila* group in the context of developing the chromosome theory, Morgan did not mention chromosomes in his statement of the theory of the gene. He emphasized that the theory was supported "from purely numerical data" of Mendelian breeding experiments (Morgan, 1926, p. 32). He then went on to provide evidence that the chromosomes and their behavior furnished the "mechanism of heredity" (Morgan, 1926, p. 32). Morgan was aware that the chromosome theory was controversial. He thus separated the Mendelian "theory of the gene," supported by evidence from breeding experiments, from the chromosome theory. Although the theory of the gene was developed by the *Drosophila* group by using interrelations to chromosomes, all the components of it were supported by breeding data containing no reference to chromosomes.

My statements of the components of the theory of the gene, which replaced the components for explaining dihybrid crosses, follow Morgan's language closely:

C6'. Assortment, linkage and crossing-over

C6.1' Genes are found in linkage groups; groups occur in corresponding pairs.

C6.2' Genes of different linkage groups assort independently.

C6.3' Usually genes in the same linkage group are inherited together; at times,

however, an orderly interchange, called "crossing-over," takes place between allelomorphs in corresponding linkage groups.

C6.4' Genes in a linkage group are arranged linearly with respect to each other.

C6.5' The frequency of crossing-over can be used to calculate the order and relative positions of the respective genes in linkage groups if such disturbing factors as double cross-overs and interference of one cross-over with another are taken into account.

These components were challenged between 1915 and 1926, but they survived the attacks to become components of the theory of the gene and part of "Mendelian" genetics today. Objections and alternatives to them proposed by Bateson and Castle will be the subject of the next two sections.

10.3 Bateson's Objections to the Chromosome Theory

As discussed in Chapter 7, before 1911 Bateson argued against Sutton's chromosome theory, as had Morgan. Morgan changed his views in 1911, but Bateson held a negative view of the theory much longer, into the 1920s, and probably harbored doubts about some aspects of the *Drosophila* group's theory until his death in 1926 (Coleman, 1970; Cock, 1983). Bateson reviewed *Mechanism* in 1916. In that review, he reiterated some of the objections that he had raised earlier to the chromosome theory (discussed in Chapter 7). Although he praised the *Drosophila* group for their extensive work with many crosses in many fruit flies, he was skeptical about their hypotheses for explaining partial coupling, especially the hypothesis of crossing-over.

Bateson emphasized that no cytological evidence existed for the actual breaking and rejoining of the pieces of the chromosomes during their observed intertwinings (Bateson, 1916, p. 455). The importance of this objection was emphasized years later by Punnett. In 1950, Punnett reflected on the reasons that he and Bateson had not accepted linkage and crossing-over as an explanation for the partial coupling, a phenomenon that they had been the first to discover. Punnett claimed that they had interpreted Boveri's work on the individuality of the chromosomes as showing that ". . . breakage and recombination was forbidden. For to break the chromosome would be to break the rules" (Punnett, 1950, p. 10). One wonders why Bateson did not cite Boveri's work in his review if it provided his evidence against the breaking of chromosomes. As discussed in Chapters 7 and 9, Boveri predicted that some sort of interchange of Mendelian factors between chromosomes might occur, and Morgan interpreted him as believing crossing-over was consistent with the evidence about chromosome individuality (Morgan, 1911c).

The Morgan group had found no crossing-over in male *Drosophila*. The lack of crossing-over in males was plausible, Bateson said, for sex-linked characters, because the X and Y chromosomes probably differed (although the visible cytological evidence was not yet adequate on that point). On the other hand, no crossing-over in non-sex (autosomal) linkage groups in the male would cast doubt on the whole hypothesis of crossing-over unless it could be shown that the homologous chromosomes in male *Drosophila* did not intertwine.

Bateson criticized *Mechanism* for providing too little actual data from the experiments to allow independent calculations about the linkage relations. He questioned whether the data were sufficient to show that only four linkage groups had been found, and he expressed skepticism that no evidence existed for linkage between factors in different groups. (This criticism is not surprising because nothing in Bateson's own hypotheses about coupling of factors and reduplication of germ cells had prepared him to think in terms of the intermediate level of organization of linkage groups of factors.) Crucial evidence for the linkage hypothesis, he suggested, would be whether the number of linkage groups correlated with the numbers of pairs of chromosomes in other species.

Bateson also leveled the charge that ad hoc modifications had been made to account for anomalies in the predicted linkage values: "The machinery for dealing with unconformable cases is extraordinarily complete" (Bateson, 1916, p. 460). After discussing such additions as the hypothesis of interference and the hypothesis that a factor had been found that modified the amount of crossing-over, Bateson asked: "Can the action of all these processes be severally traced? Can their consequences be distinguished from each other, and especially from those of multiple crossing-over?" He concluded:

> Meanwhile the suspicion is unavoidable that, given a conviction that the factors *must* be arranged in rows along four chromosomes, the various interpretations provide rather a method, or perhaps we should say alternative methods, by which the facts can be reconciled with the hypothesis, than a proof that this hypothesis is correct. (Bateson, 1916, p. 460)

He did concede that Bridges's work on non-disjunction showing the relation of abnormalities in sex-linked characters to sex chromosomes made it difficult to deny that the two showed a "very special relation" (Bateson, 1916, p. 462). Complete coupling of some characters with sex, was, he thought, the best evidence for associating particular characters with a particular chromosome. From Bateson's perspective, non-disjunction had the advantage of being a phenomenon that did not involve crossing-over, because cross-over was a hypothesis to which he strongly objected. Furthermore, Bateson himself had tried to explain characters that were coupled to sex (discussed in Chapter 9); correlations between sex and chromosomes were more plausible to him than the breaking of chromosomes during crossing-over.

Bateson's desire for some sort of view of the genes ("gens") based on physical analogies, rather than as particles on the chromosomes (discussed in Chapter 9), reappeared in his review:

> The properties of living things are in some way attached to a material basis, perhaps in some special degree to nuclear chromatin; and yet it is inconceivable that particles of chromatin or of any other substance, however complex, can possess those powers which must be assigned to our factors or gens. The supposition that particles of chromatin, indistinguishable from each other and indeed almost homogeneous under any known test, can by their material nature confer all the properties of life surpasses the range of even the most convinced materialism. Hence it may well be imagined that even if cytologists decide that in synapsis there is no anastomosis and no transference of material [i.e., the breaking and rejoining of crossing-over], the effective transference of the gens may occur. The transference may be one of "charges." Perhaps even we might profitably consider whether the chromosomes may not be thrown up, and the gens *grouped along their lines* by the interplay of the same forces. (Bateson, 1916, p. 462, my emphasis)

Trying to understand Bateson's nonmaterial, nonvitalistic view of the genes as "charges" is a fascinating historical problem, but that is not our task here (for further discussion of this point, see Coleman, 1970; Darden, 1977; Cock, 1983). Within the context of analyzing his assessment of the *Drosophila* group's extensions of the chromosome theory, the important point to note is that he objected to considering the genes to be material particles, but he had no objection to considering them as something else arranged *linearly*, or, as he said, "grouped along their lines." Although a skeptic about the chromosome theory, Bateson seemed willing to concede a possible *linear* relation among the genes. In contrast, as will be seen in the next section, Castle came to accept the chromosome theory and the existence of single cross-overs, but he questioned linearity. Castle formed alternative hypotheses to account for the anomalies that the *Drosophila* group explained by linearity, double cross-overs, and interference. The seemingly unified hypotheses of sex linkage, autosomal linkage, crossing-over between homologous chromosomes, linearity, double cross-over, and interference could be delineated and altered separately. Such modularity of the components is evident from Bateson's and Castle's different assessments and alternative hypotheses.

In sum, Bateson raised numerous problems about the adequacy of the evidence for the chromosome theory: no cytological evidence for breakage and rejoining; no crossing-over in male *Drosophila*; inadequate evidence for only four linkage groups in *Drosophila*; and the ad hoc additions of interference and double cross-overs. He concluded that alternative models for the nature of factors were still plausible. Cock (1983) suggested other reasons for Bateson's opposition, in addition to Bateson's arguments against the chromosome theory based on the evidence: Bateson's desire for a theory of heredity that would also explain embryological development; Bateson's distaste for microscopy; the lack of a cytologist in Bateson's laboratory; and Bateson's personal dislike of Morgan. An individual's assessment of a theory thus may depend on numerous criteria, only some of which have an evidential basis.

After 1916, evidence in favor of the chromosome theory mounted, especially as a result of the continued *Drosophila* work. As to the crucial evidence about the number of linkage groups, by 1919 Morgan claimed that the best evidence for that correspondence was still the four linkage groups in *Drosophila*, corresponding to its four pairs of homologous chromosomes. In addition, he was able to state that "in no other animal or plant does the number of linkage groups exceed the number of chromosome pairs" (Morgan, 1919, p. 133). By 1926, Morgan added a section of a chapter entitled "The Number of the Linkage Groups and the Number of Chromosome Pairs." In addition to *Drosophila*, he cited evidence for the correlation in sweet peas, the edible pea, Indian corn, and the snapdragon. He was still able to assert that no more independent linkage groups were known than chromosome pairs in any organism (Morgan, 1926, pp. 36-37). The genetic evidence usually progressed more rapidly than the cytological; chromosome numbers were difficult to determine accurately, especially in mammals. For Castle and Wright's guinea pigs, for example, the estimates of chromosome number by 1926 varied from eight to thirty. Only a few different Mendelian factors were known in guinea pigs and little evidence for linkage in them or in other rodents existed (Provine, 1986, pp. 120-121). The kind of work done with *Drosophila* was not easy to duplicate in other species, especially mammals.

After a visit to Morgan's laboratory in 1921, Bateson changed his views about the association of Mendelian factors with chromosomes, if not about the hypothesis of crossing-over (Cock, 1983, pp. 20-21). In a paper of 1922, Bateson discussed the Mendelian discoveries as pushing the analysis of heredity from the zygote (the fertilized egg) back to the gametes. He continued:

> We have now turned another bend in the track and behind the gametes we see the chromosomes. For the doubts—which I trust may be pardoned to one who had never seen the marvels of cytology, save as through a glass darkly—can not, as regards the main thesis of the *Drosophila* workers, be any longer maintained. The arguments of Morgan and his

colleagues, and especially the demonstrations of Bridges [of non-disjunction], must allay all scepticism as to the direct association of particular chromosomes with particular features of the zygote. The transferable characters borne by the gametes have been successfully referred to the visible details of nuclear configuration. (Bateson, 1922, p. 392)

Punnett, Bateson's student, also came to accept the chromosome theory and incorporate it into later editions of his popular book, *Mendelism*. He also included diagrams of chromosomes crossing-over and the beads on a string model for factors in the chromosomes (e.g., Punnett, 1927, ch. 8).

10.4 Castle and the Debate about Linearity

Castle came to accept the chromosome theory sooner than Bateson did. Like Bateson, Castle was an early Mendelian who did not use cytology in association with his Mendelian research. Castle worked with mammals; their chromosomes were much more difficult to study than those of *Drosophila* (Provine, 1986).

Even though he came to accept the claim that genes are parts of chromosomes, Castle continued to play his role as a critic of hypotheses proposed by the *Drosophila* group. One of his challenges was to their hypothesis of the linear arrangement of genes along chromosomes. Historical discussions of Castle's challenge to linearity usually begin with his first paper on that topic in 1919 (Castle, 1919b; for discussion, see Carlson, 1966, p. 79; Allen, 1978, p. 267; Taylor, 1983, p. 195; Wimsatt, 1987). I believe the origin of Castle's challenge goes back at least to a 1909 paper in which Castle published diagrams that looked like representations of two-dimensional organic molecules. The diagrams represented the relations among unit-characters for coat color in rabbits (see Figure 10-3). He speculated that the protoplasm contained organic molecules "built up in some such way, and that regressive variations arise by dropping off the constituent

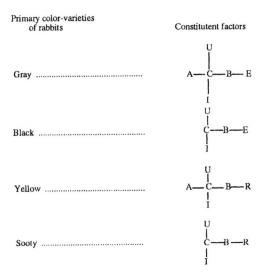

Fig.10-3. Castle's Diagrams for unit characters in rabbits.

parts of the molecule one by one" (Castle, 1909, p. 156). Castle's chemical view was endorsed by Johannsen (1913, pp. 531, 666). Both Castle and Johannsen were seeking ways to characterize the genes in terms of structural, two-dimensional, chemical models. The interrelation that Castle sought related genetics and chemistry, not genetics and cytology. His chemical models were two-dimensional, not linear.

Morgan introduced the idea of linearity in his first papers relating sex factors to the sex chromosomes. Although he speculated that the factors were "chemical substances" (Morgan, 1911a, pp. 365, 404), he did not postulate two-dimensional chemical models, as had Castle. Morgan stated that the factors were "placed lineally along the chromosome" without explicitly providing any separate argument for linearity. His focus was on arguments for the association of factors with the sex chromosomes and the twisting and splitting of chromosome threads to explain partial coupling (Morgan, 1911a p. 404). Thus, in the very first published version of his hypothesis of linkage, Morgan seemed to *assume* that the arrangement of factors in the chromosomes was linear. That the appearance of the chromosomes showed linear, threadlike bodies was perhaps the reason for this (almost implicit) assumption.

Sturtevant recounted his role in developing the first linkage maps, which were based on the linearity assumption:

> In 1909, Castle published diagrams to show the interrelations of genes affecting the color of rabbits. It seems possible now that these diagrams were intended to represent developmental interactions, but they were taken (at Columbia) as an attempt to show the spatial relations in the nucleus. In the latter part of 1911, in conversation with Morgan about this attempt—which we agreed had nothing in its favor—I suddenly realized that the variations in strength of linkage, already attributed by Morgan to differences in spatial separation of the genes, offered the possibility of determining sequences in the linear dimension of a chromosome. I went home and spent most of the night (to the neglect of my undergraduate homework) in producing the first chromosome map . . . . (Sturtevant, 1965, p. 47)

For the linearity assumption to work, and the cross-over frequencies to yield correct predictions about linear relations, the *Drosophila* group (as discussed in Section 10.2) postulated double cross-overs and interference of one cross-over with the other. Castle came to accept the Morgan group's claim that genes were located in the chromosomes and that they had definite relations to each other, but he questioned the linearity component of the theory:

> Morgan had suggested that what binds or links two characters together is the fact that their genes lie in the same body within the cell-nucleus. Such bodies he supposes are the chromosomes. The evidence for this conclusion is very strong. Morgan and his associates have demonstrated the existence in *Drosophila* of four groups of linked genes corresponding with the four pairs of chromosomes which the cell-nucleus of *Drosophila* contains. Morgan has further suggested (and has beyond doubt established the fact) that the genes within a linkage system have a very definite and constant relation to each other. He supposes their arrangements to be linear . . . .

> That the arrangement of the genes within a linkage system is strictly linear seems for a variety of reasons doubtful. (Castle, 1919b, pp. 25-26)

Castle's most strenuous objection to linearity focused on the discrepancy between predicted linkage values (based on the assumption of one cross-over) and the actual ones found. The *Drosophila* group had introduced the additional hypotheses of double cross-overs and interference to explain the anomalies. Castle was willing to allow one instance of breakage and reunion between homologous chromosomes; however, he objected to the "subsidiary hypotheses" of double cross-overs and interference to account for the deviations from predicted values. Castle proposed a model that differed from the linear model of beads on a string proposed by the Morgan group. In his published account, he seemed to be recounting steps in reasoning to the hypothesis: "If the arrangement of the gens is not linear, what then is its character? This query led me to attempt graphic presentation of the relationships indicated by the data of Morgan and Bridges but finding this method unsatisfactory I resorted to reconstruction in three dimensions" (Castle, 1919b, p. 28).

Perhaps by "graphic presentation" he meant the sort of two-dimensional diagrams that he had used in 1909 (see Figure 10-3). If so, then the strategy he seems to have employed was, first, to question linearity, then to try to develop a two-dimensional model, which failed. Next, he proposed a three-dimensional model. If genes were parts of chromosomes and had "definite and constant" relations to each other, then the three possible arrangements were linearity, two-dimensional relations in a plane, or three-dimensional relations. The linear appearance of the chromosome made plausible a linear arrangement of its parts, but other arrangements were possible; the size of the genes was unknown and no visual evidence about the genes themselves supported linearity.

Castle's alternative model was three-dimensional, with distance in three-dimensional space calculated by using cross-over frequencies. Castle analyzed Morgan and Bridges's data from *Drosophila* crosses. After criticizing their interpretation of linearity as not being consistent with their own data, he used their data to construct a three-dimensional model with rings of wire representing the genes and single wires between them representing their distance apart. The model was of "the form of a roughly crescentic plate longer than it is wide and wider than it is thick" (Castle, 1919b, p. 28) (see Figure 10-4 for Castle's diagram). Castle's model had discrepancies of its own, however, given his assumptions about crossing-over between points in the three-dimensional model. In order to account for the anomalous data, he proposed that "transverse breaks are more frequent than oblique longitudinal ones" (Castle, 1919b, p. 31).

Castle speculated about what his model actually represented:

> What, it might be asked, does this reconstruction signify? Does it show the actual shape of the chromosome, or at any rate of that part of it in which the observed genetic variations lie? Or is it only a symbolical representation of molecular forces. These questions we can not at present answer. (Castle, 1919b, p. 30)

Despite this seeming agnosticism about what the model represented, his discussions showed the continuing influence of the model of an organic molecule. One of his critiques of linearity rested on his claim that it is doubtful "whether an elaborate organic molecule ever has a simple string-like form" (Castle, 1919b, p. 26). Castle speculated that crossing-over was "like the replacement of one chemical radical with another within a complex organic molecule and it seems highly probable that such is its real nature" (Castle, 1919b, p. 32).

The *Drosophila* group nicknamed Castle's model the "rat-trap model" because of its wire cage appearance (Carlson, 1966, p. 80). They took up Castle's challenge to linearity and responded

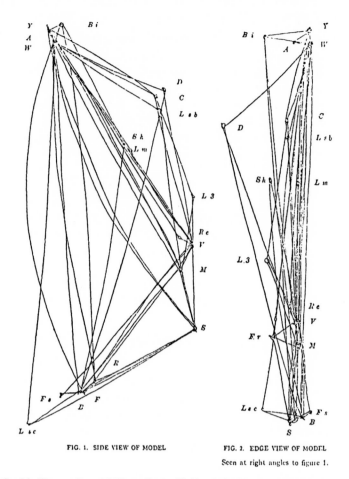

FIG. 1. SIDE VIEW OF MODEL   FIG. 2. EDGE VIEW OF MODEL

Seen at right angles to figure 1.

Fig. 10-4. Castle's "Rat-trap" model (From Castle, W. E. (1919), "Is the Arrangement of the Genes in the Chromosome Linear?" *Proceedings of the National Academy of Sciences* 5, p. 29).

in detail to his critiques in a series of papers. Sturtevant, Bridges, and Morgan (1919) criticized Castle's comparison to organic molecules. That organic molecules probably never have a stringlike form was beside the point, they said, because "*chromosomes* do have a thread-like form" (Sturtevant, Bridges, and Morgan, 1919, p. 169). Some of the discrepancies in their data were defended as within the bounds of experimental error, but they admitted to making minor errors in other values, which they corrected.

To respond to Castle's critique about the assumption of double cross-overs, they claimed that by doing experiments with mutants at loci only short distances apart, the whole X chromosome had been analyzed. The results could be explained by assuming only single cross-overs. Such data, both published and unpublished, they claimed, supported a linear interpretation and made plausible the assumption that double cross-overs were the explanation for results of crosses involving longer distances. After examining the particular three-dimensional arrangement suggested by Castle, they argued that it was approximately linear, with the few exceptions explained as being due to too few flies to be significant. Furthermore, they criticized Castle's additional (ad hoc) assumptions about the locations of breaks in his model. Their conclusion was that linearity with occasional double cross-overs was the better hypothesis.

Muller (1920) responded to Castle's attack on linearity in much more detail than the shorter paper by Sturtevant, Bridges, and Morgan (1919). Muller's lengthy paper included numerous arguments, based on consideration of a priori assumptions, additional data about the loci Castle analyzed, and difficulties with using Castle's own assumptions in constructing three-dimensional models for *Drosophila* data. Some of his critiques echoed those of Sturtevant, Bridges, and Morgan's earlier response, but others were different. Muller, for example, claimed that double cross-overs were, a priori, no less likely than single ones: if the chromosomes could break once, they could break again.

When Muller tried to construct three-dimensional models for numerous *Drosophila* genes, he found it impossible to produce consistent models. He admitted that some anomalies remained for a linear interpretation, but he explained them in various ways, such as claiming that one particular mutant factor decreased the amount of crossing-over and distorted the values. He concluded, however, that the evidence, on the whole, based on hundreds of thousands of flies, supported the hypotheses of linearity and double cross-overs.

The *Drosophila* group's critiques finally convinced Castle. By the 1924 edition of his book, *Genetics and Eugenics*, Castle said of the three-dimensional model: "This explanation has met with more difficulties than it has cleared away" (Castle, 1924, pp. 212-213). Castle's student, Dunn, said of the controversy in Castle's biography:

> Castle's failure to understand (or to accept) some of the subsidiary hypotheses of the chromosome theory caused the *Drosophila* workers a lot of trouble, but it also caused them to re-examine their own evidence, in which some minor errors were found, to carry out additional experiments, and to scrutinize their assumptions more rigorously. The linear arrangement hypothesis survived all this and was strengthened by the controversy. (Dunn, 1965b, p. 45)

In summary, Bateson and Castle showed different responses to the anomalies that arose for the predicted values based on the linearity assumption. Bateson, for these and other reasons, doubted the entire chromosome theory, but did not object to the hypothesis of a linear arrangement of genes. Castle, on the other hand, analyzed the *Drosophila* group's version of the chromosome theory into separable components, and questioned both linearity and the supposedly ad hoc hypotheses of double cross-overs and interference to save it. The various strategies of theory assessment exemplified in Bateson's and Castle's responses to the chromosome theory will be discussed in Section 10.7.

10.5 Modularity and Alternative Hypotheses

Various hypotheses developed and defended by the *Drosophila* group were systematically related. They provided a coherent account of linkage data by claiming that genes were beadlike particles arranged linearly along chromosomes that switched pieces in single crossing-over, double crossing-over, and showed the interference of one cross-over with another. The challenges of Bateson, Castle, and Goldschmidt (discussed briefly in Chapter 9), however, show that a number of modular hypotheses could be delineated and asserted or denied separately. Thus, the strategy of **delineate** separable assumptions and **alter** them independently is exemplified in these challenges.

Bateson, Castle, and Goldschmidt proposed alternatives to different hypotheses. An analysis of their challenges shows the range of plausible hypotheses that were debated. Bateson suggested retaining the linearity relations among genes, but giving up the idea of factors as material particles along the chromosomes. Unfortunately, he never developed his alternative about the transference of linearly arranged "charges" into a viable hypothesis to explain partial coupling. Goldschmidt (1917) suggested retaining the hypothesis that genes were arranged linearly along the chromosome, but he suggested an alternative to crossing-over between chromosomes. He proposed that the chromosomes dissolved between cell divisions while the genes moved into the cytoplasm to become active in carrying out cell functions. At the next cell division, the chromosomes were reconstituted, with some genes in different loci than in the previous division. Hence, "crossing-over without chiasmatypes" would occur, that is, the genes could be reshuffled into different linear linkage groups, without the chromosomes breaking and rejoining.

Castle, in contrast, came to accept the claim that factors were parts of chromosomes and that one cross-over break could occur between homologous chromosomes; but he explored alternatives to linearity, double cross-overs, and interference. Castle's early diagrams of the arrangements of the factors (prior to his acceptance of the chromosome theory) were two-dimensional. He explicitly claimed that he considered two-dimensional alternatives to linear linkage before developing his alternative three-dimensional model for the *Drosophila* cross-over data.

Thus, the historical record provides evidence that a thorough search was made to find alternative hypotheses to account for linkage data. The *Drosophila* group, as well as Goldschmidt and Bateson, all considered linear hypotheses. Castle explored two-dimensional and three-dimensional relations. Bateson denied, while the others asserted, an association of the genes with the chromosomes. The *Drosophila* group and Castle proposed a tight relation between genes and chromosomes (perhaps the genes as completely constituting the chromosomes?); Goldschmidt, in contrast, proposed a hypothesis in which the genes could move and act independently of the chromosomes. The *Drosophila* group claimed that anomalies in predicted linkage values could be explained by double cross-overs and interference, while Castle explored alternative hypotheses that lacked those mechanisms. Thus, the strategies of **delineate and alter** and **propose the opposite** (or, more generally, **deny and explore other plausible alternatives**) can be seen as exemplified in these debates. Additionally, the modularity of the theoretical components is illustrated by these various alternatives.

In addition to the strategies for generating alternative hypotheses, several theory assessment strategies are also exemplified in these debates. These include the familiar ones of **explanatory adequacy, predictive adequacy,** and **lack of ad hocness.** With mounting evidence, especially from three factor crosses showing linkage relations, the *Drosophila* group's hypotheses overcame the charges of ad hocness. Their hypotheses had faced serious challenges from rivals and survived to become integral components of the theory of the gene.

10.6 The Problem of Mutation

The *Drosophila* group's work, along with that of other geneticists, served to expand the scope of Mendelism and provide new areas of research. The one that will be discussed in this section is the problem of the mutation of genes. That history was marked by a problem, originally not part of the scope of Mendelism, which came within its domain and gave rise to new lines of research.

De Vries (1901-3) found what he called "progressive mutations" in *Oenothera*, the evening primrose. These mutations, in contrast to Mendelian varieties, did not segregate; after their appearance, they were pure-breeding, that is, they produced only the newly arisen mutant forms when self-fertilized. These mutations produced numerous changes in the primrose and thus they looked promising as a method by which new species could be formed. In addition, study of progressive mutations held the promise of allowing experimental breeding to become a method for studying evolutionary changes. De Vries's mutation theory attracted attention from other biologists interested in the possibility of studying evolutionary changes experimentally (Allen, 1969). It was one of several alternatives to Darwinian selection in the early nineteenth century (Bowler, 1984).

In the early days of Mendelism, the scope of segregation was unknown. Numerous biologists, in addition to de Vries, discussed exceptions to segregating characters (e.g., Spillman, 1902; Castle, 1903). Segregation was not, at the outset, proposed as a universal phenomenon found in all hereditary characteristics. Thus, the mutations found by de Vries, as well as other, nonsegregating variations investigated in other species, were not necessarily anomalies for Mendelism. Instead, in early Mendelism, they were considered to be outside Mendelism's limited scope. But as the scope expanded—from alternative characters to blend forms to quantitative variations—exceptions came under greater scrutiny. Moreover, de Vries's mutation theory was seen as sufficiently important as a possible explanation for the origin of new species that testing it became an important task, quite apart from its implications for Mendelism.

If de Vries had been right, then the study of hereditary units would have had a Mendelian component and a non-Mendelian one. Such a genetics would have separated the domain into two parts: segregating varieties, and new species produced by the origin of new, nonsegregating pangens. (Actually, the "nonsegregating" components might not have been considered so "non-Mendelian." Meijer, 1983, argued that Mendel himself made a similar division of hybrids into ones that showed segregation and ones that did not. As a result of the differences between his findings for *Pisum*, which showed segregation, and for *Hieracium*, which did not show segregation, Mendel doubted the universality of his "law for *Pisum*.")

If de Vries had been right that new nonsegregating mutations produced new species, then no separate field of population genetics would have been needed. Evolutionary studies and non-Mendelian genetics would have been combined. That de Vries was wrong about the appropriate level of organization for solving the problem of the origin of new species was proved by subsequent developments.

De Vries (1901-3) used the term "mutation" to refer to changes in pangens (for discussion of the history of this term, see Maull, 1977, p. 149). Morgan discovered the white-eyed male fly, as discussed in Chapter 9, while looking for species-forming mutations, such as de Vries had found in *Oenothera* (Allen, 1978, p. 148). The numerous mutations found by the *Drosophila* group were smaller scale changes than de Vriesian progressive mutations, however. The mutations found in *Drosophila* cultures did segregate and did not produce new species at one jump. The meaning of the term "mutation" changed gradually, as a result of its use by the *Drosophila* workers (see, e.g., Morgan et al., 1915, p. 35). It ceased to refer to mutations that could form species with a single change; the study of mutations came within the scope of Mendelism.

By following the discussion of mutation through the writings of Morgan and his students, especially Muller, we can trace the introduction of ideas about mutation into Mendelism. Their work clearly shows the growing importance of the problem of mutation and the growing knowledge about it. In *Mechanism*, no separate chapter was devoted to mutation. The topic was explicitly discussed only briefly in the context of whether factors could be changed by selection or whether some of the supposed effects were due to new mutations (Morgan et al., 1915, pp. 205-207).

Bateson's presence and absence hypothesis, discussed in Chapter 6, may be viewed as an attempt to explain the origin of recessive mutants by the loss of dominant factors. The presence and absence hypothesis was criticized by the Morgan group, as we have seen. Sturtevant, in criticizing the hypothesis, said: "It seems very unlikely that protoplasm (chromatin?) is such a simple substance that the only possible change in a given unit (molecule?) involves the loss of that unit" (Sturtevant, 1913a, p. 18). Thus, speculations about the chemical nature of mutation were used in the arguments against presence and absence (for more detailed discussion, see Swinburne, 1962).

By 1919, the final chapter of Morgan's *The Physical Basis of Heredity* was entitled "Mutation" and occupied twenty-eight pages. He again presented arguments against the presence and absence hypothesis and stressed that nothing was known to suggest the nature of the changes that produced different allelomorphs at a specific locus in a linkage group (Morgan, 1919, p. 251). He also drew on recent work within the Mendelian chromosome theory to explain numerous of the *Oenothera* cases as due to abnormal chromosomal arrangements, lethal genes, and crossing-over. Muller and Sturtevant's work on crossing-over in stocks of *Drosophila* with lethal mutations produced cases with ratios that corresponded closely to de Vries's data for *Oenothera* (Morgan, 1919, pp. 257–265). The *Oenothera* mutants were explained by the theory of the gene and the chromosome theory. Thus, the *Drosophila* work, as well as the extensive work done by others in Mendelian crosses, made explicable the seemingly anomalous ratios deviating far from 3:1 or 9:3:3:1. Once again, *Drosophila* proved to be an excellent model organism for genetic work; *Oenothera* proved to have been a disastrous choice for de Vries.

As a result of the debates with Castle about contamination (discussed in Chapter 8), as well as the work of Johannsen (1903) and others, genes were shown to be relatively stable, to resist contamination by hybridization and modification by selection, and to mutate occasionally. New knowledge about mutation came to be part of Mendelism.

Oenothera mutants investigated by de Vries were explained by Mendelism and abnormal chromosome behavior. This development can be analyzed either as removing anomalies or as expanding the scope of Mendelism to cover cases that were originally outside its domain. Understanding mutations that segregated became a problem within the scope of "Mendelian" genetics. New knowledge about mutation added new components to the theory of the gene that were not present, even as rudimentary, implicit parts of the components of Mendelism in its earliest days. These new findings opened new problems for research.

In *The Theory of the Gene* in 1926, Morgan discussed the origin of the new problem:

> Mendel did not consider the question of the origin or the nature of the genes. . . . The question did not arise, because the characters yellow and green, tall and short, round and wrinkled, were already present in the peas selected for the experiment. Only later, when the relation of the mutants to the wild species from which they were supposed to have come was considered, did their origin arouse interest. (Morgan, 1926, p. 72)

Morgan discussed much new information about the nature and origin of mutations. He explained almost all of de Vries's purported non-Mendelian *Oenothera* mutants as chromosomal or Mendelian abnormalities. He criticized de Vries's idea that new genes (without an allelic pair) arise in the production of new species. The mutations found in *Drosophila* were changes in one of a pair at a specific linkage locus on a chromosome (Morgan, 1926, ch. 5).

In addition to criticizing de Vries's claim that mutations arose by the addition of a new unit, Morgan (1926, ch. 6) once again criticized Bateson's claim that recessive mutants arose by the loss of genes. He also devoted several chapters to discussions about abnormal numbers of chromosomes. In his conclusions, he made clear the current usage of the term "mutation":

The preceding chapters have dealt with two main topics: with the effects following a change in the number of the chromosomes; and with the effects following a change within a chromosome (a point mutation). The theory of the gene is broad enough to cover both these kinds of changes, although its main concern is with the gene itself. The term *mutation* also has come, through usage, to include the effects produced in both these ways. (Morgan, 1926, p. 300)

Morgan's discussion of mutation in 1926 provides an excellent review both of the state of the theory of gene at the time and of how old controversies about mutation had been resolved. Muller's work on mutation during the 1920s provides impressions of new and exciting questions to be tackled in research yet to be done. Carlson recounted the state of Muller's line of research in the 1920s: after his work on lethal mutations had been used in demolishing de Vries's "spurious mutation theory," Muller faced an open field of research about the "nature of real mutations" (Carlson, 1981, p. 96). Muller discussed a host of new problems and questions about the nature of the gene and the study of mutations. In 1922, he outlined various ways of trying to understand the nature of the gene. Mutation was one of the most promising. Understanding the changes that genes undergo would provide a key to understanding the structure that permits such changes, he claimed (Muller, 1922, p. 180).

Of the forward-looking nature of Muller's work, Dunn said:

Imagining the gene as a material body was, in Muller's case at least, a creative act which resulted in a great expansion of the horizons of genetics. Far from narrowing the focus to a "bead on a string," as some biologists feared it would, it had just the opposite effect and brought the powerful force of the rigorous and imaginative thinking of the new physics into genetics. This influence was felt in many ways, chiefly in directing attention to substances which, by their response to mutation induction, might reveal the chemical and physical composition of the gene. By 1940, the substances pointed to were nucleic acids, particularly DNA, which absorbs light of the wave-length most effective in producing mutation. (Dunn, 1965a, p. 172)

Our discussion will not follow Muller's new line of research on mutations through the 1930s and 1940s because the goal is to understand the development of the theory of the gene only to 1926. By 1926, the following component of the theory of the gene had been added:
C7'. Mutation
C7.1 Genes occasionally mutate so that they produce a different character.
C7.2 Such point mutation does not alter their linear relation to other genes in the linkage group.
　　The strategy exemplified in the addition of this new theoretical component to the theory of the gene will be discussed in the next section.

10.7 Strategies: Using Interrelations and an Analog Model

Both the development of the chromosome theory in the hands of the *Drosophila* group and Muller's new research program on mutations exemplify many of the strategies discussed in previous chapters, as well as a few new ones. Strategies exemplified by their work fall into the three categories of strategies that we are discussing, namely, strategies for producing new ideas, strategies for assessing theories, and strategies for anomaly resolution. Some distinctions will be useful in this analysis. Although these were all intimately intertwined in the *Drosophila* group's work, I want to clearly distinguish (1) the components of the Mendelian theory of the gene (an intrafield theory), (2) the chromosome theory that extended the Sutton-Boveri theory (an interfield theory), and (3) the analog model of beads on a string for depicting the structure of the relations of the factors linearly along the chromosome. The strategy of **using interrelations** between genetics and cytology played a role in the development of all three.

The linkage anomalies proved to be what I call a "model anomaly," namely, an anomaly that required modification of the theory to account for it. (Model anomalies contrast with monster anomalies, discussed in Chapter 8. Monster anomalies do not require modification of theoretical components, but can instead be explained away as some sort of monstrous happening in one case or in a set of malfunctions in numerous cases. The distinction between monster and model anomalies will be discussed further in Chapter 12.) Hypotheses to resolve the model linkage anomalies involved different kinds of proposed changes to the Mendelian theoretical components, as discussed in Chapter 9. Bateson's hypotheses of coupled factors and reduplication of cells did not involve interrelations to chromosomes.

Bateson, Saunders, and Punnett's original discovery of coupled characters in sweet peas resulted from the use of Mendelian breeding techniques. Their attempted explanations ignored chromosomes. Punnett later found seven linkage groups of characters in sweet peas using genetic, breeding techniques (Morgan, 1926, p. 36). Although he found these in testing the predictions of the *Drosophila* group, it is conceivable that Bateson and his colleagues could have discovered *groups* of coupled factors independently of any hypotheses about chromosomes. Breeding data eventually confirmed the hypothesis of linkage groups, but one can imagine an alternative historical path in which the same data could have been produced by extensive breeding experiments and then patterns of groups in the coupling noted. In other words, I am suggesting a contrary-to-fact possibility that groups of coupled characters (i.e., linkage groups) could have been found using only Mendelian methods, without information about chromosomes. It was a historical, contingent fact that the discovery of linkage groups resulted from a prediction based on interrelations to cytology. Similarly, Morgan used the numbers of chromosomes in a species to predict that the number of linkage groups in a species would correspond to the number of chromosome pairs. Yet, that too could have been discovered empirically by noting patterns in breeding data on the numbers of groups of coupled characters in different species. Bateson and his colleagues had, after all, unexpectedly discovered coupling while analyzing Mendelian breeding data; they were not testing Sutton's prediction of correlated characters.

Although Mendelian techniques could have been used to discover the number of linkage groups empirically, I do not think anything in the theoretical components of Mendelism before 1911 would have *predicted* such patterns in coupling. The points raised by Bateson, a staunch Mendelian, in his review of *Mechanism*, show his questioning of evidence to support exactly those predictions. He asked whether any evidence of linkage between factors in different linkage groups had been found and he indicated that the crucial evidence for the chromosome theory would be whether the number of linkage groups corresponded to the number of homologous chromosome pairs in a given species. Nothing in his prior Mendelian work, including his own discovery of (two) coupled characters prepared him for the new theoretical idea of linkage groups of strings of factors at the level of organization between factors and germ cells. The chromosome theory, formed by the strategy of **making interrelations** between two different bodies of knowledge, played a crucial role in allowing the *prediction* of this new level of organization of Mendelian factors.

When interrelations are made between two bodies of knowledge, the two may be in various stages of development. One field might be very well developed and the other just beginning, or vice versa. Alternatively, as was the case with the cytological study of chromosomes in germ cells and Mendelism, they both might develop some knowledge independently and their interrelation provide new hypotheses for each. New hypotheses were proposed for each field, both in the original proposal of the chromosome theory in 1903 and 1904 (discussed in Chapter 7), and in the hands of the *Drosophila* group with the development of hypotheses about linkage, crossing-over, and non-disjunction. Nevertheless, it is possible to imagine that either Mendelism or cytology might have been more or less developed and benefited more or less than it did from the interrelation established between them.

After the connection was made, genetic analysis usually outstripped the cytological. In 1903, Sutton used the connection to Mendelism to predict that the maternal and paternal members of homologous chromosome pairs assorted independently; Carothers confirmed that cytologically in 1913. Morgan predicted that chromosomes broke and rejoined at crossing-over; cytologists did not confirm this visually until the 1930s. Accurate observation of chromosomes was usually more difficult to do than breeding experiments. In fact, the cytological study of chromosomes benefited enormously from the interfield connection to genetics. It is possible to imagine that the reverse could have occurred, however, if good cytological techniques had been developed first (as was the case for the history of cytology between Mendel's period of the 1860s and the rediscovery in 1900).

One point of all these exercises designed to stimulate historical imagination and to consider contrary-to-fact possibilities is to show the conceptual distinction between, on the one hand, the theoretical components of Mendelism (developed entirely using breeding experiments and postulations of factors to account for their results) and, on the other hand, the chromosome theory, bridging Mendelism and cytology. The new theoretical components for linkage and crossing-over could be stated entirely in terms of Mendelian factors without mentioning the chromosomes. In fact, in 1926, Morgan, knowing of the controversies surrounding the chromosome theory, did just that. He was careful to state the components of the theory of the gene without mentioning chromosomes, as we discussed in Section 10.2.

I followed Morgan's language closely in my statement of Components 6.1' through 6.5', with no mention of chromosomes. As discussed, Morgan stressed that all the claims he made in his statement of the theory of the gene were based on data from breeding experiments and were thus supported by evidence independent of the chromosome theory. Despite their simultaneous development by the *Drosophila* group, the theory of the gene, supported by genetic data, should be distinguished from the historically more controversial chromosome theory. The theory of the gene was an *intrafield* theory and was a modified version of the Mendelian theory of 1900-1903 (which I stated in Chapter 5) with new components about linkage and mutation. The chromosome theory was an *interfield* theory that extended the claim that Mendelian factors were carried by the germ cells (Component 3 in Table 5.2) to the claim that the factors were parts of the chromosomes.

A reason for making the distinction is to show how the strategy of **using interrelations** to cytology proved to be a very fruitful method for developing the theory of the gene, as well as for simultaneously developing the interfield chromosome theory. Interrelations to cytology helped to develop the new theoretical components of Mendelism, that resolved the linkage anomalies. If geneticists had tried to resolve the linkage anomalies without the connections to chromosomes—if, for example, Bateson's reduplication hypothesis had had evidence in its favor—then, the theoretical components of Mendelism would have been very different in 1926.

The interrelations between genetics and cytology did more than merely resolve the linkage anomalies found in the Mendelian breeding data, however. In the hands of the *Drosophila* group, the development of the interrelations served to extend the Sutton-Boveri chromosome theory. In addition to the further development of the interfield chromosome theory, the interrelations provided new techniques of genetic analysis, such as mapping, that allowed new kinds of Mendelian predictions to be made, namely, relative linkage values. The chromosome theory also provided a way of using two different techniques to provide two independent sources of evidence for resolving new anomalies, such as the genetic and correlated chromosomal anomalies of non-disjunction.

Furthermore, the use of interrelations to cytology and the development of the chromosome theory allowed the *Drosophila* group to develop a hypothetical *analog model* for picturing the relations of the factors along the chromosome—the model of beads along a string. Crossing-over occurred between "beads" and resulted in the shifting of the pieces of the strands. (It is interesting to ask whether the model was like pearls strung along a thread or whether it was like the beads I played with as a child that we called "pop" beads. Pop beads had no thread but each bead had a hole at one end and a protrusion at the other so that each bead attached to the next one; they could be popped apart to make necklaces and bracelets of different lengths. One wonders if the chromosomes were thought to contain a thread of material not identified with the factors, or if the totality of the beads made up the entire chromosome. The pictures, such as in Figure 10.2, do not allow a decision to be made between these two analog models.)

Little historical, notebook evidence exists about the formation of the hypotheses of cross-overs, double cross-overs, and interference. Some notebook evidence is extant in the Sturtevant papers. His laboratory notebooks recording the cross-over data have been preserved and show his usage of lines to show crossing-over, much as his published paper did (Sturtevant, unpublished; Sturtevant, 1913b). But this evidence does not aid in understanding the development of the hypotheses; it only provides evidence that Sturtevant used diagrams of threads crossing to set up the categories in which he recorded genetic data. His notebooks thus provide evidence that he "saw" the character combinations in fruit flies as being a result of crossing-over.

Despite the lack of notebook evidence about steps in their thinking, the exposition of the hypotheses in the published work of the *Drosophila* group easily lends itself to speculations about how the analog model may have aided in hypothesis formation. In his published work, Morgan cited Janssens's diagrams of intertwining chromosomes in support of crossing-over (see Chapter 9). Seeing those diagrams may well have played a role in suggesting breaking threads. The fine structure of the chromosome was unobservable during this period; the giant salivary chromosomes in *Drosophila*, showing banding patterns, familiar in genetics books today, were not discovered until the 1930s (Painter, 1933). Before 1930, the image of visible intertwining threads, observable in stained chromosome specimens, was extended to the unobservable level hypothetically. The analog model of beads on a string provided the hypothetical, theoretical model for depicting Mendelian factors along the chromosome. Crossing-over provided the theoretical explanation for the data of incomplete linkage.

The analog model can be seen as playing the role that Boyd (1979) proposed for what he called "theory-constitutive metaphors." The use of familiar analogs, according to Boyd, aids in the introduction of new scientific hypotheses and terminology. The visual image of chromosomes as intertwining threads and the hypothetical beads on a string model may have supplied the terminology of "crossing-over." "Crossing-over" was used both to designate the predicted switching of pieces of the chromosomes (theoretical) and to designate "cross-over classes" of organisms in breeding experiments that exhibited the incompletely coupled characters (data). The *Drosophila* group thus used their theory-laden terminology for labeling their breeding data, in the light of the chromosome theory. Bateson would never have designated the data, showing the small classes in partial coupling, the cross-overs.

When the anomalies arose for predicted linkage values, based on linearity and the equal likelihood of only one cross-over, then the analog model may have supplied new ideas for forming new hypotheses to resolve those anomalies, as well as providing language used in the new hypotheses. Hesse (1966) discussed the way analogies used in original theory formation can serve in forming additional hypotheses to resolve anomalies that subsequently arise for the theory and also how analogies provide meaning for new theoretical terms. The language of "double cross-overs" gains meaning from the analogy to threads crossing and recrossing each other. Similarly "interference" likely was named because of the analogy to one twist interfering with the formation of another (see quotation above in Section 10.2). The idea of one break relieving tension in the twisted thread and thus preventing another break from occurring nearby reflects the use of the visual analogy in exposition and may indicate its role in hypothesis formation.

By employing such language as "double crossing-over" and "interference," the meanings from ordinary usage were carried over into genetics and provided geneticists with an understanding of the phenomena via a mechanical analogy. The confusing numbers of classes of character combinations in breeding experiments was made orderly from the perspective of the much simpler, visual, mechanical images of threads crossing, recrossing, breaking, joining, and interfering. The analog model supplied visual and mechanical images that were constitutive of the new hypotheses and supplied the theoretical language in which they were expressed. (For further discussion of the role of metaphors and analogies in giving meaning to new terms, see Boyd, 1979; Hesse, 1966; and Lakoff and Johnson, 1980.) In sum, the analog model functioned in providing terminology for the original hypothesis of crossing-over. It also served in resolving anomalies that arose for the hypothesis of single cross-overs and supplied theoretical language for the new hypotheses that explained the anomalies.

The model of beads along a string became a "public" metaphor, as Boyd (1979) says of theory constitutive metaphors in science; it was a model used in explication of the *Drosophila* group's hypotheses and developed by others as genetics developed. The model of genes arranged linearly along the chromosome is the source of the visual ideas taught in genetics courses today; such ideas aid students in explaining data about fruit flies and character combinations in the breeding experiments done in laboratory exercises. The analog model eventually became part of textbook accounts of Mendelism.

Thus, the strategy of **using interrelations** between genetics and cytology in this case was intimately tied to the strategy of **develop and manipulate an analog model**. The analogy provided a picture representing the structural relations among the postulated, unobservable, theoretical, Mendelian factors and the visible chromosome threads. The strategy of **using interrelations** was enormously fruitful in the hands of the *Drosophila* group: producing new theoretical components for the intrafield theory of the gene; furthering the development of the interfield chromosome theory; and supplying a new analog model, new techniques, and new predictions in a new research program. The strategies of **using interrelations** and of **developing and extending an analog model** served both as strategies for producing new ideas and strategies for resolving anomalies.

The *linear* model was not the only possible analog model, as Castle's exploration of two-dimensional relations and his alternative three-dimensional model clearly indicated. Given how

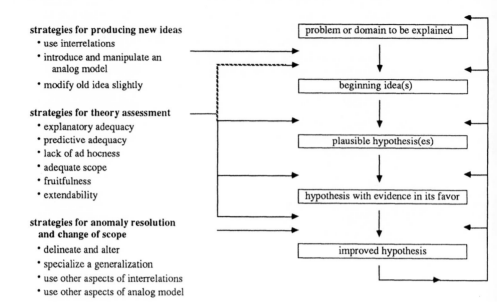

strategies for producing new ideas
- use interrelations
- introduce and manipulate an
  analog model
- modify old idea slightly

strategies for theory assessment
- explanatory adequacy
- predictive adequacy
- lack of ad hocness
- adequate scope
- fruitfulness
- extendability

**strategies for anomaly resolution
and change of scope**
- delineate and alter
- specialize a generalization
- use other aspects of interrelations
- use other aspects of analog model

problem or domain to be explained

beginning idea(s)

plausible hypothesis(es)

hypothesis with evidence in its favor

improved hypothesis

Figure 10-5. Refined stages and strategies, VII.

little was known about the possible size of the factors and how little was known about the internal structure of the chromosomes, it was reasonable to assume that the factors were at a much lower level of organization within the chromosome than the *Drosophila* group seems to have assumed. That the visible two-dimensional chromosome threads contained an elaborate three-dimensional internal structure was not an unreasonable hypothesis for Castle to explore. Castle did pursue his own three-dimensional, wire, analog model, while Bateson toyed with vague analogies to physical forces.

It is interesting to speculate how much of a role Bateson's and Castle's alternative analog models might have played in their assessments of the *Drosophila* group's hypotheses. They both came to the new hypotheses already somewhat committed to models that were different from the linear, material beads along a string. Such prior commitments may have played a role in their unwillingness to accept the new one immediately. Suppose someone had formed a mental picture for conceiving the hypothetical Mendelian factors, such as a picture of forces attracting each other or a picture of the two- or three-dimensional chemical structure of an organic molecule. Further suppose that that someone was then presented with a very different model. Resistance to the new model would not be surprising. Still, this is only speculation. Biographers are in a better position to describe the prior, often implicit, commitments to models or to more general methodological or metaphysical assumptions that influence individual scientists' judgments in assessing a new theory. The public record only provides hints about such commitments.

Bateson's concern that a theory of heredity also be a theory of embryological development illustrates another theory assessment strategy. Debate can occur about the scope of the domain that a theory must explain to be adequate. Bateson believed that problems of heredity and development were inextricably related. A theory of linked material units that showed no promise for explaining how characters were produced during development was inadequate, from his perspective (Coleman, 1970; Darden, 1977; Cock, 1983). Thus, another strategy in theory assessment is to consider the **scope of the domain** that it explains, namely, the number of kinds

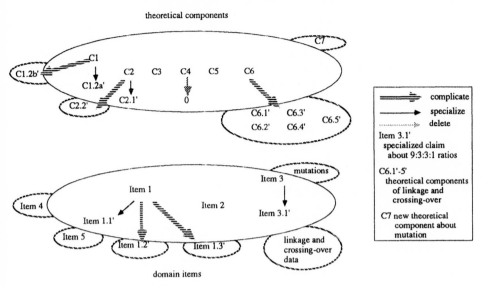

Fig. 10-6. Changes in theoretical components (C6, C7) and domain.

of domain items that it explains. If a kind of domain item is not explained and the theory shows no promise for explaining it, then that will be an inadequacy of the theory. A proponent for the theory could respond to this criticism by saying that the kind of item in question must be explained by another theory. Resolution of such a dispute would have to await further theory development.

Finally, the development of the new problem of mutation shows the strategy of **taking an old idea outside the theory and modifying it quantitatively** to form a new theoretical component. Progressive mutation was a theoretical idea developed by de Vries outside of Mendelism. The *Drosophila* group appropriated the term and changed its meaning in the light of the phenomena in their Mendelian experiments. Eventually they were able to explain away de Vries's supposed instances of progressive mutations with the newly expanded theoretical components of the theory of the gene. Thus, the development of the new idea of mutation can be seen as exemplifying the strategy of **quantitatively altering an old idea** to account for new, related phenomena. The result of adding the new components about mutation was yet another expansion of the scope of the domain of Mendelism, making it appear more and more like a very general theory for solving the old problem of explaining hereditary phenomena. An important new research program developed to explore the nature and size of the gene by finding out what caused it to mutate.

Figure 10-5 summarizes the strategies for producing new ideas, for theory assessment, and for anomaly resolution exemplified in the development of the new Mendelian components of linkage, crossing-over, and mutation. Figure 10-6 uses the representation developed in Chapter 6 to show the addition to the domain of the new items of linkage, crossing-over and mutation, and the corresponding, extensive changes in the theoretical components.

CHAPTER 11

Unit-characters to Factors to Genes

11.1 Introduction

During the development of the theory of the gene, the need for postulating the existence of a hypothetical entity, the gene, gradually emerged. No one person discovered the gene. The distinction between the character and the unit carried by the germ cells gradually entered the literature. The factor hypothesis emerged between about 1906 and 1910, and with it came debate about the nature of the unit factor, in contrast to the unit-character. Johannsen introduced the term "gene" in 1909, although determining his meaning of that term, as well as of his other terms, "genotype" and "phenotype," has proved difficult for historians. The general usage of "gene" to refer to the theoretical entity postulated by Mendelism can be traced to the adoption of that term by Morgan and others in the *Drosophila* group. In 1917, Morgan wrote a paper defending the introduction of the term "genetic factor" to refer to an entity distinct from the character, and defending the postulated entity against the charge of being "imaginary " (Morgan, 1917, p. 517). By the 1920s, the main theoretical claims of Mendelism had been labeled by Morgan the "theory of the gene."

My analysis of the components of the theory shows these changes. Component 1 from the period 1900-1903 was stated in the following way:

As of 1900-1903:
Component 1. Unit-characters
> C1.1 An organism is to be viewed as composed of separable unit-characters.

As discussed in Chapter 8, by about 1910 the factor hypothesis had developed:

As of about 1910:
Component 1'. Factors and characters
> C1.1' Characters are produced by factors.
> C1.2a' One factor may produce one character or
> C1.2b' Multiple factors may interact in the production of one character.

168

By 1926 these components had developed further and "genes" had replaced "factors." The components in 1926 may be stated as:

C1'. Genes and characters
    C1.1' Genes cause characters.
    C1.2a' One gene may cause one character or
    C1.2b' Multiple factors (genes at different loci in linkage groups) may interact in causing one character or
    C1.2c' One gene may affect many characters.

My statement of these components in 1926 is a bit more precise and expressed more explicitly in terms of "causes" than was Morgan's 1926 statement, as will be discussed more fully later.

This chapter will trace the conceptual developments producing these changes in components of Mendelian theory from unit-character to factor to gene. Previous chapters have focused primarily on changes in theoretical components that resulted from resolving empirical anomalies and expanding the scope of the domain explained by the theory of the gene. Although much of the genetics literature dealt with reporting new empirical findings, proposing new hypotheses to explain the data, and testing those hypotheses, occasional discussions dealt with less empirical matters. A number of specific issues were discussed by geneticists, or their critics, that I classify as "conceptual problems," rather than empirical anomalies. Resolution of these conceptual problems led to the development of these new theoretical components by the 1920s. This chapter, thus, shifts focus from historical developments produced by resolving empirical anomalies to developments resulting from solving conceptual problems; also, strategies for finding and solving conceptual problems will be proposed.

The kinds of conceptual problems that I will be discussing focused on several issues: symbolism, terminology, and the defense of postulating a new theoretical entity. Issues about Mendelian symbolism were raised. To what did the symbols, such as *A* and *a*, refer? How should the symbolism be changed in the light of new hypotheses? As the factor hypothesis developed during the period between 1906 and 1910, papers appeared questioning the nature of the factors. Geneticists came to realize that their terminology was in need of standardization. After 1915, with the successes of the *Drosophila* work, questions were raised about the claim that a new theoretical entity, the gene, existed. Debates occurred about the properties to be attributed to genes. How were gene and character to be distinguished? Was there really a stable entity that was not contaminated in hybridization experiments? Was it an imaginary entity? Arguments for the adequacy of the theory of the gene involved answering such questions about the nature of genes, as well as showing the adequacy of the theory in resolving empirical anomalies.

Section 11.2 discusses the distinction between empirical anomalies and conceptual problems. This section compares my usage of "conceptual problem" with other uses in philosophy of science. Section 11.3 discusses changes in mathematical symbols and structural models and the important role played by symbolism in the development of Mendelian theory. Section 11.4 discusses changes in terminology, from discussion of "characters" carried by germ cells, to the introduction of terms to refer explicitly to a new theoretical entity, especially the terms "factor" and "gene." Section 11.5 discusses Morgan's 1917 defense of the need for postulating a new theoretical entity, the gene. This section also traces some of the properties ascribed to the newly postulated entity. Finally, Section 11.6 proposes strategies for finding and solving conceptual problems. The strategies exemplified in the conceptual changes in the theory of the gene include introducing and manipulating symbolic representations, standardizing and clarifying terminology, as well as postulating a new theoretical entity at a new level of organization and adding to the list of properties ascribed to it.

11.2 Conceptual Problems

A loose distinction can be made between empirical problems and conceptual problems. Empirical problems have been labeled "anomalies" in previous chapters. They are problems that arise because of anomalous data or data outside the current explanatory scope of the theory. New hypotheses proposed to resolve empirical anomalies often lead to new questions about the theoretical entities or processes that they postulate. Such conceptual problems arise, not directly from problems related to data, but from the nature of the claims in the hypothesis. Thus, conceptual problems arise from questions about the theoretical components themselves. In this category of conceptual problems, I will include issues about the adequacy of symbolic representations, the introduction of new theoretical terms, and disputes about the need for and properties ascribed to a new theoretical entity. This does not aim to be a complete list of types of conceptual problems; as with other distinctions introduced in this book, it is a list useful for analyzing this particular case.

In general, a sharp distinction may not exist between empirical and conceptual problems. Resolving empirical problems may involve satisfactorily answering questions about the nature of theoretical entities. Furthermore, relations between what is considered to be empirical and what is considered to be theoretical change over time (Shapere, 1984, pp. 342-351). The cell was a theoretical entity in the mid-nineteenth century, but would probably be considered an observational entity throughout much of the twentieth century. A thorough analysis of the nature of such changes has yet to be provided by philosophers of science and none will be attempted here.

Despite the difficulties of, in general, distinguishing empirical and conceptual problems, the distinction is useful for a number of reasons. First, the contrast between empirical anomalies and conceptual problems will be useful for discussing strategies about how to find and solve conceptual problems. Empirical anomalies result, as we have seen in numerous examples in previous chapters, from data that cannot currently be explained by the theory. The origin of empirical anomalies is not especially problematic. They often result from a failed prediction. The nature of conceptual problems and how they arise is less straightforward, however. I suspect scientists find it easier to recognize an empirical finding that has not been adequately explained than to recognize a conceptual problem in need of resolution. Only occasionally did geneticists specifically address issues I am labeling conceptual problems, in contrast to the numerous debates about empirical anomalies, which have been the focus of previous chapters. Thus, this chapter will discuss strategies for *finding*, as well as resolving, conceptual problems.

Another reason that I need to distinguish empirical anomalies and conceptual problems stems from the nature of some of the historical developments. One goal of this analysis is to develop strategies for theory change based on the actual changes that occurred during the development of the theory of the gene. Not all changes occurred in response to empirical anomalies. Discussing only strategies for resolving empirical anomalies (as well as strategies for theory assessment) would be incomplete. The historical concerns of geneticists to be discussed in this chapter should help to make clear why I need this category of conceptual problems.

Some other philosophers of science have used the term "conceptual problem" or have seen the need for such a category of problems, other than empirical ones. Toulmin (1972, p. 176) used this term much more broadly than I do. He did not use it to distinguish empirical from conceptual problems. As types of conceptual problems he included explaining phenomena, improving techniques, interdisciplinary interactions, and the resolution of conflicts between scientific and nonscientific ideas. My usage of "conceptual problem" will be much more narrowly focused on issues about the adequacy of symbolic representations, terminology, and the nature of new theoretical entities. Toulmin (1972, p.161) did mention several similar issues to be separated and analyzed in scientific change, although he did not specifically label them conceptual problems. In his list, he included analyzing the "language," and the "representation techniques" (such as mathematical formalisms and diagrams). Of terminology and symbolic representations, he said: "all such [theoretical] terms, images and/or models are alternative and more-or-less adequate means of expressing or symbolizing the collective concepts . . . of a scientific discipline" (Toulmin, 1972, pp. 198-199).

My usage of "conceptual problems" is closer to Shapere's (1974b, pp. 558-565) category of "theoretical inadequacies," which he characterized as problems concerning the adequacy of the theory itself. These include issues such as whether a theoretical claim is oversimplified. Shapere, however, included in that category an incompleteness in the theory, namely a domain item for which the theory cannot account. I would be more likely to classify such an incompleteness as an empirical problem for the theory, but also distinguish it from an anomaly arising from a failed prediction; the distinction is between "incomplete" and "incorrect" with respect to a given domain.

Laudan's (1977) usage of "internal conceptual problem" is narrower than the way I will use the term. Such problems may arise, he said, from internal inconsistencies or conceptual ambiguity or circularity within the theory. So far as I know, the theory of the gene did not face the criticism of being inconsistent or circular; however, debate did occur about the need for the terms "factor" and "gene." Laudan noted that scientists often engage in debates centered on issues about the nature of theoretical claims, such as whether bodies can act on each other at a distance. Removing ambiguities and specifying the meaning of scientific terms are ways for resolving conceptual problems, Laudan said (1977, p. 50). The development of the theory of the gene was marked by such developments. My category of conceptual problems includes more than Laudan's internal conceptual problems; it includes issues about the adequacy of Mendelian symbolism and debates about the need for postulating a new theoretical entity. Laudan was correct in claiming that philosophers of science have concentrated on empirical problems and done less analysis of conceptual ones. (Laudan also discussed "external conceptual problems," which arise from questions about the compatibility of the theory with other accepted theories. Those problems will be discussed in terms of successful or failed interfield relations in Chapter 13.

This historical case of theory change shows that resolution of conceptual problems was important in the development of the theory of the gene. Strategies for finding and solving conceptual problems thus need to be devised. This chapter takes a step along that road.

11.3 Symbolic Representations

During the development of the theory, changes occurred in diagrams, mathematical symbols, and analog models (e.g., chromosomes as strings) presented by geneticists in their publications. The depictions of the theoretical entities of Mendelism between 1900 and 1926 show the change from letters, used to designate types of germ cells in hybrids, to round beads on a string. These representations graphically show the changes from the early hypothesis about the purity of the gametes to hypotheses about material units located linearly along chromosomes.

In general, the introduction of symbols (at first perhaps only to designate classes of data) and the conceptual manipulation of such symbols can be seen as important first steps in moving away from the data itself toward an explanation of that data. Such steps may result in the postulation of a new theoretical entity. A hypothetical path to the discovery of segregation was discussed in Chapter 3. In that imaginary episode of problem solving, I suggested that 3:1 ratios might lead someone with knowledge of simple algebraic formulas to introduce the symbolic notation: $(A + a)(A + a) = 1AA + 2Aa + 1aa$. Once the data has been depicted in this way, the question can be asked: what do the A and a represent? Then, the designation of A and a as representing types of pollen and egg cells is a next step. Because most pollen grains look the same and egg cells are difficult to see even with a microscope, the symbols refer to unobservable differences. The "theoretical" symbols can be manipulated to suggest tests of the adequacy of such a representation. For example, a backcross (Aa x aa) yields the prediction of 50 percent dominant and 50 percent recessive forms; such a cross can be done to confirm the prediction. Multiplication may be designated by x and hybrid crosses were also designated by x. Multiplication is a mathematical manipulation of symbols that can be used for depicting hybrids. The results of multiplying the symbol can be used to predict the outcome of such crosses.

Symbols free one from the data to think in terms of an unobservable level of organization. The symbols can be used to make predictions of observable phenomena to test the adequacy of the symbol system. Then new conceptual problems arise when questions are asked about the symbols, such as what do A and a represent? Symbols open new possibilities of "conceptual manipulation." One can imagine, in an abstract way, what happens when certain kinds of symbols are multiplied. One can imagine the results of a test cross before doing it. Playing with symbols in the "mind's eye" is a creative way of imagining what may be occurring at unobservable levels of organization. Moreover, the use of symbols open the path for new questions to be asked about what the symbols represent. (I thank Edward Manukian for calling my attention to this role of symbol manipulation in the development of new theories.)

The previous paragraphs and Chapter 3 speculated about conceptual play. Turning to the actual history as discussed in Chapters 4 to 10, I believe it is reasonable to claim that, historically, the introduction of symbols to designate types of germ cells was the first step toward the postulation of the existence of a new theoretical entity, the gene. The symbols and analog models for representing the unobservable "something" in the germ cells changed as Mendelism developed. The genetics literature contained discussions about appropriate symbolism, problems introduced by the use of one symbol system rather than another, and discussions about what the symbols represented. This section will discuss some of those issues and show how the symbols and analog models changed during the development of the theory. The extent to which the symbolism suggested new ideas, as opposed to being changed following a discovery made without such guidance, is an interesting question that I see no way of answering with extant historical evidence about this case. The role of symbols and analog models in strategies for producing new ideas is a fascinating question about which I wish I had more historical evidence. What the historical record does show is the way theoretical claims were represented graphically in publications.

As Figure 11-1 shows, Mendel's diagrams used capital and small letters to designate different types of "pollen cells" (*Pollenzellen*) and "germinal cells" (*Keimzellen*) that were crossed during breeding. Arrows designated combinations during fertilization. The result of the fertilization

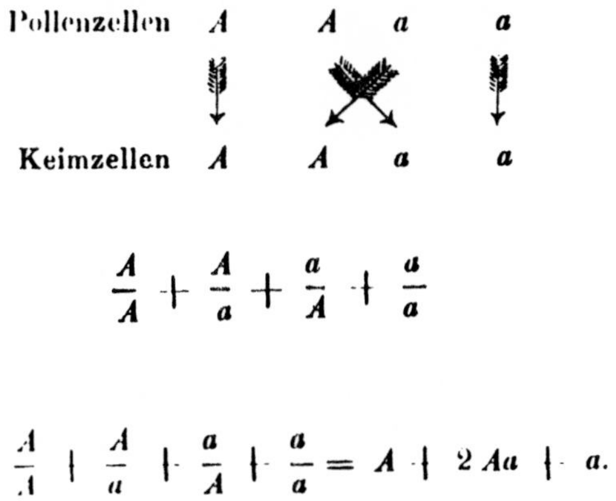

$$\text{Pollenzellen} \quad A \quad\quad A \quad a \quad\quad a$$

$$\text{Keimzellen} \quad A \quad\quad A \quad a \quad\quad a$$

$$\frac{A}{A} + \frac{A}{a} + \frac{a}{A} + \frac{a}{a}$$

$$\frac{A}{A} + \frac{A}{a} + \frac{a}{A} + \frac{a}{a} = A + 2\,Aa + a.$$

Fig. 11-1. Mendel's diagrams for hybrid cross (From Mendel, Gregor (1865), "Versuche über Pflanzen-Hybriden," *Verhandlungen des naturforschenden Vereines in Brünn* 4, p. 30)

(shown at the bottom of Figure 11-1) was $A + 2Aa + a$. Much has been made of the fact that Mendel used A and a, not AA and aa for the homozygous forms (discussed in Chapter 4). Thus, whether Mendel would have claimed that nonhybrid forms had a double (diploid) nature is unclear. His symbols referred to types of germ cells. Whether he thought of them as referring to a lower level of organization, that is, as something carried by the germ cells, is not clear from his statements or his symbols (discussed in Chapter 4). Olby argued, partly on the basis of Mendel's symbols for *Phaseolus* (bean) crosses, that Mendel "could conceive of *three* contrasted characters *which can exist together* in the F_1 *hybrid*, but which are mutually exclusive in the germ cells" (Olby, 1985, pp. 246-247). Mendel's symbols for such hybrids were, for example, $A_1 A_2 a$.

Interestingly, de Vries, who had an explicit material particle view of pangens, used algebraic symbolism for crosses and dropped the squared superscript in his representation as shown in Figure 11-2. One wonders what he meant by the "individuals d and d^2." How did d and d^2 individuals differ? As of 1900, he may not have had a view of paired pangens in homozygous forms, although his theory of pangenesis makes it a reasonable expectation that he thought one was contributed by each parent.

Before the formation of the chromosome theory and the factor hypothesis, and before a clear separation of genotype and phenotype, Mendelism only demanded that the phenotype of the hybrid (the heterozygous form) be represented by double (or more) different symbols. A cross of a pure yellow pea with a pure yellow pea yielded pure yellow peas, which could be symbolized as $A \times A = A$. In contrast, for a hybrid produced by a cross between yellow and green, Aa was needed to designate the nature of the hybrid yellows. Pure yellows designated by A were different from hybrid yellows, Aa. Without any clear claim that symbols referred to paired, unobservable factors, when varieties with pure characters (i.e., nonhybrids) were crossed, there was no need to designate the homozygous individual with double symbols, for example, AA and aa.

Thus with a factor hypothesis, but no chromosome theory, there was no theoretical reason that three different factors could not exist together in hybrids resulting from complex crosses. Bateson, in fact, proposed such as an explanation for a complex cross in 1903. Before abandoning this early view of multiple allelomorphs in a single hybrid organism, Bateson suggested that, in a cross of poultry with different combs, "imperfect segregation" occurred: hybrids were

50% dom. + 50% rec. pollen grains, and
50% dom. + 50% rec. ovules.

If dominating is designated by d and recessive by r, fertilization yields:

$$(d + r) (d + r) = d^2 + 2 \, dr + r^2$$
or
$$25\% \, d + 50\% \, dr + 25\% \, r.$$

Fig. 11-2. DeVries's description of hybrid cross results.

occasionally produced having three different allelomorphs for the same character, which he designated as R, P, and S (Bateson, 1903). Bateson, as we have discussed, argued against paired chromosomes as the location of Mendelian factors.

The chromosome theory, with its paired homologous chromosomes, allowed only two different allelomorphs to exist in any one normal organism (although within a population multiple allelomorphs, all mapping to the same locus on a genetic map, could be found, as discussed in Chapter 6). The factor hypothesis and the chromosome theory thus demanded double symbolic representations for allelomorphs in both heterozygotes and homozygotes (not single letters or three letters). By 1926, it was still debatable whether two allelomorphs existed for every hereditary characteristic in an organism, including those for which no mutants had been found and thus no genetic crosses could be done.

The use of a capital letter for the dominant form and a small letter for the recessive gained wide acceptance. Bateson used 1AA:2Aa:1aa in his exposition of Mendel's work without even noting that Mendel did not use a double basis for the homozygotes (e.g., Bateson, 1902). The symbolism 1AA:2Aa:1aa contained a difference between AA and Aa that was easily exploited to symbolize a blend (or otherwise different) form in the heterozygote. Thus, the symbolism was originally used in the explanation of 3:1 ratios, but it contained an unused distinction that allowed easy extendability to explain a difference in phenotypic appearance between the homozygous dominant (AA) and the heterozygote (Aa).

With the development of Bateson's presence and absence hypothesis, symbolism changed. A capital letter represented the present, dominant factor but a zero was used for the absent recessive, e.g., AO. This symbolism was adopted by those who accepted Bateson's hypothesis. With the discovery of multiple allelomorphs, the binary symbolism of capital and small letter, or capitals and zeros, became inadequate. Morgan, who had accepted the presence and absence hypothesis, abandoned it with the discovery of multiple allelomorphs in *Drosophila*. He specifically argued that the symbolism based on that hypothesis was inadequate (Morgan, 1913).

Sturtevant (1965, pp. 52-53) recounted his 1914 development of an explanation for coat colors in mice using multiple allelomorphs. With hindsight, he claimed, he was able to see that Cuénot had formed the same explanation in 1907 in a paper that Sturtevant had read. Cuénot, however, had used different capital letters for each alternative allelomorph, such as C, M, G, F, and U. Sturtevant said, "Our failure to realize that Cuénot had understood and explained the situation can only be excused by his use of unorthodox symbols and by the fact that he apparently felt there was nothing about the relation that called for elaboration or emphasis" (Sturtevant, 1965, p. 53).

Sturtevant continued his discussion of symbolism by noting that with the abandonment of the presence and absence hypothesis, the system of gene symbols associated with it broke down. Of the new developments, he said:

According to that [AO] scheme, each pair of genes was named for the somatic effect of the dominant allele, but with the accumulation of many recessive mutant genes, including, for example, a dozen or so for eye-color, it became necessary to name each one for the mutant allele, usually the recessive. The wild type was considered as a standard of reference, usually symbolized as "+." This system, as gradually developed is now universally used in the *Drosophila* literature, and essentially the same scheme is applied to microorganisms. The older scheme, or some compromise between it and that used for *Drosophila*, is still usual for most other higher organisms—a difference in language that sometimes makes for confusion and misunderstanding. (Sturtevant, 1965, p. 53)

These changes illustrate the relation between theory development and changes in symbolism. The symbolism changed as new hypotheses were proposed. Problems with hypotheses raised problems about symbolic notation based on them. The use of symbolic notation that did not clearly reflect a new underlying hypothesis might result in others not understanding the hypothesis.

New problems arose about what Mendelian symbols represented when a distinction began to be made between the unit-character and whatever accounted for the difference between the germ cells, represented originally by A and a. Much of the discussion of this issue focused on the nature of the "unit factors," with the development of the factor hypothesis between about 1906 and 1910. Changes in terminology from "unit-character" to "factor" to "gene" will be discussed in the next section. The rest of this discussion about changes in symbolic representations will focus on the shift from the use of letters in algebraic manipulations to their use in more graphic illustrations. Hypotheses about structural relations were reflected in Castle's and Johannsen's diagrams and discussions of two- and three-dimensional representations, as well as in the *Drosophila* group's diagrams of linkage maps and beads along a string.

Castle's early diagrams (1909) were like organic chemical molecules. His representations for different color varieties in rabbits were discussed in Chapter 10 and reproduced in Figure 10-3. He explicitly claimed to be using diagrams like those in organic chemistry:

> To aid in expressing the interrelationship of the factors I think it useful to imitate the organic chemist and employ diagrams. Thus a diagram might be constructed as follows to express the relations of the six factors in a reproductive cell transmitting the color characters of a gray rabbit:—

$$
\begin{array}{c}
U \\
| \\
A\text{—}C\text{—}B\text{—}E \\
| \\
I
\end{array}
$$

> It is possible that in protoplasm we have organic molecules built up in some such way, and that regressive variations arise by dropping off the constituent parts of the molecule one by one. If A drops out, we have a black rabbit instead of a gray; if E is replaced by R, a yellow one is produced . . . if C is lost, we have an albino, whose breeding capacity varies with the number of other invisible factors which remain. (Castle, 1909, p. 156)

Castle might have viewed contamination (or modification, as he usually called it) of factors by hybridization as resulting from slight changes in parts of an organic molecule or in other sorts of chemical changes in substances within the cells (see, e.g., Castle, 1912, pp. 361-362).

Johannsen (1911, 1913) endorsed Castle's organic chemistrylike diagrams and also used this form of representation in speculating about possible structures (see, e.g., Johannsen, 1913, pp. 531, 666). Of a possible analogy to organic chemistry, Johannsen said:

> The genotypes may then be characterized as something fixed and may be, to a certain degree, parallelized with the most complicated molecules of organic chemistry consisting of "nuclei" with a multitude of "side-chains." Continuing for a moment such a metaphor, we may even suggest that the genes may be looked upon as analogs of the "radicals" or "side-chains." All such ideas may as yet be premature; but they are highly favored by the recent researches . . . . (Johannsen, 1911, p. 143)

These attempts by Castle and Johannsen to use a form of representation like that of organic chemists may well indicate that they were speculating about genes as parts of organic molecules, using the chemistry of the period. Castle's diagrams show how the Mendelian letter symbols could be incorporated into structural, analog models in two or three dimensions, possibly representing the arrangements of atoms or groups of molecules into a larger molecule. They show the admittedly speculative attempts of geneticists to move from the algebraic manipulation of symbols to a more realistic depiction of chemical structures in the germ cells.

Neither Castle nor Johannsen developed this form of symbolism extensively in later work, perhaps because they realized its limitations and because no methods from chemistry were available to do experiments relating chemistry and genetics during this period. Castle's further development of the idea that variable characters indicated the lack of constant, unvarying factors in the germ cells caused him to question the adequacy of the symbolism of Mendelian unit characters in terms of letters:

> In our descriptions we call these characters A, B, C, etc., and the recombinations are AB, BC, AC, etc. In our formulae A is always A, and B is always B, but it is an open question whether in our living animals the characteristics or qualities designated by these symbols are from generation to generation as constant and changeless as the symbols. (Castle, 1912, p. 353)

In his "The Mendelian Notation as a Description of Physiological Facts," East (1912) defended the use of constant symbols to represent constant unit factors, even though the ultimate nature of the factors was not known. He compared the Mendelian notation to formulas in algebra and chemistry and defended its usefulness in predicting the outcome of experiments. The next section will discuss his defense in more detail and quote his definition for unit factors.

Castle's later rat-trap model (discussed in Chapter 10 and reproduced in Figure 10-4) was an attempt at representing the genes as parts of a three-dimensional chromosome. Once he had abandoned the contamination hypothesis and accepted the stability of the genes, he no longer objected to letters symbolizing genes. The rat-trap model may have been a further development of the idea of genes as parts of a single molecule. As we discussed in Chapter 10, Castle objected to the *Drosophila* group's hypothesis of linearity because he considered it doubtful that organic molecules ever have a simple stringlike form (Castle, 1919b). His own symbolic representation located letters representing genes in a three-dimensional space, based on his hypotheses of single cross-overs (but not double cross-overs) and no linearity of relations. Castle's model received extensive criticism by the *Drosophila* group (see Chapter 10).

The linkage maps developed by the *Drosophila* group located names or the first letter of the name of mutant allelomorphs along lines (see Figure 10-1). They claimed that the diagrams, based on cross-over data from breeding experiments, represented the linear arrangements of factors or genes along the chromosomes. Sometimes the letters were replaced by beads (see Figure 10-2) and in other diagrams, the chromosomes had a two-dimensional structure and the genes were represented by letters inside the chromosomal representations (e.g., Morgan, 1926, p. 60, reproduced in Figure 12-3). These graphic representations brought order to the confusing ratios in the data for linked genes that occasionally crossed over. They illustrated in a dramatic way the new hypotheses proposed by the *Drosophila* group, based on relations between genes and chromosomes. The linkage diagrams, unlike Castle's organic chemistry ones, also illustrated fruitful interrelations between two levels of organization that could be investigated with techniques from two different fields.

The symbolic representations of letters along lines or beads along a string showed significant theoretical developments. The symbols used in the early days of Mendelism such as *A* and *a* designated some unobservable difference between germ cells in hybrids produced by crossing varieties with alternative characters. What accounted for that difference was unknown. Thus, the field began with symbols that represented germ cells as if they were black boxes, differing in some unknown way. (See Shapere, 1974b, p. 562, for further discussion of black-box incompleteness as a theoretical inadequacy.) The black boxes were filled in as the theory of the gene developed. Debate occurred about the adequacy of various symbolic representations for the genes and their relations in the germ cells, just as it did about the adequacy of various explanations for empirical anomalies.

By 1926 the theory of the gene dispelled some of the mystery about what made the germ cells different and, thus, about what the early symbols had represented. According to the theory, germ cells of hybrids, originally symbolized by *A* and *a*, differed in the beadlike genes located at corresponding loci on homologous chromosomes, now symbolized by the first letter of the name of the mutant character and located on a map with other symbols for other genes on the same chromosome.

The symbolism of beads along a string raised new conceptual problems. The new mystery was what the beads themselves represented. Did they represent single material particles or something else? What were the beads made of? Did they have internal structure? Were genes independent or did their positions along the strings affect their action? These general unanswered questions will be discussed more fully in the next section. They illustrate how conceptual problems arise about the theory itself and about the symbols used to represent the theoretical claims.

These developments show that geneticists were not content merely to manipulate symbols to make predictions without trying to push to lower levels of organization and answer questions about what the symbols for different types of germ cells represented. Two- and three-dimensional representations, embedding the Mendelian letters within them, were introduced and debated as to their adequacy for representing structures in the germ cells or developmental interactions. The representations were often based on hypotheses about interrelations between Mendelian factors and other entities at other levels of organization, such as Castle's and Johannsen's organic molecule diagrams, and depictions of genes along chromosomes.

This section has shown how the symbolism developed along with other developments in Mendelism. Changing hypotheses raised issues about the adequacy of various symbolic representations and necessitated changes. New symbols and models were devised to solve the conceptual problems raised by the older, inadequate representations.

11.4 Terminology

Changes in terminology sometimes signal important theoretical developments. At other times, a new term may be substituted for one or more old ones without any significant change in theory. I will argue in this section that the change in terminology from "unit-character" to "factor" indicated an important theoretical change, but the adoption of "gene" was less important. "Factor" seems to have gradually come into usage and I have been unable to find anything that looks like its original introduction, accompanied by an explicit definition. In fact, the genetics literature contains few explicit definitions of its key theoretical terms. Some scientists were given to coining new terms, while others rarely did so. Bateson and Johannsen, for example, introduced numerous new terms, Castle and Morgan very few. It is not an easy historical task to determine the meaning of a term at a given time; different individuals often used the same terms differently or used different terms to mean the same thing. As is not surprising, the meaning of terms usually must be determined from the contexts in which they were used and in the light of knowledge about the scientist's position on various issues.

As discussed in earlier chapters, much of the work in the period between 1900 and 1910 did not distinguish between what was later called the phenotype (the visible characters in an organism) and the genotype (all the genes in an organism). Early usage of the term "unit-character" sometimes referred to the phenotypic character, such as the yellow color of peas, but at other times was used to designate a type of germ cell that later produced plants with that character. Bateson's introduction of the term "allelomorph" was originally at the phenotypic level:

> Each such character, which is capable of being dissociated or replaced by its contrary, must henceforth be conceived of as a distinct *unit-character*; and as we know that the several unit-characters are of such a nature that any one of them is capable of independently displacing or being displaced by one or more alternative characters taken singly, we may recognize this fact by naming such unit-characters *allelomorphs*. (Bateson, 1902, p. 22)

In 1903, Sutton used Bateson's term "allelomorph" to refer to the Mendelian units that he hypothesized were part of the chromosomes. Gradually, "allelomorphs" came to refer to alternative factors for the same trait, such as yellow or green color of peas. "Multiple allelomorphs" referred to more than two alternatives for the same trait. In 1903, Bateson speculated that more than two allelomorphs for a given trait might be found. After developing his presence and absence hypothesis in 1905, however, Bateson opposed the hypothesis of multiple allelomorphs, which was advocated by the *Drosophila* group. "Allelomorph" continued to be used in the period between 1910 and 1926 for referring to one of a pair of characters (e.g., Castle, 1916a, p. 101; 1924, p. 139) or one of a pair of factors (e.g., Morgan, 1919, p. 23) or one of a pair of genes (e.g., Morgan, 1926, p. 93). By the 1930s, "allelomorph" was shortened to "allele" and used to refer to the alternative states of a gene found at the same locus in a linkage group (see, e.g., Shull, 1935). Normally, only two alleles are found in any one organism, carried by the two homologous chromosomes; however, in a population, more than two alleles for a trait may exist. The multiple allelomorph hypothesis became a theoretical component, as discussed in Chapter 6.

Sometime in the period between 1903 and 1905, Bateson began using the terms "unit factor" and "factor." It was a distinction he needed in proposing the presence and absence hypothesis (e.g., Bateson and Punnett, 1905, pp. 136-137). Dominant characters were determined by a present factor, but recessive characters were represented by a loss of that factor. Castle critiqued the "factor hypothesis" in the context of criticizing Bateson's presence and absence hypothesis in 1906 (Castle, 1906).

Bateson gradually began using "factor" outside the context of the presence and absence hypothesis, in discussing Mendelism more generally: "We know first the fact deduced from Mendel's original experiments with peas, that the bodily characters may result from the transmission of distinct unit-factors" (Bateson, 1907, p. 166). In 1909, Morgan, in his critique of the factor hypothesis (discussed in Chapter 6) indicated that the factors were "sometimes referred to as the actual characters themselves—unit-characters" (Morgan, 1909d, p. 365). In his 1909 book, Bateson's terminology was not consistent. He often used "character" for what was found in the gametes, but sometimes caught himself. In discussing Mendel's work, for example, he said:

> If the germ-cells which that zygote eventually forms are bearers of *either* tallness *or* dwarfness, there must at some stage in the process of germ-formation be a separation of the two characters, or rather of the ultimate factors which cause those characters to be developed in the plants. (Bateson, 1909, p. 11)

By 1911, Punnett explicitly stated the distinction: the "factor" was the term for the "something" in the gamete that "corresponds" to the unit-character (Punnett, 1911, p. 31).

In his 1911 review article, Spillman discussed current terminology. He criticized the unit-character conception that one unit in the germplasm was responsible for the development of a definite portion of the organism. He attributed this view to de Vries, but not to Mendel. He suggested that one character was probably the result of many factors and one factor probably was related to many characters. After introducing usage of interacting factors (and their absences) to account for various characters in his breeding experiments, he claimed, "It would be a serious matter to convince any one who has watched the shifting and recombining of these factors that they are not real things" (Spillman, 1912, p. 765).

In 1912, East discussed the Mendelian symbols and the nature of unit factors. East had introduced the multiple factor hypothesis, and thus he needed an explicit distinction between character and factor:

> The facts of heredity that one describes in the higher organisms are the actual somatic characters, variable things indeed, but still things concrete. Their *potentialities* are transmitted to a new generation by the germ cells. We know nothing of this germ cell beyond a few superficial facts, but since a short description of the breeding facts demands a unit of description, the term unit factor has been coined. As I hope to show, a factor,* not being a biological reality but a descriptive term, must be fixed and unchangeable.
>
> [* He attached the following note:]
> I hope this statement is not confusing. The term factor represents in way a biological reality of whose nature we are ignorant just as a structural molecular formula represents fundamentally a reality, yet both as they are used mathematically are concepts. (East, 1912, p. 634, my emphasis)

Obviously East was struggling with the issue of what to claim about the nature of factors, since so little was known.

In his book, *The Gene: A Critical History*, Carlson (1966) documented the change in terminology from unit-character to factor to gene. He emphasized Castle's "fallacy" of confusing the unit with the character or assuming that, if the character showed increased variability, then the unit must also have varied (Carlson, 1966, ch. 4). Carlson discussed Castle's critique of the hypothesis that pure, nonvariable factors existed (Carlson, 1966, p. 26). An important conceptual shift occurred with the conceptual severing of the unit from the character. It was a conceptual shift that Castle resisted, as discussed in Chapter 8.

In 1903, Johannsen published work on "pure lines" in beans. Johannsen was working on the problem of the efficacy of selection in producing new forms. Johannsen's work has not been discussed in previous chapters because it was more within the context of problems related to the evolution of new varieties and species than specifically to problems about heredity. His work on selection of beans for size characters contributed to prevalent arguments at the time that selection was not efficacious in producing new varieties, but could only be used to sort out stable "genotypes" already present within a population. Johannsen was not doing Mendelian hybridization experiments and the characters he studied in beans were not tested for segregation. Nonetheless, his work on stable genotypes was interpreted by some Mendelians as indicating that the Mendelian factors were stable and not variable (e.g., Muller, 1914).

As discussed in Chapter 8, Castle's contamination hypothesis was an alternative to the claim of stable, noncontaminating factors. Castle opposed Johannsen's views about the inefficacy of selection in producing new types and pointed out that the characters in beans studied by Johannsen did not show Mendelian segregation. Castle argued against the "purity of the gametes," and against assumptions about "pure genes" (Castle and Phillips, 1914, pp. 5-6).

Johannsen had introduced the term "gene" in his 1909 book, *Elemente der Exakten Erblichkeitslehre*. He coined the term by, he claimed, using a part of Darwin's term "pangene" (Johannsen, 1909, p. 124). Actually Darwin (1868) had used the term "gemmule" for the hereditary units he postulated in his "provisional hypothesis of pangenesis," and de Vries (1889) had changed the name to "pangen" when he provided his own altered version of Darwin's theory in his theory of intracellular pangenesis.

Johannsen argued for using his new term "gene" as a substitute for the German "*Anlagen,*" "*Zustande,*" and "*Grundlagen*" (Johannsen, 1909, p. 124) and for the English terms "unit-factors," "elements," or "allelomorphs" (Johannsen, 1911, p. 132). He claimed it was free of all hypotheses and stood for the "something" in the gametes that "determined" (*bedingt*) or "has some effect on" (*mitbestimmt*) a character in the developing organism (Johannsen, 1909, p. 124). Johannsen's usage seems to have added little to the concept of Mendelian factor already in use. If anything, it removed some of the connotations; for example, Johannsen's term avoided de Vries's numerous claims about pangens. In addition to his claim that paired, antagonistic pangens showed Mendelian segregation, de Vries had claimed that pangens were atomlike units in the nucleus that were carried by chromosomes, reproduced, and grew into cell parts.

Morgan (1911c) used the term "gene" in the published version of his address at a 1910 symposium on Johannsen's work, in which Johannsen introduced the term in English (Johannsen, 1911). Yet, Morgan did not use "gene" consistently until 1917. Allen (1978, pp. 209-210) argued that the Morgan group did not adopt Johannsen's term in *Mechanism* in 1915 because Johannsen claimed that the gene was to be an abstract concept, consciously dissociated from any theories of hereditary particles carried by chromosomes. Since the *Drosophila* group advocated a material view of units located on chromosomes, they chose not to use Johannsen's term in the early years. Interestingly, Bateson, who also opposed the chromosome theory, used "factors or gens" in his review of *Mechanism* (Bateson, 1916, p. 462). Muller, however, who was an early advocate of the material nature of the Mendelian factors, also used the term "gene" in 1914. Thus, Allen's interpretation of why the *Drosophila* group did not use "gene" in *Mechanism* perhaps did not apply to Muller.

In contrast to Allen's interpretation, Coleman suggested that Morgan's later willingness to adopt Johannsen's term "gene" perhaps reflected Morgan's sharing with Johannsen some skepticism about knowing the physical nature of the Mendelian factors (Coleman, 1970, p. 239). Yet, Coleman's interpretation would not explain Muller's early usage (Muller, 1914) because Muller advocated their material nature. Of course, neither Morgan nor Muller shared Johannsen's skepticism about the location of the genes on the chromosomes or his skepticism about the generality of the theory of the gene for solving the problem of heredity. Thus, the changing usages of the terms "factor" and "gene" by the *Drosophila* group are difficult to use in drawing uncontroversial historical conclusions.

Johannsen opposed attributing to the gene a material particle nature (Winge, 1958). In addition, in 1911, Johannsen opposed the chromosome theory (Johannsen, 1911). Even by 1923 he only admitted that the *mutants* the Morgan group had found were associated with the chromosomes. He speculated that the segregating characters were "rather superficial in comparison with the fundamental Specific or Generic nature of the organism." He continued:

> We are very far from the ideal of enthusiastic Mendelians, viz. the possibility of dissolving the genotypes into relatively small units, be they called genes, allelomorphs, factors or something else. Personally I believe in a great central something as yet not divisible into separate factors. (Johannsen, 1923, p. 137)

He then suggested that this "central something" was "perhaps only provisional." This passage shows the difficulty in determining exactly what meaning Johannsen attached to his term "gene." Johannsen scholars have, in fact, debated the extent to which Johannsen's terms "gene," "genotype," and "phenotype" changed throughout his career and differed from the modern usage (Churchill, 1974; Wanscher, 1975; Roll-Hansen, 1978b). "Phenotype" and "genotype" may have referred to statistical attributes of a *population*, not to the characters and total genes making up one *individual* organism, as the terms are used today (Winge, 1958; Churchill, 1974; Roll-Hansen, 1978b, Mayr, 1982, p. 782). If Johannsen had not been so given to coining new terms and if his term "gene" had not gained wide acceptance, I suspect his work would not have required discussion in a history of the development of the theory of the gene.

Carlson (1966) entitled one of his chapters "From Factor to Gene Through Three Ph.D. Theses." In that chapter he discussed the work of Morgan's students: Bridges, Sturtevant, and Muller. Carlson documented the theoretical changes that resulted from Bridges's work on non-disjunction, Sturtevant's on mapping, and Muller's on the mechanism of crossing-over (discussed in Chapter 10). Carlson implied that the change from "factor" to "gene" was to be attributed to the *Drosophila* group and involved the claim that the gene was located on the chromosome (Carlson, 1966, ch. 9). Carlson concluded his chapter with an apt quotation from Morgan's 1917 article entitled "The Theory of the Gene": "The germ plasm must, therefore, be made up of independent elements of some kind. It is these elements that we call genetic factors or more briefly genes" (Morgan, 1917, p. 515; quoted in Carlson, 1966, p. 76).

In his 1919 book, Morgan used "gene" instead of "factor," but in the revised edition of *Mechanism* in 1922, the *Drosophila* group retained their usage of "factor" from the first edition. Although Morgan eliminated "factor" in his 1926 *Theory of the Gene*, he stated the theory in terms of "paired elements (genes)" (Morgan, 1926, p. 25).

In his 1927 edition of *Mendelism*, Punnett continued to use "factor." In his discussion of Morgan's work, however, he made a remark that supports Carlson's interpretation that the usage of "gene" had come to be associated with the *Drosophila* group's interpretations: "In place of the older term 'factor' most Americans prefer to use 'gene,' a word presumably derived from the Greek" (Punnett, 1927, p. 149, footnote 1). Punnett's only mention of Johannsen was of the 1903 pure line work. He, thus, associated the term "gene" with the *Drosophila* group's work, not with Johannsen's introduction of it. It is reasonable to assume that the *Drosophila* group's eventual adoption of Johannsen's term "gene" established its general usage.

Thus, the terms "factor" and "gene" seem to have been used somewhat interchangeably between 1911 and 1926. Those terms referred to the units (or perhaps something less definite than an individuatable unit) that were found in the gametes and showed Mendelian segregation. Each usage, however, must be carefully read to see whether its user assumed the units were localized on the chromosomes. It would have been less ambiguous if the *Drosophila* group had coined their own term for the material unit that segregated *and* was located on the chromosome. On the other hand, by using the terms that other Mendelians who opposed the chromosome theory used, they allied their theory with the successes of Mendelism and left open the question of the success of the chromosome theory. Just as with mutation, the *Drosophila* group did not coin a new term for the concept that they changed in significant ways. They might well have considered Johannsen's reason for coining new terminology:

> It is a well-established fact that language is not only our servant, when we wish to express—or even to conceal—our thoughts, but that it may also be our master, overpowering us by means of the notions attached to the current words. This fact is the reason why it is desirable to create a new terminology in all cases where new or revised conceptions are being developed. Old terms are mostly compromised by their application in antiquated or erroneous theories and systems, from which they carry splinters of inadequate ideas, not always harmless to the developing insight.

> Therefore I have proposed the terms "gene" and "genotype" and some further terms, as "phenotype" . . . . (Johannsen, 1911, p. 132)

In retrospect, one might have expected to find explicit definitions of "factor" or "gene" as an entity that *causes* characters. In retrospect, it sounds correct to claim that genes *cause* hereditary characters, although the relation is not one cause, one effect. It seems plausible to say that the ratios of hereditary characters produced by breeding experiments were explained by postulating the existence of causal factors, with specific properties and behaviors. With hindsight, it seems geneticists were postulating an underlying, unobservable causal factor to explain their data.

Causal language, however, was used surprisingly infrequently in the literature. Biologists were cautious. The data demanded that some difference between types of germ cells in hybrids be postulated; how that difference was to be more definitely characterized was unclear. Only gradually did biologists realize that to speak of yellow and green *characters* in the germ cells of peas was not appropriate. The change from unit-character to the factor hypothesis was gradual, evoked debate, and created problems about the nature of the factors. Johannsen defined the gene as the something in the germ cells that "determined" or had "some effect" on a character. In 1911, Punnett defined the factor as the something in the gamete that "corresponded" to the unit-character. Even in 1926 in his statement of the theory of the gene, Morgan did not state that genes *caused* characters. Instead, he vaguely said, "The theory states that the characters are *referable* to paired elements (genes) in the germinal material . . ." (Morgan, 1926, p. 25, my emphasis).

Mendelians, perhaps wary of the speculative claims about hereditary particles in nineteenth-century theories of heredity, moved beyond their breeding data very carefully. Only those properties of the elements in the germ cells demanded by the breeding data were ascribed to genes. The move to a symbolic representation was an important move away from the specific details in the data. Considering the *A* and *a* as "factors," rather than "unit-characters" was another important step away from the data. But going all the way to a claim that factors or genes *cause* characters was perhaps just too explicit, based on what was needed to account for breeding data. The *Drosophila* group was willing to consider the genes to be beadlike, material particles in the chromosomes, but they were still wary of making hypotheses about gene action. Thus, explicit definitions of genes as causal factors were not given, although discussion of the nature of the "effects" of genes (e.g., Morgan, 1926, pp. 300-307) makes it plausible to claim that the theory of the gene contained the component: genes cause characters. The need for postulating a new theoretical entity, variously called "factor" or "gene" will be discussed in the next section.

11.5 A New Theoretical Entity and Its Properties

Theories of heredity as far back as Aristotle's had postulated a transfer of something between male and female and between parent and offspring. That something was unobservable to the unaided senses and accounted for the resemblance of offspring to one or both parents and to grandparents and to more distant ancestors. It was not always postulated to be a material stuff (see, e.g., Stubbe, 1972). Particulate theories of heredity abounded in the late nineteenth century, explaining the resemblance of offspring to parent by postulating unobservable hereditary particles. The phenomenon of reversion, that is, an offspring resembling a grandparent or some more distant ancestor, also often evoked hypotheses about units, dormant in the parents, but passed on to the offspring (e.g., Darwin, 1868; de Vries, 1889). Mendelians were often at pains to argue that properties attributed to Mendelian genes were less speculative than the nineteenth-century theories and based more firmly on experimental evidence from breeding experiments (e.g., Morgan et al., 1915, pp. 223-227).

   Although debate occurred about its physical nature and location on the chromosomes, Mendelians did not doubt that some unobservable difference existed between types of the F_1 hybrid's germ cells and that that difference was correlated with characters produced in the generation. The data demanded some sort of unobservable factor to explain hybrid crosses in which, for example, the color green was found in one of the original parents, disappeared in the hybrid, and reappeared in the F_2 offspring in a form that would continue to breed true. Recessives appeared too often and with too much regularity for a theory of spontaneous mutation to adequately account for them.

   It is reasonable to claim, based on the historical evidence, that most Mendelians did not doubt that something, not yet observable with microscopes or analyzable with chemical or physical techniques, accounted for the postulated differences between germ cells in the hybrid. Postulating theoretical entities at ever lower levels of organization was a common strategy in late nineteenth-century biology, as evidenced by the successes of the cell and germ theories of disease.

   De Vries argued that pangens were material particles analogous to atoms in chemistry. Others speculated that Mendelian factors were chemical substances. Considerable debate occurred as to what constituted the difference between the germ cells of hybrids. Many biologists moved cautiously beyond claims about ratios of characters to hypotheses about the causes of the ratios, as discussed above. Symbols for types of germ cells were one step away from observable characters. Another step was the introduction of terms to designate what was carried by the germ cells (because the characters themselves were not carried). With the postulation of the factor hypothesis, debates occurred about the need to postulate a new theoretical entity and about what properties should be ascribed to it.

In 1917, Morgan wrote "The Theory of the Gene" to defend the theory against various criticisms and present evidence for the real existence of genes. He succinctly stated the problem:

> So far I have spoken of the genetic factor as a unit in the germ plasm whose presence there is inferred from the character itself. Why, it may be asked, is it not simpler to deal with the characters themselves, as in fact Mendel did, rather than introduce an imaginary entity, the gene.
>
> There are several reasons why we need the conception of the gene. (Morgan, 1917, p. 517)

He then listed the reasons and discussed each one. They included (1) the manifold effects of each gene; (2) that the variability of the character is not due to the corresponding variability of the gene; (3) that characters that are indistinguishable may be the product of different genes; (4) the inference that each character is the product of many genes; and (5) evidence that genes have a real basis in the germplasm, namely, that they are located in the chromosomes. These reasons show the extent to which the underlying unit had been severed from the observable character.

Morgan continued his defense of the existence of genes and the chromosome theory. The "coincidence" that four groups of linked genes are associated with the four chromosome pairs in *Drosophila*, he said, "adds one more link to the chain of evidence convincing a few of us that the gene in Mendelian inheritance has a real existence" (Morgan, 1917, p. 520).

By this period, the clear need was present to separate the underlying, unobservable something in the germ cell from the character whose development it affected. Thus, a new theoretical component, explicitly postulating the existence of genes in the germplasm, was added to the theory. The existence of a given gene was inferred from the nature of and behavior of mutant characters. The number of properties ascribed to genes was growing, and many new questions could be raised about these new theoretical entities.

Morgan indicated that by 1917 Mendelism had developed considerably since its early beginnings, and new properties could be ascribed to genes: "[Since 1865] much water has run under the Mendelian mill. In consequence we can now add certain further attributes to the rather formal characterization of the gene as deducible from Mendel's law alone" (Morgan, 1917, p. 516). He did not explicitly list the "postulated attributes" of genes, but used genes to explain numerous results of breeding experiments (as well as defending the theory against various criticisms). In general, in the historical literature, one does not find explicit lists of properties attributed to the gene at various times. Nonetheless, it is possible to analyze the historical claims in order to construct such a list.

A growing list of properties for a theoretical entity indicates that it was becoming less "imaginary" and better understood. An analysis to produce such a list aids in showing the theoretical changes that occurred.

Nersessian (1984) in her analysis of the development of claims about electromagnetic fields constructed a list of general properties and traced their change over time. She provided a helpful list of questions that may be used, in general, for tracing changes in claims about a new theoretical entity. Such questions include: What does it do? How does it do it? What is its function? What effects does it produce? What kind of "stuff" is it? How can it be located? The general properties that she traced over time were "stuff," "function," "structure," and "causal power" (see

Nersessian, 1984, pp. 153-159 for more detail).

Work in AI to develop techniques for representing knowledge so that it can be manipulated by computer programs has also developed a method of representing an entity in what are called "frames." A frame consists of a name and a list of associated general properties and relations, which are filled in with specific values for the specific entity, or left blank, if no information yet exists about the value of that general property. More generally: "A Frame is a collection of questions to be asked about a hypothetical situation: it specifies issues to be raised and methods to be used in dealing with them" (Minsky, 1981, p. 109). (For more discussion of frames, see Minsky, 1981; Thagard, 1988.)

It is instructive to represent the changing properties ascribed to genes in this way. Frames for

---

A. General types of properties
    1. is a kind of
    2. has part
    3. part of
    4. number in {value of part of}
    5. causes
    6. structure
    7. association to others of same kind
    8. behavior during a process

B. Allelomorph, 1903
    1. is a kind of—pure, individual, hypothetical hereditary entity
    2. has part—unknown
    3. part of—germ cell
    4. number in germ cell—many
    5. causes (or correlated with)—one independently variable, hereditary character
    6. structure—unknown
    7. association to others of same kind—pairs
    8. behavior of all in organism during gametogenesis
        a. one of each pair to each germ cell (during segregation)
        b. maternal and paternal pairs randomly mix

C. Chromosome, 1903
    1. is a kind of—pure, individual, microscopically observable entity
    2. has part—chromatin
    (has part—nucleoproteids)
    3. part of—nucleus
    (part of—cells)
    4. number in cell—few
    5. causes—(hypothesis—groups of hereditary characters)
    6. structure—threadlike
    7. association to others of same kind—pairs
    8. behavior of all in organism during gametogenesis
        a. one of each pair to each germ cell (during reducing division)
        b. maternal and paternal pairs do not randomly mix
    (predicted to be wrong on the basis of the claim that allelomorphs
    were part of chromosomes; predicted—mix randomly; confirmed in
    1913 by Carothers)

Table 11.1. Frames for allelomorph and chromosome, 1903

representing the claims about the theoretical entity in Mendelism at various time periods can be constructed. Table 11-1 shows a list of general properties in (A) and fills in values in (B) for allelomorph in 1903 and in (C) for chromosome in 1903. The values of the general properties are based on Sutton's 1903 discussion (analyzed in Chapter 7). I will not pause here to justify the choice of the general slots or the particular values for them, as represented in Table 11-1. Alternative representations are possible and might represent Sutton's claims more or less well.

The frames for allelomorph and chromosome in Table 11-1 are refined versions of frames constructed for use in an AI computer program. Roy Rada and I have experimented to determine the adequacy of versions of frames for representing allelomorph and chromosome, by constructing an AI system named SUTTON. The name of the system follows the convention of Langley, Simon, and their colleagues, whose discovery systems have such names as BACON, GLAUBER, and STAHL (see Langley et al., 1987). SUTTON simulates the discovery of the chromosome

---

A. Factor, about 1910
    1. is a kind of—individual hereditary entity
    2. has part—unknown
    3. part of—germ cell (hypothesis—chromosome)
    4. number in germ cell—many, some estimates for actual numbers in some species
    5. causes (or correlated with)—independently variable hereditary character
        a. one-one, or
        b. many-one (multiple factor hypothesis)
    6. structure—unknown
    7. association to others of same kind—pairs
        a. (hypothesis—presence paired with absence in heterozygotes)
    8. behavior during hybridization experiments
        a. segregation
        b. occasional exceptions to segregation with coupled factors

B. Gene, 1926
    1. is a kind of—individual, stable hereditary entity
    2. has part—unknown
    3. part of—chromosome
    4. number in chromosome—many, some estimates for some species and some chromosomes
    5. causes (or correlated with)—independently mutable hereditary character
        a. one-one
        b. one-many
        c. many-one
        d. many-many
    6. structure—beadlike or unknown
    7. association to others of same kind
        a. allelomorphic pairs in any one organism
        b. multiple allelomorphs may exist in a population
        c. linearly arranged in linkage groups on chromosome
        d. allelomorphic pairs at corresponding loci in corresponding linkage groups
        e. number of corresponding linkage groups equals number of homologous chromosome pairs
    8. behavior during hybridization experiments
        a. segregation
        b. independent assortment of linkage groups
        c. sometimes linkage groups cross over

Table 11-2. Frames for factor and gene.

theory of heredity, the claim that genes are parts of chromosomes. (A full discussion of SUTTON would take us too far afield from our purposes here. In addition, I regard that work as a report of preliminary experiments in implementing discovery strategies. Here, I only wish to discuss the frame representations and, in the next section, strategies associated with them. I will not discuss the specific rules and procedures for reasoning about parts and wholes that simulate the discovery of new hypotheses in SUTTON. For more detail, see Darden and Rada, 1988a; 1988b.)

Finding good lists of the general properties to include in such a representation of a new theoretical entity is an interesting philosophical problem. Nersessian's list and my general property list are two examples. Philosophers could no doubt add others from general metaphysical considerations. I will not pause here to pursue that issue further. (See Shapere, 1969, pp. 155-157 for discussion of several general properties attributed to new theoretical entities.) The general properties that I have listed (see Table 11-1.A) for genes were created to try to balance the needs of representing historical claims and the need to specify general types of properties that could apply to other entities, such as chromosomes. The general properties in Table 11-1.A are, I think, minimally adequate for representing some of the important properties.

Representing the properties of allelomorphs and chromosomes (as of 1903) in this way shows the numerous values of general properties they shared and indicates why it was reasonable to postulate the chromosome theory of heredity. The frame representation of allelomorph can also be used to represent the changes in the properties ascribed to the gene over time. Nersessian provided a similar sketch of the changes in the general properties ascribed to electromagnetic fields at various historical times (Nersessian, 1984, p. 158).

I have not created a frame for the theoretical unit of Mendelism in 1900, because the discussions of unit-characters that contain the seeds from which claims about genes developed were too vague. The character and the unit associated with it in the germ cells were confused, as discussed in Chapter 6. Just setting up a frame structure to represent the unit and its short list of

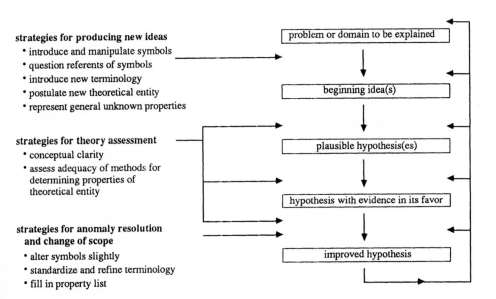

Fig. 11-3. Refined stages and strategies, VIII

properties as of 1900 introduces more clarity than the historical record shows.

Table 11-2 fills in the values that represent properties attributed to Mendelian factors, as of about 1910, and genes, as of about 1926. Examining the similarities and differences between the properties of allelomorph in 1903, factor in 1910, and gene in 1926, is a way of tracking the changes in properties over time. It is difficult to force all the theoretical components of Mendelism, which have been represented in sentences in previous chapters, into this form of representation of properties ascribed to the theoretical entity. Furthermore, implicit claims must be made more explicit than they were in the historical literature. In constructing such a representation, one must be careful not to use too much hindsight. Despite its limitations, this form of representation does graphically show the growing list of properties for the gene as the theory developed.

Additionally, the unknown values of general properties show what issues had not been settled at particular times and were candidates for more research, if adequate experimental techniques could be found to tackle them. What the structure of the gene was and what chemical parts composed it were obvious unanswered questions, for example. The method of representing general properties and missing values is thus a method for finding conceptual problems, as will be discussed in the next section.

Gradually, the need for postulating a new theoretical entity in Mendelism became clear. A new theoretical component had been added that I am characterizing as "genes cause characters." The list of properties ascribed to genes grew as the theory developed. The hypothetical unit and the observable character had been separated.

11.6 Strategies for Finding and Solving Conceptual Problems

This chapter has discussed changes in the theoretical components of Mendelism that resulted from finding and solving conceptual problems. Several strategies are exemplified in these developments, related to the introduction of new symbols, new terms, and the postulation of a new theoretical entity. The strategies discussed in this section are listed in Figure 11-3.

The introduction of symbols to refer to the hypothetical differences between types of germ cells led to the new conceptual problem—what do the symbols represent? Thus, one way of finding conceptual problems is to introduce and manipulate symbolic representations and then ask, what do the symbols represent?

In order to solve this problem, techniques may be needed from fields other than the field in which the problem arose. Alternatively, the symbols may refer to entities, or structures, or processes, at new levels of organization requiring new techniques that have not yet been developed. The strategy exemplified in such problem-solving attempts is the strategy of **finding the referent of the symbols.**

Another question that arose about symbolic representations was the issue of the compatibility of the symbols with the hypotheses. Some changes in hypotheses necessitated a change in symbolism, such as the presence and absence theory leading to the change of *Aa* to *AO*. Thus, another way of finding conceptual problems is to ask—do the symbols adequately represent current hypotheses? (For discussion of such an issue in evolutionary biology, see Provine's critique of the adequacy of Wright's adaptive landscape diagrams for representing Wright's hypotheses, in Provine, 1986, pp. 283-287.)

A way of solving such a conceptual problem about symbolism is to modify the symbolism. Various strategies for producing new ideas can be used in modifying symbolism, such as changing the old one slightly, using analogies, and importing symbolic representations from other fields. The failure to provide an adequate symbolic representation can result in a new hypothesis not being understood by others. Such a misunderstanding occurred in the case of Cuénot's representation of multiple alleles. It is also important to find good analog models, not only for exposition, but for the development of new hypotheses in the face of anomalies. In Chapter 10, I suggested that the analog model to beads along threads may have played such a role in suggesting the ideas of crossing-over and double crossing-over. Castle's ideas about slight modifications in genes may have been related to his idea that genes were parts of organic molecules that could be slightly modified with changes in one or more atoms or "radicals" composing the molecule.

The discussion of new terms indicates that conceptual problems may be lurking in a field when a plethora of different terms are being used for the same thing. As each scientist sees the need for new terminology and postulates new terms, a confusing proliferation may result. At some point, the field should stop and take stock of the various terms, their meanings, and their usage. Standardization and common agreement is a way of solving a terminology problem. Sometimes deeper conceptual disagreements may lie behind terminological differences. Sorting out the terms may aid in uncovering areas of disagreement.

The issue arises as to when to use the strategy of **introduce new terminology**. The *Drosophila* group can perhaps be faulted for coining too few new terms, given the significant changes they made in the ones they adopted, such as "mutation" and "gene." Yet, when the meaning changes gradually over time, it may be difficult to see when a new term is needed. Also, as mentioned earlier, using an old term that is associated with successes is a way of "buying into" the previous success. The *Drosophila* group bought into the successes of Mendelism by using the Mendelian "factor," and then added their claim that the factors were parts of chromosomes. As we have seen, although Johannsen's term "gene" was adopted, the meaning he himself attached to it is not easy to determine. Scientists should perhaps be wary of adopting a new term coined by someone else; it may have misleading connotations.

There is a possible strategy associated with new term introduction that I will not advocate: the strategy of **explicitly stating and defining a new term**. Sometimes that is a useful approach. In general, however, I am not enamored of the method of requiring necessary and sufficient conditions be stated before inquiry can proceed. New scientific concepts, such as the unit-character, are often fuzzy in their early stages. Too much rigor in the early stages may not be possible or desirable. I believe new terms should sometimes be allowed to "float" in their early stages, that is, they should be used in one way and then another, until their area of applicability becomes clear.

After a new theory is well-developed and well-supported by evidence, then precise definitions of its terms may be easier to devise; however, there may be no need to do so. East also had a rather negative view of the demand for explicit definitions: "The first thing one does if he wishes to oppose the idea of a unit character is to ask for a definition. A perfect definition of a unit character is as difficult to formulate as for a flower, yet one can obtain an idea of a flower by proper application" (East, 1912, p. 645). Geneticists rarely gave explicit definitions. Thus, in developing strategies for theory change related to changing terminology, I do not suggest introducing new terms with explicit definitions as the best method in the early stages. After usage has developed, then scientists may find that pausing to ponder definitions and standardize terminology are useful methods for producing clarifications in the field.

The most interesting of the conceptual issues discussed in this chapter is the issue of when to postulate a new theoretical entity and how to determine its properties. The gene filled a gap between parents and offspring; something had to be transmitted to account for heredity. Some difference in types of germ cells could be inferred from character ratios. It need not have been a discrete material entity. Minimally, what was needed was something with the property of segregating in a pure way and later being associated with the development of a character in the offspring. Filling in the gaps in steps of processes is a reason to postulate a new theoretical entity. This idea will be developed further in the next chapter as we discuss the way hypotheses about genes filled in the gaps in pedigree diagrams for relations between parents and offspring.

I would like to be able to say that the case exemplifies the strategy of **postulating an underlying causal factor** to account for observable regularities; however, as noted in Section 11.4, explicit causal claims about genes were rare. The nature of gene action was unknown. Some nineteenth-century theories had postulated that hereditary particles *grew* into cells or cell parts (e.g., Darwin, 1868). Although there were speculations about the relations between genes and enzymes, as will be discussed in Chapter 13, the claim was not made that genes *cause* enzymes. Although I am stating the theoretical component as "genes cause characters," geneticists were usually not so explicit. Thus, I am not claiming that the case explicitly exemplifies the strategy of **postulating an underlying causal factor**.

The strategy of representing general properties of a newly postulated theoretical entity is a method for generating conceptual problems. The blanks in the representation point to problems to be solved. Finding methods for filling in the blanks may lead to important theoretical advances. Sometimes, finding a property, such as what the parts of the new entity are, will involve pursuits at a different level of organization, perhaps using techniques from a different field.

Thus, a general method for finding conceptual problems is to devise a representation that is incomplete, either a diagrammatic representation (see the next chapter) or an explicit list of types of properties (such as the frame representation discussed in Section 11.5). Filling in what is known and examining what is missing shows problems to be solved. Many different strategies may be involved in solving conceptual problems generated in this way. Numerous of the strategies discussed thus far are relevant, as are the strategies for diagnosing and fixing faults in the theory (discussed in the next chapter).

Strategies for theory assessment that are especially relevant to conceptual problems include determining that claims about a new theoretical entity are clear and that its postulated properties can be adequately investigated (listed in Figure 11-3). Strategies for anomaly resolution relevant to conceptual problems include **altering symbols slightly, standardizing and refining terminology**, and **filling in missing items in a property list**. These are strategies for incremental improvement in a theory (also listed in Figure 11-3).

In summary, this chapter has discussed various conceptual problems with the theory of the gene, related to changing symbolism, terminology, and the postulation of a new theoretical entity. The resolution of these problems resulted in new theoretical components in the theory of the gene, components in which a new theoretical entity was postulated. Strategies for finding and solving conceptual problems have been suggested, based on these historical changes. These strategies include **introducing and manipulating a new symbolic representation, standardizing and clarifying terminology, postulating a new theoretical entity at a new level of organization, as well as developing a list of general properties and filling in the blanks** for the new theoretical entity. They are depicted within the context of stages of theory development in Figure 11-3.

CHAPTER 12

Exemplars, Diagrams, and Diagnosis

12.1 Introduction

Throughout this analysis, I have emphasized that I have often stated claims more explicitly than was found in the historical literature. Theoretical components stated in sentences have been useful for my purpose of tracing the development of the theory of the gene. This lack of fidelity to the actual published record, however, raises a question: if geneticists usually did not explicitly state their theories in sentences or explicitly define their theoretical terms, how were theories usually represented in the literature?

This chapter will explore one possible answer to that question. Morgan's 1926 book, *The Theory of the Gene*, will be examined to see how the theory was conveyed, other than in the statement of the theory in sentences that Morgan provided (Morgan, 1926, p. 25). An analysis of his exposition suggests an important role for examples and diagrams of typical hybrid crosses as a means of representing theoretical claims. Diagrams illustrated the roles that genes were claimed to play in typical hybrid crosses. The examples and diagrams illustrating the role of genes served as exemplars of how to explain hybridization data by depicting behaviors of unobservable genes. Examples of typical hybrid crosses illustrated the explanatory repertoire of the Mendelian geneticist in 1926.

Section 12.2 discusses Morgan's exposition of the 1926 theory, both in terms of examples and through his summary of the components of the theory in sentences. Section 12.3 discusses the representation of the theory of the gene by exemplars and diagrams. Section 12.4 considers various issues in the philosophy of science, given this view of the theory as showing the role of theoretical entities in typical steps of hybrid crosses. The issues include the relations between data and theory, the nature of the explanation that the theory provides, the role of exemplars as concrete problem solutions, and the role that exemplars can play in providing abstract problem-solving schema for use in similar problems. Section 12.5 recasts the discussion of types of anomalies in the light of this different representation of the theory. Monster anomalies will be analyzed as problems showing where typical steps fail in unusual cases. Model anomalies will be analyzed as showing the need for a new exemplar.

Finally, Section 12.6 will explore an analogy between diagnostic reasoning and anomaly resolution. Just as a repairperson diagnoses faults in a device or a physician diagnoses a disease in the human body, a scientist engaged in resolving an anomaly for a theory is trying to locate the fault in the theory. Providing a new hypothesis to resolve the anomaly is sometimes like fixing a faulty module in a device or providing treatment for a disease. Considering the theory as it is depicted in a series of normal steps in hybrid crosses provides a "flow diagram" for a normal process that can be used in diagnosing the fault in a particular anomalous case.

This chapter, then, introduces and makes use of an alternative, diagrammatic representation of the theory of gene. The theory is represented as a set of exemplary problem solutions showing how to use genes in explaining typical hereditary phenomena.

12.2 Morgan's Exposition of the Theory of the Gene

Prior to presenting the explicit statement of the theory of the gene, Morgan (1926) discussed typical hybrid crosses and their results. These included the usual Mendelian pea crosses: tall crossed with short yielded 3 tall to 1 short in the F_2 generation; yellow and round peas crossed with green and wrinkled ones yielded 9:3:3:1 ratios in the F_2. In addition to these typical crosses, Morgan added others that had resulted from the expansion of the scope of Mendelism. First, incomplete dominance was depicted for flower color in four o'clocks. Figure 12-1 reproduces Morgan's figure showing the phenotypic results of 1 red to 2 pink to 1 white. Dominance was not universal and incompletely dominant forms provided a good illustration of the nature of the F_2 hybrids (i.e., how pure reds differ from pink hybrids was not obscured in the phenotype of hybrids). Then, Morgan presented the role that the genes played in the steps of the hybrid cross. His diagram is shown in Figure 12-2. Beadlike particles were depicted. Two of the same type were present in the parents; one of each type was represented in the genes of the F_1 hybrid. Morgan illustrated the role of genes in the hybrid during gametogenesis and fertilization using circles and

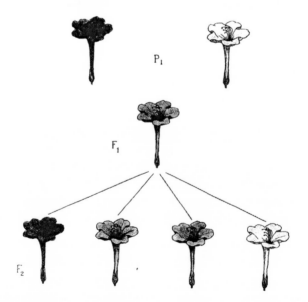

Figure 12-1. Morgan's figure for phenotypes of four o'clocks (From Morgan, Thomas H. (1926), *The Theory of the Gene*. New Haven: Yale University Press, p. 6) .

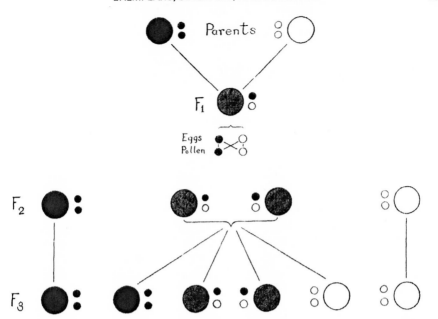

Figure 12-2. Morgan's Illustration of genes in four o'clocks (From Morgan, Thomas H. (1926), *The Theory of the Gene*. New Haven: Yale University Press, p. 7).

a cross between them. Finally, the circles in the F_2 and F_3 generations graphically showed the genes associated with the 1:2:1 ratios in the phenotypic characters.

The diagram represented numerous implicit assumptions: genes were beadlike; they occurred in pairs in homozygous and heterozygous forms; they segregated in a pure form during the formation of gametes in hybrids (with no contamination between red and white); they occurred in equal numbers; they combined randomly; and so forth. A picture is worth a thousand words. I will not try to state them all.

Morgan continued his examples of typical genetic phenomena by providing pictures of the phenotypes of sweet peas and fruit flies showing incomplete and complete linkage. Then data from *Drosophila* crosses with incomplete linkage and crossing-over were presented along with pictures of phenotypes. The cross-over data were used to calculate relative distances on genetic maps. After presenting pictures of the four *Drosophila* maps, showing relative positions of dozens of genes, he stated the theory of the gene in sentences. It is worth quoting him again here; note that he did not yet mention chromosomes in this explicit statement of the theory of the gene in sentences. After discussing these examples, Morgan said:

We are now in a position to formulate the theory of the gene. The theory states that the characters of the individual are referable to paired elements (genes) in the germinal material that are held together in a definite number of linkage groups; it states that the members of each pair of genes separate when the germ-cells mature in accordance with Mendel's first law, and in consequence each germ-cell comes to contain one set only; it states that the members belonging to different linkage groups assort independently in accordance with Mendel's second law; it states that an orderly interchange—crossing-over—also takes place, at times, between the elements in corresponding linkage groups; and it states that the frequency of crossing-over furnishes evidence of the linear order to the elements in each linkage group and of the relative position of the elements with respect to each other.

These principles, which, taken together, I have ventured to call the theory of the gene, enable us to handle problems of genetics on a strictly numerical basis, and allow us to predict, with a great deal of precision, what will occur in any given situation. In these respects the theory fulfills the requirements of a scientific theory in the fullest sense. (Morgan, 1926, p. 25, italics omitted)

This statement of the theory provided an abstract characterization of the "typical" and "occasional" (i.e., crossing-over) behavior of the unobservable genes during germ cell formation. The *general* verbal statement summarized all the cases that the separate diagrams illustrated in *concrete* examples. The theory focused on the step between parents and offspring during the formation of germ cells. Although illustrated by hybrid crosses between parents with differing characters, Morgan's abstract statement generalized the findings for *all characters* of an individual organism. The theory made *general* claims about the nature of unobservable details in normal hereditary processes. Morgan stressed that all the claims made in the theory of the gene about the unobservable "elements" in the germ cells were based on data from hybridization

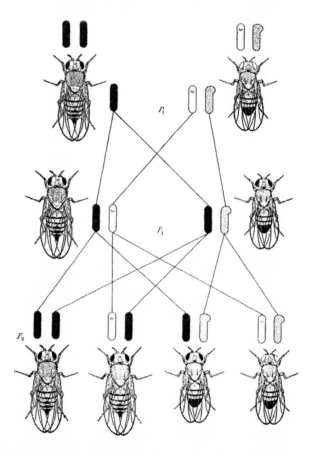

Figure 12-3. Morgan's illustration of sex linkage (From Morgan, Thomas H. (1926), *The Theory of the Gene*. New Haven: Yale University Press, p. 60).

experiments. Furthermore, he emphasized that the theory made predictions that were supported by the results of additional hybridization experiments, such as various test crosses.

Not until a subsequent chapter did he introduce discussion of the chromosomes passing "through a series of manoeuvres that go far toward supplying a mechanism for the theory of heredity," as well as presenting "the view that the chromosomes are the bearers of the hereditary elements or genes" (Morgan, 1926, p. 45). He did not label this "the chromosome theory of heredity," as he had labeled the "theory of the gene," but much of the rest of the book was devoted, not to discussion of Mendelian breeding experiments in support of the well-established Mendelian theory of the gene, but to evidence for the chromosome theory.

After introducing the chromosome theory, he then was able to add depictions of chromosomes to his diagrams, with genes designated by letters on the chromosome. Figure 12-3 shows Morgan's figure illustrating the crosses with the white-eyed mutant and the explanation of the results in terms of inheritance of sex chromosomes, with one X carrying the gene for white eye. It graphically depicts the relationships between an observable phenotypic character (eye color), microscopically observable sex chromosomes, and hypothetical genes. It also illustrates the genetic and chromosomal combinations during hybrid crosses and the steps leading to the results in the F_1 and F_2 generations.

The various diagrams of typical hybrid crosses and their explanation in terms of the behavior of genes and chromosomes can be viewed as "exemplars," namely, concrete examples that served as model cases. Morgan's 1926 book was the written version of a series of general lectures and it was used as a textbook in courses. Segregation, independent assortment, linkage, and crossing-over constituted the explanatory repertoire of the Mendelian geneticist, as of 1926. Morgan taught his readers about the theory and how to give Mendelian explanations by providing the concrete examples of each type of process, illustrated by diagrams. The theory did not have to be stated in sentences; the theoretical claims were illustrated in the examples of typical crosses. The diagrams graphically portrayed the role of genes and chromosomes in accounting for the ratios of inherited characters.

The examples supplied model cases. Similar results of similar hybrid crosses could be explained by invoking similar steps and filling in the details about the characters in the specific cross. Students could be taught how to apply the theory of the gene and the chromosome theory by examining the roles attributed to genes and chromosomes in these typical hybrid crosses with peas, four o'clocks, and fruit flies. Then, those patterns, depicted in the typical steps of typical hybrid crosses, could be used to explain similar results in other hybrid crosses showing the same patterns in the data.

Mendelian genetics is taught in a similar way today, beginning with an exemplary explanation of 3:1 ratios (often in Mendel's yellow and green peas) in terms of segregating genes (e.g., Moore, 1972; Strickberger, 1985). Other cases of 3:1 ratios, the student learns, are to be explained in a similar way. Then, patterns for independent assortment, linkage, and crossing-over are gradually added. Becoming an expert entails learning the patterns and learning when to apply them. When experienced geneticists see a 3:1 ratio, they know that those alleles are undergoing a typical Mendelian segregation process. Morgan's exposition of the theory served to teach such patterns to his readers in 1926.

12.3 Exemplars and Diagrams

This perspective provides an alternative way of thinking about the development of the theory of the gene. I have been analyzing its development in previous chapters in terms of changing statements of the components of the theory. Instead, this view of the importance of exemplars suggests looking at that history in terms of an expansion of the explanatory repertoire of the theory by adding to its set of typical examples. The example of the role of Mendelian factors in segregating so as to explain 3:1 ratios became part of a much larger set of exemplars, including examples of independent assortment, sex linkage, and crossing-over. The exemplars depict details in typical hereditary steps between parents and offspring, steps that can be represented diagrammatically.

From this perspective, the history of the problem of heredity can be traced by showing how the details of the steps from parents to offspring were filled in. The nature of the connection between parents and offspring had been a gap in the diagram, a black box to be filled. By looking back at the early history of heredity from this perspective, it is possible to view attempts to solve the problem of heredity as attempts to fill in the gaps in pedigree diagrams. Such diagrams of hereditary trees showed characters that were passed from parents to offspring, but provided no depiction of the hereditary material passed along. The discovery of cells enabled the problem to be pushed to the level of the germ cells that united at fertilization.

Mendelian genetics allowed inferences to unobservable differences between germ cells in hybrids to take the analysis to yet a lower level of organization. From observable steps (e.g., 3:1 ratios in the F_2), inferences were made about unobservable entities and their activities at previous steps between parents and offspring. The unseen details about the typical behavior of genes in the formation of germ cells constituted the theoretical components of Mendelism. Ratios of characters in the F_2 and following generations were to be explained by the existence and assortment of particlelike genes. Typical Mendelian ratios were explained by fitting the ratios into a diagram of typical steps in which pure genes were depicted as segregating during gametogenesis and recombining at fertilization.

With the formation of the chromosome theory, chromosomes were added to the diagrams. Additional independent evidence for the claims about the hypothetical genes was supplied by locating them on the observable chromosomes and showing the parallels between the behavior of chromosomes and genes. Segregation, independent assortment, linkage, and crossing-over were graphically depicted by locating genes on chromosomes. The diagrams of crosses fitted the pieces together, the pieces being the parents of the original cross, their chromosomes, and their genes. The chromosomes and genes passed to the offspring and accounted for the characters found in children and grandchildren. All the pieces fitted together in consistent steps in diagrams representing typical hereditary processes. Fitting data into typical steps explained that data.

Thus, a way of viewing the theory of the gene and the chromosome theory is as follows: the theory is represented by stepwise diagrams, depicting structures at different levels of organization; the structures play specific functional roles in typical hereditary processes. The chromosome and gene theories were often represented, not in specific statements, but by diagrams showing the steps in which the theoretical entities were claimed to function.

12.4 Exemplars and Explanation

This view of the theory of the gene and the chromosome theory sheds light on a number of issues of concern to philosophers of science. The issues to be discussed in this section include the role of exemplars, the relation of data and theory, and the nature of explanation. This view also allows me to recast methods of anomaly resolution developed in previous chapters. That task will be taken up in the next two sections of this chapter.

Kuhn (1970; 1974) discussed the importance of "exemplars," concrete problem solutions in which a formalism is applied and given empirical grounding. He discussed how problems presented in textbooks are a way that exemplars are taught. The student learns to see other cases as similar to the exemplary problems and learns to use the pattern again to solve a new puzzle in a similar way. (This point is also made in Kitcher, 1984; Schaffner, 1986a.) The genetics case exemplifies the role of shared examples, rather than abstract statements of formulas, as a means of representing the theory. Providing examples of Mendel's pea crosses and showing how segregating genes explain the 3:1 ratios was a typical way of communicating Mendelian theory. After about 1910, finding another instance of 3:1 ratios and applying Mendelian factors to explain them was like puzzle solving; no new theoretical development resulted from applying the theory in yet another instance of the same type. Such application of an old pattern is importantly different from adding additional exemplars. Additional exemplars applied to new types of problems (new types of data to be explained), such as quantitative characters, sex-linked characters, and incompletely linked characters. Adding new exemplars constituted theory development, not just the application of the theory to another instance of the same kind.

Kitcher (1981; 1989) analyzed theories, saying that they provided one or more abstract argument patterns that are invoked, in particular cases, by instantiating the abstract patterns, which he called "schemas." According to Kitcher, a scientific theory explains a particular case by showing that the case is an instance of one of its schemas. The domain that the theory explains consists of all the cases in which the argument pattern can be instantiated properly. Kitcher's and Kuhn's insights may be combined. Exemplars may serve as the source for constructing one or more abstract patterns, in this case, abstract steps in abstract diagrams of typical hereditary processes. Or, alternatively, no abstraction may actually be formed; geneticists may substitute new variables, such as different characteristics or different chromosomes, when applying a concrete diagram to a new case. The set of exemplars representing the standard cases of segregation, independent assortment, linkage, and crossing-over provide the set of abstract patterns of steps in typical hereditary processes. (Kitcher is interested in philosophy of language and hence talks in terms of providing "arguments." The idea of an abstract argument schema, however, can be applied to abstract steps representing hereditary *processes*, rather than representing *arguments*.) Thagard (1988, pp. 41-43) has also discussed the role that Kuhn's exemplary problem solutions can play in providing good abstract schemas for problem solving, such as the schemas discussed by Kitcher. Explanation then involves instantiating a particular schema in an instance in order to show how it can be subsumed under the theory, the theory being considered as a set of problem-solving schemas (among other things). (For a similar view of explanation, see Schank, 1986.)

One analysis of explanation is that it involves providing an argument; the initial conditions and a general law are premises and the data to be explained are the derived conclusion (Hempel, 1965). The view of the theory of the gene as providing a general pattern of typical steps in a hereditary process allows a different analysis of explanation for this case. The data to be explained are fitted into a diagram of typical steps in a hereditary process. Figure 12-2, for example, shows a typical hybrid cross in which the parents differ with respect to one characteristic. Initial conditions, data, and theory are combined in the diagram. The kind of organisms used in a cross are the initial conditions, the role of the genes in germ cell formation is the theory, and the resulting 1:2:1 ratios are the data. When such a cross is made and the segregation mechanism (represented by the "beads" in the diagram) is functioning, then the 1:2:1 data result. Explaining the 1:2:1 data involves fitting that data into this series of typical steps for an instance of segregation of one pair of alleles.

The diagrams not only represent the theoretical steps, but combine the theoretical and the observational. Various steps in the diagram represent observable phenomena; their nature aids in inferences to the nature of the unobservable steps. Since white was found in the parental generation, disappeared in the F_1, and reappeared in the F_2, it was a very reasonable inference that something was present in a dormant (or partially active) state in the F_1 step, correlated with the pink flowers. The 1:2:1 data serve as a means of inferring the nature of the unobserved process of segregation of genes during the formation of the germ cells in the F_1 hybrid. Thus the 1:2:1 data are a part of the diagram as an end result and are used in making inferences to earlier steps, which, then, explain those data.

Within the context of the steps in the process that are represented by the diagram, theoretical entities have the same ontological status as observable ones. The theoretical ones are just smaller and knowledge of their nature and behavior less certain until adequate evidence has established that the "theoretical" steps in the diagram are correct. At an earlier stage in the nineteenth century, that the material passed from parent to offspring was cellular was uncertain. Various developments, such as the discovery of the mammalian egg and the role of the sperm in fertilization, served to confirm the existence of germ cells in sexually reproducing organisms. As a result, by the early twentieth century, it was not controversial to include them in a diagram of steps between parents and offspring. Cell "theory" thus became less "theoretical," in the sense of being uncertain or hypothetical.

The term "theory" is used sometimes only to indicate uncertainty, but, at other times, for a well-confirmed generalization. Thus, by the early twentieth century, cell theory was still a theory in the sense of being of a very general claim that allowed the inclusion of a depiction of germ cells in all representations of the stages between parents and offspring. The theory, however, was no longer hypothetical. The general claim that organisms are composed of cells is now widely accepted. Cells are microscopically observable and the sequence of stages in germ cell formation are no longer in any doubt.

The 3:1 ratios were explained by placing them as the *end* step in a process that produced them. Other types of data might be obtained about steps prior to the end in order to confirm the adequacy of the diagram as representing typical steps. Once different types of germ cells had been postulated by Mendelism, it was suggested that visible differences might be found between different types of pollen cells. In fact, in a few instances viable differences in the pollen cells were found, although in most hybrids, the different alleles produce no visible difference in the different types of germ cells. Furthermore, additional data resulted from new experiments to test the adequacy of the postulated steps. Test crosses (e.g., *Aa* x *aa*) were done to test the adequacy of the representation of the genotypes. Thus, the steps of the diagram, once it is drawn, can be examined to see where an experiment or observation might be made to confirm a given step. If additional data can be "fitted into" the diagram, then such a fit serves in explaining the data and serves to provide evidence for the correctness of the diagram. Predictions made by manipulating the symbolic representations in the diagram, once confirmed, provide important evidence in favor of the diagram's accuracy in representation.

When the theory of the gene was used to explain additional instances of segregation, it did so by showing that they were instances in which the typical diagram could be used. Similarly, the other types of cases—independent assortment, linkage, crossing-over—provided additional abstract explanatory patterns that could be invoked in given cases. The theory was thus a repertoire of explanation patterns. Invoking the segregation pattern, to explain yet another instance of segregation in yet another species, ceased to be important for theory development.

It was just another application of the same pattern. What was important in the development of the theory was a case in which the typical pattern was inadequate. As discussed in previous chapters, anomalies often drove theory development. Exceptions to 9:3:3:1 ratios resulted in the discovery of linkage and crossing-over. New theoretical components were added to the theory and typical examples of these became part of the explanatory repertoire of exemplary problem solutions.

The next two sections discuss how viewing the theory as representing typical steps in typical hereditary processes allows a recasting of the discussion of types of anomalies and strategies for anomaly resolution.

12.5 Monster and Model Anomalies

When anomalies arise, then the problem becomes to localize the purported "typical" step or steps where something has gone wrong. The problem of localizing anomalies in the steps of typical crosses will be discussed in the next section. Here, I want to recast the distinction between (what I have called) "monster and model anomalies." A monster anomaly does not require a change in the set of patterns for normal, well-functioning cases. In contrast, a model anomaly does requires such a change, either the alteration of a typical pattern or the addition of one or more new patterns to the set. Three examples of anomalies will be reconsidered here. First, the monster anomaly of 2:1 ratios discovered by Cuénot in mice (discussed in Chapter 8) and the monster anomaly of non-disjunction discovered by Bridges (discussed in Chapter 10) will be discussed. Then, the model linkage anomalies (discussed in Chapters 9 and 10) will show how the set of exemplars expanded.

When Cuénot discovered 2:1 ratios rather than 3:1, they represented an anomaly that was judged to be a challenge to the the normal steps in segregation. As discussed in Chapter 8, different biologists localized their alternatives in different theoretical components, now considered to be different steps in a typical segregation process. Debate occurred as to which step was in need of modification. Morgan proposed that the germ cells of the hybrid were not pure and thus challenged the step of pure segregation by asserting its opposite. Cuénot suggested that pure germ cells formed but did not combine randomly in fertilization; his alternative was selective fertilization, a process localized in the step subsequent both to pure segregation and to the formation of the types of pure germ cells in equal numbers. The alternative that was eventually confirmed was that of Castle and Little, who proposed that the homozygous dominant form (AA) was not viable. In the lethal hypothesis, all steps in segregation functioned normally until after fertilization. The malfunction occurred sometime during embryological development (the exact stage was debated for decades, e.g., Eaton and Green, 1962). Thus, in the Castle and Little hypothesis, all the normal steps of pure segregation and random fertilization were preserved; a new, previously implicit assumption was delineated—equal viability of all zygotes.

Thus, disagreement occurred as to whether the 2:1 anomaly was a model anomaly that required a change in theory, or whether it was just a quirky, monstrous case. Morgan viewed it as an indication that no step of pure segregation ever occurred after hybrid crosses. No exemplar of pure segregation (and perhaps no Mendelian theory at all) would have been been used in explaining hybrid crosses had Morgan's hypothesis been confirmed. Had Cuénot's hypothesis been correct, then another exemplar would have become part of Mendelism—cases showing selective fertilization. In addition to the exemplar of yellow and green peas producing 3:1 ratios, the set would have included yellow and non-yellow mice producing 2:1 ratios, with the genes represented by Y and Y not combining with each other. Yet no change in the exemplary set resulted. The Castle and Little hypothesis of homozygous dominant lethals showed that the case was an example of a monstrous happening, after all the typical steps had occurred. Homozygous lethals themselves became a known malfunction class, in other words, a monster anomaly of a known type. When 2:1 ratios were found in additional cases, a possible explanation pattern was to suggest that they were yet another instance of homozygous lethals. This example shows that Mendelism expanded not only by changing the theory, but also by expanding the set of known malfunctions—the typical theoretical steps were known to apply but some other quirky process prevented the normal result from being found. The anomaly was thus resolved without altering the claim that the segregation steps were general; the theory did not have to be modified in the face of Cuénot's monster anomaly.

Non-disjunction was another anomaly for typical genetic steps that produced monstrous results; however, its role in theory development was somewhat different from the 2:1 case. It served to show that an anomaly in genetic data could be explained by a correlated (a somewhat stronger word would be better) anomaly in the chromosomes. Abnormal genetic data were explained by the non-disjunction of chromosomes. Finding corresponding anomalies in genetic data and in the observed chromosomes of that strain of fruit flies served as "proof" of the chromosome theory, Bridges claimed (see Chapter 10). What would have been a problematic anomaly for the theory of the gene was removed as an anomaly for it. Finding correlated anomalies in genetic data and chromosomes served as evidence in favor of the correctness of depicting hypothetical genes along the observable chromosomes in nonanomalous cases.

Thus, some anomalies get resolved by showing that they are monsters. The typical process has just not worked in their case; one or more of the typical steps has been localized as failing. No fundamental modification of the theory is required; the steps involved are still considered "typical." Normally genes segregate. Normally chromosomes in homologous pairs disassociate in germ cell formation.

But other anomalies become "model" anomalies and are seen as requiring a change in the theory in order to account for them. Some changes are small modifications, such as changing the assumption of paired alleles to allow for multiple alleles in populations. That change was so small it was not illustrated in Morgan's set of exemplars nor discussed in his abstract statement of the theory. Others, such as the anomalies to 9:3:3:1 ratios, required more extensive changes in the theory. The most important modification to Mendelism occurred as a result of the discovery of linkage. The "typical" case of independent assortment was clearly delineated from segregation, in the process of showing that independent assortment was not so typical after all, as discussed in Chapter 9. Consequently, in localizing the problem of exceptions to 9:3:3:1 ratios, the steps producing segregation (and 3:1 ratios) were for the first time clearly distinguished from the steps of independent assortment. Moreover, a new exemplar was added for "typical" linkage cases. And Morgan, in his 1926 exposition of the theory, provided alternative examples and diagrams for linkage in *Drosophila* to accompany the usual Mendelian ones in peas that did not show linkage. Model anomalies thus result in an expansion of the set of exemplars, as well as requiring modifications in the theoretical components.

Extending the scope of the theory may occur, not in response to an anomaly, but simply by showing that additional types of cases can be explained. Additional exemplars are added. The repertoire of typical cases expands. Extension of the scope of Mendelism to include quantitative characters is almost such a case, but the theory did actually change to include the multiple factor hypothesis in order to explain such ratios as 15:1 (see Chapter 6).

In summary, the discussion of types of anomalies has been recast in the light of the analysis of the theory as a repertoire of exemplary explanation patterns. Monster anomalies are cases in which an exemplar fails in unusual cases; monster anomalies may be either unique cases or examples of malfunction classes. Lethal genes are an example of a class of monster anomalies. In contrast, model anomalies show the need for a new exemplar; they turn out not to be quirks (or monsters), but examples of a typically normal pattern that had not been included in the previous stage of theory development. Linkage and crossing-over are examples of model anomalies.

12.6 Diagnosing and Fixing Faults in the Theory

This section will explore an analogy between (1) the theory, represented by steps in diagrams of typical steps in hereditary processes, and (2) structural or functional diagrams, which depict components of a device or stages in a process. After discussing this analogy, anomaly resolution will be compared to diagnosing and fixing faults in a device.

Diagrams are often useful for representing the structural components of a system or the functional stages in a process. Examples include electric circuit diagrams used to represent an artifact of human construction, and diagrams of the circulatory or immune system, representing natural systems. When repairpersons or doctors diagnose a fault or disease, they may try to localize the problem within a part of the normally functioning system, by referring to the part as represented by a component of the diagram. Such diagrams can be at various levels of detail, just as theories can. One can imagine going up or down in levels of organization and in the amount of detail at any one level within a diagram. "Black boxes" can occupy places in the diagram, when no more detail about that step is known or when such detail is irrelevant to the purpose served by the diagram.

Viewing the problem of heredity as the need to fill in details in a diagram of steps between parents and offspring allows an analogy to be made to such schematic diagrams. Seeing that a blank exists between parents and offspring as to what is passed from one to the other is a way of finding the scientific problem of heredity. Thus, scientific research problems can be found by considering steps in natural processes that have not been filled in. Hypotheses can be made as to how to fill the blanks. Not all the black boxes need be filled in at once. It was possible to use pedigree diagrams to represent hereditary characters as being passed from generation to generation, even before it was known what exactly was passed from parent to offspring. Thus, an incompleteness in the theory (as opposed to an incorrectness indicated by a model anomaly) can be represented by a black box, a place holder, in the diagram of typical steps.

The theory of the gene can be viewed as filling in unobservable details in steps in typical hereditary processes. The abstract steps in a typical segregation process can be diagrammed. One way of constructing such a diagram is to drop out the details in a diagram of a specific cross, such as the cross of differently colored four o'clock flowers. Figure 12-2 is such an abstract pattern. Numerous of the theoretical components that have been stated in sentences in previous chapters are depicted in that abstract diagram, including assumptions about the existence of genes, that they occur in pairs, that the pairs separate in hybrids so that the germ cells are pure, and so on. Symbols for chromosomes could be added to that diagram to represent a higher level of organization of the genes. Varying amounts of detail can, thus, be depicted in such a diagrammatic representation of typical steps in a segregation process.

The theory of the gene did not fill in all the details of the steps and structures that were used in its representation. Alternative allelomorphs, for example, were represented by different symbols, even though it was unknown how they differed. Such "promissory notes" in the early stages of theory formation are easily represented by some difference in the symbols (e.g., A and a, or black and white beads). They provide problems of incompleteness to be solved at later stages of theory formation, perhaps by filling in details at lower levels of organization.

When a repairperson is diagnosing a fault in an electrical device, knowledge of the normal functioning may be important information. A schematic diagram of the circuits shows which components should be present and provides possible sites of malfunction. The modularity of the system is important. Separable components can be considered as the locations of sites of failure. The role that components play in normal functioning may aid in localization, and the nature of the malfunction can be used to focus attention on some parts of the system rather than others as the most likely sites for failure. Similarly, physicians go through long training in the normal functioning of systems in the human body, such as the circulatory and immune systems. Symptoms of disease allow localization in one or more malfunctioning body parts. (Other methods for diagnostic reasoning exist besides appeal to functioning of a normal system; see, e.g., Buchanan and Shortliffe, 1984; Reggia, Nau, and Young, 1983; Chandrasekaran and Mittal, 1983; Sembugamoorthy and Chandrasekaran, 1986.)

An analogy can be made to localizing anomalies. A correct theory of heredity fills in details of diagrammatic steps of normal hereditary processes. When anomalies arise, then a task is to localize the fault in one or more of the theoretical steps, namely, the modular components of the theory. This is the task of localization in anomaly resolution discussed in previous chapters. (Localization was one of the steps in anomaly resolution listed in Table 8-1). Each step of a normal process that occurs prior to the location in the diagram representing the anomalous data may be a candidate for being the problematic site. The nature of the anomaly itself is very important in generating plausible, problematic steps; for example, when the anomaly of $2Aa:1aa$ ratios was found, then the focus was on the steps in segregation involving the production of the AA forms that were missing (as discussed in Chapter 8). The steps in which the hypotheses were localized were (1) the formation of pure A type germ cells (Morgan's impurity hypothesis), (2) the formation of both types in equal numbers (my hypothesis), (3) the random combination of all types (Cuénot's selective fertilization hypothesis), and (4) the development of the AA zygote into an organism (Castle and Little's lethal hypothesis). Those are the sequential steps in the normal process of producing 3:1 ratios. A diagram of normal steps in segregation, showing the modular steps, aids in the localization of potential sites of the fault.

Finding implicit assumptions, such as the equal viability assumption that is subverted in the case of lethals, involves realizing that important steps have been omitted from the diagram. Morgan's segregation figure (Figure 12-2), for example, does not show the step of normal embryological development after fertilization but before the adult phenotypic characters form. Filling in more details in the diagram about unrepresented steps in the process is a method for uncovering implicit assumptions during anomaly resolution. Resolving some anomalies or removing nonanomalous incompleteness in the theory may not be possible at a given time, if no methods for investigating another level of organization exist. Resolving some problems or filling in some blanks may have to wait for new techniques or new conceptual developments in other fields.

Monster anomalies are like a malfunction in a device or a failure in a part of the body; the normal process is captured by the schematic and the monster anomaly is a failure in a part. For quirky, unusual cases the failure is unlike other known failures. For a quirky case, diagnosing the failure may be a difficult problem for the repairperson or physician who has not seen that kind of failure or disease before. Similarly, it took geneticists some time to unravel all the details of the quirky chromosomal mechanisms in de Vries's anomalous *Oenothera* mutants.

For malfunction classes, such as lethal gene combinations turned out to be, the monstrous anomaly is like a known kind of failure in a device or a disease. Seeing a problem as an example of a well-known malfunction class is a routine, puzzle-solving task. Finding a new malfunction class is more interesting and adds to the knowledge about how things can go wrong. Scientists often receive credit for discovering the first instance of a malfunction class. Baur is credited with discovering the first case of lethals (Dunn, 1965a, p. 102). As discussed in Chapter 8, Castle and Little (1910) credited him with having found a similar anomaly; Baur's finding served as evidence in their arguments that their hypothesis was not ad hoc but was a kind of anomaly for Mendelism found in other cases. Similarly, medical researchers are credited with the discovery of new diseases. Technicians are less likely to be rewarded for finding a malfunction that causes a device to be recalled, but engineers may be grateful to learn where their design is routinely failing.

After localizing the problem posed by an anomaly, fixing the fault requires different responses, depending on whether the anomaly is a monster or a model anomaly. Monster anomalies are resolved by showing what went wrong in the normal process. The theory is saved. The claimed typical steps are typical; the anomaly is just a malfunction. Model anomalies, however, require a change in the claims about what is normal (or general) in hereditary processes. To continue the analogy, the schematic diagram of normal processes is shown to be incorrect by model anomalies.

A model anomaly serves as a model for typical processes that had not been known to be typical prior to the discovery of the anomaly. Bateson, Saunders, and Punnett (1906) are credited with the first discovery of the incomplete linkage of Mendelian factors, although they used different terminology to label the phenomena (coupling) and provided explanations that differed from the hypothesis of linkage of genes on chromosomes (coupling and reduplication), as discussed in Chapter 9. Linkage data eventually were explained by limiting the generality of the pattern of "segregation." A second pattern for independent assortment had to be clearly differentiated for dihybrid crosses and its generality limited to apply to linkage groups associated with different chromosomes (see Morgan's statement earlier about the applicability of "Mendel's second law," as Morgan named it). Thus, the generality of segregation to apply to both 3:1 and 9:3:3:1 ratios was limited. Numerous new details in the steps of typical processes, related to linkage and crossing-over, were added to the diagrams for typical steps. Model or exemplary cases of these were added to the set of typical diagrams; linkage was a model of a typical process. Thus, model anomalies are resolved by changing the theory so that they become models of typical cases. The assumptions about what is normal are fixed, so that diagrams of the previously anomalous case are included in the representation of the normal system.

Analyzing the theory as representing typical, modular steps that can be failing in a particular case (a monster) or failing in numerous cases and thus requiring a modification in design (such as model anomalies requiring linkage components to be added)—such an analysis makes use of the analogy to schematic diagrams and diagnostic reasoning. Documenting the ways faults were localized and fixed in the modules of the theory is a way of analyzing theory development. The strategy that emerges from this analogy is to represent the theoretical steps diagrammatically and use such a diagram as an aid in diagnosing the location of anomalies. Consider whether the anomaly is just a malfunction of the normal steps or whether the diagram itself is in need of change. Then, strategies for producing new hypotheses, discussed in previous chapters, can be invoked to construct the new component, such as changing the old one slightly by specializing or generalizing it (discussed in Chapter 6). Once the new component is added, it should function as a step in typical processes to produce the expected end result. (My thanks to John Josephson for helpful comments on these ideas.)

In summary, this chapter has suggested a representation of the problem of heredity as finding the steps between parents and offspring. The theory of the gene and the chromosome theory are viewed as supplying components of the series of steps involved in gametogenesis and fertilization. Thus, theory construction is viewed as filling in the blanks for typical steps in hereditary processes. Examples of using the theory to account for typical cases serve as exemplars for a repertoire of explanation patterns provided by the theory. Explaining unproblematic new cases involves instantiating one of the patterns in that case.

A diagrammatic representation of the typical steps supplied by the theory allows depiction of the theory at various levels of detail and allows incomplete portions to be represented by boxes yet to be filled. Such an explicit representation of steps aids (and is sometimes driven by) anomaly resolution. Faults can, at least sometimes, be localized in one or more of the modular steps. Anomalies may either be shown to be monsters not requiring a change in the "typical" steps provided by the exemplars or they may be more serious. Model anomalies require a change in one or more modules (steps) of the theory and provide new exemplars. Some anomalies (or incompletely filled blanks) may point to the need for new theories at other levels of organization in order to resolve them.

The physical nature of the gene, how genes reproduced, and how they functioned to produce characters were unsolved problems for the theory of the gene; that is, they were blanks not filled in for the details of the steps between parent and offspring. Biologists in the early twentieth century attempted to resolve them without much success by appealing to other theories at other levels of organization. The next chapter discusses some of the early, failed interfield relations to resolve these problems and indicates how molecular biology eventually filled in these blanks at a lower level of organization. In retrospect, we can see that they were not fatal unsolved problems for the theory of the gene, although some critics at the time saw them as such.

CHAPTER 13

Genetics and Other Fields

13.1 Introduction

In 1926, the theory of the gene and the chromosome theory had much evidence in their favor, but unsolved problems remained. These included problems about gene action during development, about the nature of the gene, and about the role of genes in evolutionary change. For some, the theory of the gene became the foundation from which new problems in these areas were tackled. For others, however, problems not solved by the theory were judged to be serious difficulties for it. Those biologists concerned with embryological development, the chemical nature of the gene, and evolutionary problems assessed the theory of the gene differently.

Embryologists did not find the theory of the gene promising for solving the problem of the differentiation of cells during development. Numerous hypotheses were made about the nature of gene action, the study of which was called "physiological genetics" (see Dunn, 1965a, ch. 18). Nonetheless, no widely accepted general theory was constructed to explain gene action.

Numerous speculations about the chemical nature of the gene were made, based on the chemistry of the early twentieth century. Few successful interfield relations were established between genetics and biochemistry during the period between 1920 and 1940. Among the views which looked promising at the time, but ultimately proved misleading to genetics, were the colloidal theory, the view that enzymes were not proteins, and the "tetranucleotide" hypothesis that DNA was a simple repeating polymer and thus too simple to be identified as the genetic material. Molecular biology developed in the 1950s. By that time, biochemists had discarded colloidal particles in favor of macromolecules, and spurred by new genetic findings, had re-examined the chemical nature of DNA, using new and improved techniques.

In contrast to embryology and biochemistry, the theory of the gene and the study of chromosomal aberrations provided fresh new approaches to investigations of evolutionary change in the 1920s and 1930s. New work began in theoretical population genetics, as well as field studies of the genetics of natural populations. By the late 1930s and early 1940s, that work resulted in the "synthetic" theory of evolution, combining Mendelian genetics and Darwinian selection (as well as other components).

This chapter will briefly discuss interactions between genetics and these other fields. The interactions illustrate the strategy of **forming** (or attempting to form) **interrelations** between, on the one hand, a theory with successes and, on the other, problems or results in other fields. Some interrelations were successful during the first half of the twentieth century, but others were not. No general histories of the relation between genetics and embryology or between genetics and colloid chemistry exist. In contrast, an abundance of literature exists about the evolutionary synthesis. The brief discussions in this chapter do not do justice to these important topics. They do, however, indicate directions for future historical work needed to better understand interfield interactions between genetics and other fields. I include them here because they will aid in further developing the strategy of **making interrelations**. Not all attempts to make interrelations succeed as well as the chromosome theory did. Furthermore, these interrelations illustrate the different ways in which the theory of the gene was assessed, given different interfield perspectives.

Section 13.2 briefly recapitulates the developments of the theory of the gene discussed thus far. Some problems within Mendelism were solved by the chromosome theory, but others remained unsolved and new ones were raised. Section 13.3 discusses the assessment of the theory of the gene from the perspective of its lack of explanation for gene action during embryological development. Section 13.4 discusses several hypotheses about the chemical nature of the gene, including the relation between genes and enzymes, the autocatalytic view of the gene based on the colloid theory, and the hypothesis that genes were proteins. Section 13.5 briefly discusses the demise of de Vries's mutation theory; it also indicates some of the components of the synthetic theory of evolution supplied by the theory of the gene and the chromosome theory. Section 13.6 develops the discussion, begun in Chapters 7, 9, and 10, of the strategy of **making interrelations** between two bodies of knowledge. The interrelations discussed in this chapter provide a basis for expanding that strategy. A brief conclusion includes quotations from geneticists about the successful relations between genetics and other fields.

13.2 Solved and Unsolved Problems in 1926

Nineteenth-century theories of heredity had very large domains of phenomena. The theories were often not only theories of heredity (explaining why offspring resemble parents), but also theories of embryological development, normal growth, and evolutionary change (Darden, 1976; Robinson, 1979). Some twentieth-century biologists also desired a single unified theory for evolutionary change, heredity, variations, sex determination, and embryological development (Sapp, 1987). In contrast, Mendelism began by explaining a very narrow domain, namely, the results of hybridization experiments between varieties differing in qualitatively different characters. Thus, a feature of the emergence of genetics as a field was a narrow focus on phenomena of a very specialized kind. From its very circumscribed beginnings, the scope of Mendelism was expanded, until, by 1926, the theory of the gene was formulated as a general theory of the transmission of hereditary characters. It is somewhat surprising that studies of qualitatively different characters in hybrid organisms provided the starting point from which such a general theory developed, but narrow, specialized investigations in science can sometimes blossom into lines of research leading to very general theories.

Between 1900 and 1910, the developments of Mendelism were focused on testing the generality of segregation and resolving anomalies generated by breeding experiments. Some of those anomalies were resolved with the development of the chromosome theory. The development of the theory of the gene by the *Drosophila* group between 1910 and 1926 resulted from the successful interrelations between genes and chromosomes. The chromosome theory solved some of the questions that were raised by Mendelism, but were not solvable with the technique of artificial breeding. It provided, as Morgan said, the "mechanism" of Mendelian heredity. Locating the genes on the chromosomes explained why paired genes segregated: the paired chromosomes separated during germ cell formation. Sex-linked inheritance was explained in terms of linkage of characters on the X chromosomes. Mendelism was thus made compatible with the cytological explanation that sex determination was dependent on the complement of X and Y chromosomes.

The anomalies of linkage resulted in the recognition of the need to separate segregation from independent assortment. And anomalies for independent assortment were resolved by the new theoretical components of incomplete linkage and crossing-over. Gene mapping techniques were developed, based on the evidence for the linearity of the genes along the chromosomes. The correlation of Mendelian findings with cytological results was a powerful method, as illustrated by Bridges's work on non-disjunction. Similar abnormalities in the genetic characters and chromosomes provided what Bridges called a "proof" of the chromosome theory (discussed in Chapter 10). The interrelation of two bodies of knowledge, developed by independent techniques, and the subsequent development of both, hand-in-hand, proved a powerful means of theory construction for the theory of the gene and the chromosome theory, as discussed in previous chapters.

Nevertheless, the theory of the gene and the chromosome theory were not without problems in the 1920s. Some problems arose from results within Mendelism itself, such as the issue of whether the position of genes in the chromosomes affected their production of mutant characters (see Carlson, 1966, chs. 13, 14). Because the case study being discussed in this book ends about 1930, we will not continue the historical discussion of intrafield anomalies for the theory of the gene produced by continued application of genetic and cytological methods. Developments such as those that we have traced so far continued, with new anomalies and new hypotheses. The 1930s saw further development of cytological techniques that made the interfield relations between genetics and cytology even more fruitful; for example, with the discovery of giant salivary gland chromosomes in *Drosophila*, cytological observations became much easier (Painter, 1933). Our focus in this chapter, which will conclude the historical discussion, will be on interfield relations other than the interrelations between genes and chromosomes.

Unsolved problems that required both new techniques and investigations at different levels of organization were numerous. What was the chemical composition of the genes? What was their structure? What was their size? How could the number of genes on a given chromosome or in a given cell or in a given species be estimated? Are all genes alike in all cells of the body? Could methods for producing mutations provide clues as to the chemical nature of the gene? How did genes act during embryological development to produce characters? What relation did mutations, modifying genes, and chromosomal changes have to evolutionary change? Solving these problems required going beyond the techniques of artificial breeding and cytological examinations. Studies using other techniques to investigate other levels of organization were required to solve them. New interfield relations were needed. The theory of the gene had developed from investigations of a very narrow domain. After its successes, the next stage was, once again, to broaden the perspective to consider how the new theory did or did not provide a basis for tackling some of the problems set aside at its beginnings.

From 1900 to 1930, attempts were made to form various interrelations between genetics and other bodies of knowledge; however, no other successful interfield theories were developed to rival or supplement the chromosome theory, despite these attempts. Chapter 9 discussed Bateson's analogies to forces; he searched for alternatives to the chromosome theory that would simultaneously explain heredity, development, and the origin of discontinuous variations important for evolutionary change. In another line of research, Castle proposed and abandoned chemical models of Mendelian factors to account for the small gradual variations that he believed were important in evolutionary change. He called his work "experimental evolution" (Castle, 1907). Both Bateson and Castle had objected to the material particle view of the genes along the chromosomes. They both believed that it was inadequate to solve the problems that they faced; their interfield perspectives did not include cytology, but did include the problem of evolution.

Others wanted a single theory to explain a larger domain, a domain that included heredity, variation, sex determination, embryological development, and evolution. They did not judge the theory of the gene and/or the chromosome theory to be a success. It was too narrowly focused and did not seem to provide a basis from which a general theory for biology could be constructed (see, e.g., Saha, 1984; Harwood, 1984; Sapp, 1987). The next three sections will briefly discuss the assessments of the theory of the gene, and the attempts to solve various unsolved problems, from the perspectives of embryological development, the chemical nature of the gene, and the nature of evolutionary change.

13.3 Genetics and Embryology

In the very process of solving the problem of gene transmission, the Morgan group showed that transmission was a problem separate from the problems of gene action and evolution. Allen analyzed this development:

> Morgan and all of his coworkers always regarded the gene as a functional unit. However, in 1915 as well as later, they began by focusing their attention largely on the transmissional and structural aspects of heredity (the relations of genes to chromosomes, the interactions among chromosomes, etc.). There was a reason for this approach, and in it perhaps lies Morgan's genius. Study of transmission of genes and their physical relation to chromosomes could be approached experimentally and quantitatively. Consequently, hypotheses could be developed and tested and the structure of the germ plasm and its transmissional qualities worked out in predictive ways. The study of gene function, in a biochemical or embryological sense, could not be approached experimentally in 1915. By focusing on what could be studied experimentally, Morgan and his group were able to develop a highly elaborate but consistent theory. They postponed consideration of functional problems, although they realized that ultimately any theory of heredity had to account for how genes functioned and how they controlled the highly complex processes of embryonic development . . . . (Allen, 1978, p. 212)

Many of the early geneticists had been trained in embryology, including Morgan himself (Gilbert, 1978). In 1900, the problem of the *transmission* of heredity characters was not clearly delineated from the problem of explaining how those characters developed during ontogeny (Sandler and Sandler, 1985; Maienschein, 1987). Although some scientific episodes can be analyzed from a perspective in which problems were clearly formulated and then their solutions actively sought, the problem of hereditary transmission does not lend itself to such an analysis. Mendelism established a newly delineated problem in the very process of beginning to solve that problem. Geneticists separated the problem of transmission of hereditary characters from the development of those characters embryologically. For those still primarily concerned with how the characters developed embryologically, the genes and chromosomes (the agents of transmission) did not look like promising candidates for controlling differentiation of cells during the development of different tissues. It was not clear to embryologists that transmission could be treated as a separate problem from development.

Attempts to relate genetics and embryology were many and varied. Morgan himself was an experimental embryologist, but was able to provide little connection between the two subject areas (Allen, 1985). The segregation of genes during germ cell formation was often thought to be closely related to (if not the same process as) the sorting out of different hereditary units in the differentiation of cells during development. With the observations that most somatic cells contained all the chromosomes, the component of Weismann's theory in which he postulated the qualitative division of chromosomes during development was disproved. Morgan (1919, pp. 242-243) argued that the two processes were not the same: segregation during germ cell formation was not the same process as whatever controlled differentiation.

Challenges to the theory of the gene from those advocating a stronger role for the cytoplasm in heredity were sometimes based on a belief that the egg cytoplasm provided organization important during embryological development (Pauly, 1987a; Sapp, 1987). The seeming lack of any organization among the individual genes affecting similar characters (they were often scattered among different chromosomes) made it difficult to see how the organization of an embryo could result from these individual particles (Russell, 1930). Further, many of the mutations studied by geneticists seemed to be minor differences, such as the color of eyes. Despite the fact that geneticists tried to counter these critiques (e.g., Morgan, 1917), embryologists went their own way with their own research programs, paying little heed to the results in genetics (see Allen, 1975, ch. 5).

Sex determination proved to be more complicated than the simple *XX* and *XY* relations proposed before 1910. Exactly how the sex chromosomes and the total numbers of chromosomes determined sex was a lively area of research and debate (Farley, 1982). The cytological explanation of sex in terms of the presence or absence of particular chromosomes could be viewed as an oversimplified explanation of the complex process of sex determination during embryological development.

What was needed was a good theory of gene action. Physiological genetics tried, unsuccessfully, to solve that problem with numerous different hypotheses. The speculation that genes either were or produced enzymes was common. Although that idea sounds very modern, hypotheses about the nature of enzymes then were very different from the current view that they are macromolecular proteins. An idea that eventually proved quite fruitful was the idea that the genes somehow controlled enzymes responsible for catalyzing chemical reactions. Those reactions ultimately produced a phenotypic character, such as the reactions that controlled chemical pigment formation, which determined color characters in plants and animals (Dunn, 1966a, ch. 18; Provine, 1986, pp. 112-125). Physiological genetics, however, needed good information from biochemistry about enzymes, proteins, the chemical nature of the chromosomes, and details about metabolic processes. Lacking adequate information in biochemistry on which to found interrelations, physiological genetics did not coalesce into a successful field with a widely accepted theory in the 1920s and 1930s, as did Mendelian transmission genetics. (For the lack of disciplinary coherence in general physiology in this period, see Pauly, 1987b.) In the next section on hypotheses about the chemical nature of the genes, we turn to some of the attempts to establish interrelations between genetics and biochemistry .

13.4 The Chemical Nature of the Gene

The chemical or physical nature of the gene was often the subject of speculation. Bateson's analogies to forces and Castle's speculations about chemical structures have already been discussed. As the importance of the role of enzymes in physiological processes became known (Kohler, 1973), speculations were often made either that genes were to be identified with enzymes or that they produced enzymes (e.g., Bateson, 1913, pp. 266-269; skeptical remarks by Morgan, 1919, pp. 244-255). But at the time, it was not known that enzymes were proteins (Fruton, 1972, p.157; Leicester, 1974, p. 183).

Debate raged during the 1920s and early 1930s about the importance of colloids to problems of physiology (Pauly, 1987a). Protoplasm was shown to be a colloidal solution and some biochemists believed the study of colloids to be the path to understanding the nature of life. Colloidal particles were thought to be aggregates of smaller molecules and to occupy the size range between small molecules and cell components visible in the light microscope, about 2 to 100 millimicrons (Alexander and Bridges, 1928). Those who advocated the existence of macromolecules, instead of the more loosely associated colloidal particles, were in the minority (Fruton, 1972, pp. 131-148). Colloidal particles were believed to have surfaces that could "adsorb" other substances. By uniting two different substances on their surfaces, colloids could act as catalysts in facilitating the reactions between the two substances (Bayliss, 1923). One view of enzymes was that they were colloidal particles, composed of various types of molecules, often with no protein component (Fruton, 1972, pp. 131-148). The chemical composition of colloidal particles was less important than the structural relations of the smaller molecules composing them; one factor critical to their reactivity was the amount of their surface area.

In 1917, the Harvard biologist Troland proposed that Mendelian factors were enzymes, which he characterized as colloidal particles able to catalyze both production of more particles like themselves (autocatalysis) and catalyze other reactions in the cell (heterocatalysis) (see Raven, 1977, for further discussion of this distinction). Troland's (1917) idea of the genes as autocatalytic chemicals was influential. Muller (1922) expanded the idea by claiming that being autocatalytic itself was not as remarkable as the fact that after mutating, the gene could reliably

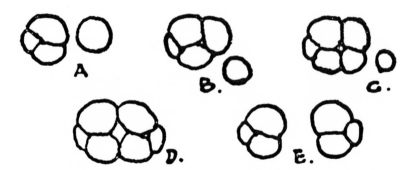

Fig. 13-1. Genes as autocatalytic colloidal particles (From Alexander, Jerome, and Calvin B. Bridges (1929), "Some Physico-Chemical Aspects of Life, Mutation and Evolution," *Science* 70, p. 509. Permission for use granted by the American Association for the Advancement of Science.)

reproduce itself in the mutated form. He emphasized, "What the general principle of gene construction is, that permits this phenomenon of mutable autocatalysis, is the most fundamental question of genetics" (Muller, 1922, p. 177). Muller, however, did not identify Mendelian genes with enzymes; he merely claimed that genes "determined the existence" of various enzymes or other proteins (Muller, 1922, p. 175).

Alexander and Bridges endorsed the autocatalytic view in 1928, and suggested that genes were beadlike colloidal particles adsorbing other substances on their surfaces and splitting apart to reproduce themselves. They provided diagrams of beadlike colloidal particles undergoing autocatalysis. Their diagram for this colloidal view of gene reproduction is reproduced in Figure 13-1. A complex colloidal particle, made up of three smaller subunits, "adsorbs" other particles on its surface until three new subunits are in place (steps A, B, C, and D). Then the daughter particle separates from the mother (step E) in this process of "reproductive catalysis." The idea of autocatalysis thus seemed a promising approach to the problem of gene reproduction (Alexander and Bridges, 1928; 1929).

Debate occurred as to whether any new laws, other than those of noncolloidal chemistry, were needed to account for these unusual properties of the colloidal solutions found in living things (Pauly, 1987a). At issue was the question of the nature of a level of organization between that of small molecules and that of cell structures visible in the light microscope. This colloidal level of organization was viewed by some biologists at the time as the key to understanding physiological and genetic phenomena. In the 1930s and 1940s, a different level of organization was discovered to be more important—the level of the macromolecule. Colloid chemistry gave way to early molecular biology with the discovery of macromolecules and investigation of their crystalloid, rather than colloidal, nature. Important in this change was the crystallization of enzymes and viruses, the discovery of large molecular weight molecules in the ultracentrifuge, and new theories of chemical bonding proposed by Pauling and others (Fruton, 1972; Olby, 1974, 1979; Kay, 1986).

Chemical analysis of chromosomes in the nineteenth century had shown that they were composed of "nuclein" (nucleoproteins, containing both nucleic acids and proteins), and early speculations had been that nuclein was the hereditary material (Wilson, 1896). As chemical techniques improved, nucleic acids were clearly identified as separate from proteins. The four nucleotides contained in nucleic acids were thought to occur in equal proportions and to form a simple repeating structure. This "tetranucleotide hypothesis" for the structure of nucleic acids, as well as increasing knowledge about the complexity of proteins, focused attention on proteins, rather than nucleic acids, as likely candidates for the role of hereditary material (Fruton, 1972, pp. 183-204; Olby, 1974).

Kay (1986) argued that a new "protein theory of life" developed in the period between 1930 and 1950. Proponents of the theory sought a unified account of the structures of enzymes, genes, viruses, and antibodies, all of which were thought to be crystalline proteins. The new field of molecular biology arose with the study of the structural properties of macromolecular proteins and the application of that structural information to problems about living phenomena, including the chemical structure and function of the gene. Surprisingly, from this perspective, the nucleic acids, rather than proteins, were shown to be the genetic material in the 1940s and 1950s. In the 1940s, biochemists, inspired by new genetic findings (Avery et al., 1944), re-examined the tetranucleotide hypothesis and showed that the four bases did not occur in four equal proportions after all (Chargaff, 1950). With the discovery of the double helix structure of DNA in 1953, the problem of gene reproduction received a plausible solution: the helix splits apart and its constituents provide the template for molecules to line up to produce a copy (Watson and Crick, 1953; Watson, 1968). At least part of the problem of gene action was solved during the late 1950s and 1960s with the cracking of the genetic code and the discovery of the details of protein synthesis (Judson, 1979). How genes control embryological development is still an active area of research. Numerous problems about gene action still remain unsolved.

The chromosome theory of Mendelian heredity explained Mendelian segregation, independent assortment, and linkage in terms of the behavior of chromosomes. It did not solve the problem of gene action or the physical nature of a gene that could reproduce, mutate, and control the production of characters. These problems required a new level of organization, whose existence was only suspected before 1930—the level of the macromolecule (Olby, 1979). Molecular biology, working at that level of organization, eventually solved the problems. The solutions did not, however, produce the (hoped for) general protein theory of life; nucleic acids play a major role also.

With hindsight, the attempts to relate genetics to chemistry in the period between 1900 and the 1930s can be viewed in different ways. On the one hand, they can be viewed as failed interrelations because of the mistaken importance placed, first on colloids and then on crystalline proteins. Alternatively, this period can be viewed as a stage during which important distinctions were made that ultimately led to the molecular biology of the gene (Raven, 1977). At any rate, the early attempts to interrelate genetics and biochemistry did not result in a successful interfield theory by 1926.

13.5 Genetics and Evolution

The relation of genetics to evolutionary questions changed dramatically in the period between 1900 and the 1940s. The development of the theory of the gene played an important role in this dramatic shift. In 1900, debates about evolutionary mechanisms were occurring between positions that may be termed "saltationism" versus "gradualism," and "mutation" versus "selection." Darwin (1859) advocated gradual development of new species, based on the selection of small, individual differences among organisms; he did not attribute an important role to "sports" or saltations, the occasional variants that were very different from other organisms. Biometricians in the late nineteenth century considered themselves to be Darwinians because they advocated the importance of small individual differences as the raw material on which selection acted. They were opposed by others, such as Bateson and de Vries, who advocated the importance of "discontinuous variations" or mutations in evolutionary change and accorded selection the minor role of eliminating the most unfit. With the beginnings of Mendelism, the debate came to be characterized as a dispute between Darwinism or biometry, on the one hand, and Mendelism or mutationism, on the other (Provine, 1971; Allen, 1978, p. 124). Thus, in its early days, Mendelism and mutationism were seen as alternatives to a Darwinian explanation of the origin of species in terms of selection of small scale variations.

By the 1920s, Mendelism and selection had been reconciled (Provine, 1971), and by the 1940s, the "synthetic" theory of evolution had been formed. The synthetic theory had components from Mendelism, from Darwin's theory of natural selection, and from other sources. The synthetic theory served to explain the origin of new species (Mayr and Provine, 1980; Mayr, 1982; Darden, 1986b). One factor in this dramatic change was the development of Mendelism from its early beginnings to the much more elaborate theory of the gene and the chromosome theory by the

1920s. As discussed in Chapter 10, the development of explanations for mutations by the theory of the gene served to discredit de Vries's mutation theory. In addition, the development of the multiple factor hypothesis (see Chapter 6) and other information about chromosomal abnormalities provided new views of the origin of variations. These new views about the nature of variation were more easily interrelated to Darwinian selection than the components of Mendelism in 1900 (Darden, 1986b).

De Vries (1901-3) based his mutation theory on findings in the evening primrose, *Oenothera*, which proved to be a very poor choice of model organism. He observed dramatic changes from one generation to the next that would breed true, rather than segregate, in the offspring of the new mutant. Such variants he called "progressive mutations." His mutation theory claimed that the origin of such mutants provided a mechanism for the origin of new species: new species were produced by a progressive mutation in a single generation. The mutations exhibited by varieties, which he called "digressive mutations," showed Mendelian segregation but did not give rise to new species. Bateson saw Mendelism as a more promising approach for evolutionary problems than did de Vries (Darden, 1977), who had said to Bateson: "Don't stop at Mendel" (letter of 1901, quoted in Provine, 1971, p. 68). Thus Bateson, but not de Vries, believed that Mendelian segregating variations were important in evolutionary change.

Despite this disagreement about the importance of Mendelism, Bateson and de Vries did agree that the biometric view was not a promising approach. De Vries argued against the adequacy of selection acting on small individual differences, which he called "fluctuating variability," for producing new species. Such small variations were produced by environmental influences and were not hereditary mutations, he claimed. Thus, Mendelism did not hold the key to understanding the origin of species, nor did biometry, from de Vries's perspective.

De Vries's theory held the promise that experiments would be relevant to the question of the origin of species, an approach that appealed to numerous biologists in the early twentieth century (Allen, 1969). Morgan began work with *Drosophila*, looking for large scale mutations (affecting numerous characters), such as de Vries had found. The *Drosophila* mutations, however, were smaller scaled and did show Mendelian segregation. *Drosophila* did not produce true breeding, new species in a single generation. De Vries's mutation theory and his sharp distinctions between individual variation, digressive, and progressive mutations did not receive support in the period after 1910. With the development of the chromosome theory and increasing knowledge about smaller scale mutations, de Vries's purported pure breeding, large scale mutations in *Oenothera* were "explained away" as due to smaller scale mutations and unusual chromosomal behavior in that genus (Morgan, 1926). The concepts of genotype and phenotype, as well as the multiple factor hypothesis, helped in sorting out the issues about hereditary versus environmental influences on variability. In addition, the multiple factor hypothesis accounted for the seemingly small scale differences and aided in the abandonment of a sharp distinction between individual differences and "discontinuous" variations.

Had de Vries's mutation theory been confirmed, then two levels of organization would have been critical to the origin of new species—the level of the pangens that produced new mutations and the level of the species, the pure breeding forms constituting the offspring of the mutants. When mutations were found to be smaller scaled and not productive of a new species in a single generation, however, these two levels alone were shown to be inadequate for solving the problem of the origin of new species. If de Vries's theory had served to guide studies of evolution, then genetics and evolutionary studies would have been much more closely related than they became in the period of the 1910s and 1920s. If the study of variation alone had produced the mechanism for the origin of species, then the problem that remained would have been to explore the nature of the mutants separating different species. Genetics would have been "experimental evolution," with two sets of theoretical components: (1) Mendelian components to explain heredity in crosses between varieties, and (2) de Vriesian components for explaining the origin of new species by progressive mutations. The situation was not so simple, however. Additional work had to be done to unravel the complex processes and additional levels of organization involved in the origin of species. De Vries had not expected that Mendelism would hold keys to understanding the origin of species. Nonetheless, the theory of the gene and the chromosome theory did play important roles in the developments leading to the synthetic theory of evolution formulated in the 1930s and 1940s.

Many of the early Mendelians were concerned with the question of whether selection could be effective in producing evolutionary change. Castle's experiments with rats, discussed in Chapter 8, had been directed to the question of whether selection could change characters, and by inference, the underlying factors. Johannsen's (1903) work on pure lines had shown that selection, with no new mutation, was ineffective in producing new forms, once the variability in a population had been separated into homozygous pure strains. As discussed in Chapters 6 and 8, the simplifying assumption with which Mendelism began—the unit-character concept—was abandoned with the development of the multiple, modifying factor hypothesis. This hypothesis served to resolve the debate with Castle: genes did not contaminate each other; also, selection of small differences was selection of modifying genes. The conflict between mutation and selection was, in part, resolved once selection could be seen to act on heterogeneous populations to select different combinations of stable factors. The sharp distinction between individual differences and mutations had dissolved into a continuum (Mayr, 1982, p. 551). The change in the one unit-one character component of Mendelism to the multiple factor component was important to this development.

By the 1920s, Darwinian selection and Mendelian heredity were reconciled (Provine, 1971), although all the components of the synthetic theory had not been clearly formulated in that period. Following the numerous diverse views leading to the evolutionary synthesis in the 1930s and 1940s is beyond the scope of this book. A definitive historical account of the synthesis has yet to be written and competing interpretations are common (e.g., Mayr and Provine, 1980; Mayr, 1982; Bowler, 1984; Provine, 1986; Burian, 1988). Instead of trying to provide an overview of the many historical strands leading to the synthesis, I will analyze the development in terms of the levels of organization interrelated by the synthetic theory and briefly indicate some of the components contributed by the theory of the gene and the chromosome theory on each level.

Again, the importance of finding good information at appropriate levels of organization was crucial for solving an old problem. Just as the macromolecular level was crucial for the development of molecular biology, important to the synthesis was the clear separation of three levels: gene and chromosomal variations, populations, and species (Darden, 1986b). Mayr has especially emphasized the importance of "populational thinking" and the study of isolating mechanisms for producing noninterbreeding species (Mayr, 1982). In my terminology, those ideas involved the recognition of the populational level and the species level.

Dobzhansky's 1937 book, *Genetics and the Origin of Species*, is acknowledged as the key publication in summarizing and promulgating the new synthetic theory (Mayr and Provine, 1980; Provine, 1986). Dobzhansky was influenced by several diverse traditions that contributed to his ability to produce such a definitive publication of the synthesis. Important influences included his early work in Russia on theoretical population genetics and its application to the genetics of natural populations (Adams, 1980), his work with Morgan and Sturtevant in experimental genetics, the work of Wright in theoretical population genetics, and his interest in studying natural populations in the wild in order to solve evolutionary problems (Dobzhansky, 1982, Gould, 1982; Provine, 1986).

In his 1937 book, Dobzhansky presented what he called the "mechanisms of evolution as seen by a geneticist" (Dobzhansky, 1937, p. 12), which came to be called the "synthetic theory." He explicitly stated the theory as having components at three "levels." The first "stage" or level of the evolutionary process, according to Dobzhansky, consisted of mutations and chromosomal changes. Once produced, those mutations were "injected into the genetic composition of the population." Then, a "mutation may be lost or increased in frequency in generations immediately following its origin, and this (in the case of recessive mutation) without regard to the beneficial or deleterious effects of the mutation." Next, the "genetic structure" of populations was molded into new shapes by the influences of selection, migration, and geographical isolation, "in conformity" with the environment and the "ecology, especially the breeding habits, of the species. This is the second level of the evolutionary process, wherein the impact of the environment produces historical changes in the living population." Finally, the "third level is a realm of fixation of the diversity already attained on the preceding two levels," with the establishment of noninterbreeding populations. Dobzhansky argued that an important problem was the study of "isolating mechanisms" at this level, such as "ecological isolation, sexual isolation, hybrid sterility, and others" (Dobzhansky, 1937, pp. 13-14).

Mendelian genetics and the chromosome theory helped to fill in details about processes at all three levels. At the first level, they provided new theories about the nature of variations, which function as the raw material for evolutionary change. Especially important was the new view of genes as stable (noncontaminating or blending) hereditary units. Darwin and numerous other nineteenth-century biologists had believed that variations blended in hybrid crosses; thus unusual sports would not be passed on to offspring in a pure form. Mendelism's purity of the gametes was a surprise and its development into the theory of the gene eliminated blending theories. That mutants could be maintained in recessive form was a new component relevant to issues about the genetic variability in populations. Issues about the frequency of dominance and its evolution were new problems opened up by the theory of the gene (see, e.g., Provine, 1986); also, multiple effects of one gene, as well as linkage, supplied mechanisms for explaining the existence of nonadaptive characters. Additionally, unequal crossing-over supplied a mechanism for the origin of new genes, another way in which the new findings provided new hypotheses relevant to the nature of the raw material in evolutionary change (Mayr, 1982, p. 567). Good theories about processes at the first level of the evolutionary process proved very fruitful to development of hypotheses about processes at the next level.

At the populational level, Mendelism provided a basis for developing theoretical models for sexually breeding populations. Those theoretical developments gave rise to theoretical population genetics with the work of Chetverikov (Adams, 1980), as well as Fisher, Haldane, and Wright (Provine, 1971; 1986). One example of considering the effects of selection on Mendelian breeding populations was theoretical work that showed that a small selective advantage of a gene in a population would be sufficient for the gene to spread surprisingly rapidly in a population (Provine, 1971). Thus, the theoretical considerations of the implications of Mendelian breeding systems served to correct ideas based on less rigorous analysis and provided a host of new hypotheses to be tested in experimental and field studies of the genetics of populations. Chetverikov investigated natural populations of *Drosophila* and demonstrated that mutations found in laboratory populations were also found in natural populations (Adams, 1980). Provine (1986) argued that theoretical controversies between Wright and Fisher stimulated fruitful work. Dobzhansky was among numerous biologists who actively did experimental and field studies of the genetics of populations testing the implications of theoretical population genetics (Provine, 1986).

At the species level, Mayr (1982) argued, genetics played an important negative role. Genetics served to eliminate various theories of hereditary mechanisms as possibilities for being mechanisms for producing new species. Such theories of variability that were incompatible with Mendelism included several different theories. One was de Vries's mutation theory, which was a saltational theory, proposing large rapid jumps to new species. Another group of evolutionary theories consisted of neo-Lamarckian theories that depended on some mechanism for the inheritance of acquired characters. A third type has been labeled "autogenetic" (Mayr, 1982) or "orthogenetic" (Bowler, 1983). This third type attributed species change to some sort of cause internal to organisms that drove change along a particular path (Mayr, 1982, p. 607). Thus, at the species level, genetics played the negative role of eliminating rival theories for speciation that were based on some sort of genetic mechanism incompatible with Mendelism. Bowler summarized this result of the synthesis: "The new Darwinism was based on the assumption that speciation did not require special genetic mechanisms; given geographic isolation, natural selection alone can separate one species into another" (Bowler, 1984, pp. 298-299).

In a more positive sense, genetics and the chromosome theory served to suggest new mechanisms for the origin of new species, such as the doubling of the entire set of chromosomes (polyploidy) or other chromosome abnormalities; however, these were not universal mechanisms. In his development of the synthetic theory, Mayr (1942) argued that geographic isolation was a more important factor in speciation than any genetic mechanism. Mayr (1982), from his perspective as one of the naturalists who contributed to the study of geographic isolation in speciation, has claimed in retrospect that genetics was not important in the development of views about speciation. Instead, he claimed, the naturalist and systematist traditions supplied ideas crucial for developing the components of the synthetic theory at the species level, such as the importance of geographic races (Mayr, 1982, pp. 561-566). Provine, on the other hand, has emphasized that, in the development of Wright's hypotheses about speciation, an important role was played by Wright's study of the genetics of practical breeding, such as inbred populations (Provine, 1986). At any rate, theoretical generalizations, based on the consequences of Mendelian breeding at the populational level, did not alone supply all the components for a universal mechanism of speciation. Additional analogies or interfield relations supplied components of the synthetic theory at the species level.

The synthetic theory explained much speciation as being a result of processes occurring in isolated populations; the most important of these processes was mutation, followed by selection. The synthetic theory provided the basis for arguing that no additional mechanisms were required to account for evolutionary changes above the species level (which came to be called "macroevolutionary changes") (Simpson, 1944; Mayr, 1982, p. 607).

Thus, I argue that one of the important aspects of the synthesis was the recognition of which levels of organization were important to solve the problem of the origin of species. The level of the production and inheritance of variations had long been recognized as crucial. The populational level had been emphasized by Darwin, but neglected by early geneticists in their attempts to equate the production of some new variations with the production of new species. Even more importantly, the synthesis involved the "discovery" of the need for the species level. Darwin

(1859) had not sharply distinguished varieties and noninterbreeding species; he argued for a continuum between varieties and species. Genetics helped to show the need for reproductive isolation in the production and maintenance of the diversity that characterized the origin of new species, but Mendelian genetics alone did not indicate the causes of the development of the inability to interbreed. The synthetic theory focused new attention on the study of "isolating mechanisms," as Dobzhansky called them. Finally, the synthetic theory also indicated that additional, new processes proposed to account for even higher levels of organization (the macroevolutionary level) were not needed. Whether additional levels, above the species level, are needed to account for trends in the paleontological record is currently a matter of debate (e.g., Stanley, 1979). Thus, the synthetic theory was a multilevel theory (Darden, 1986b), which postulated a temporal sequence of processes at the levels of variation, the population, and the species, as adequate for accounting for all evolutionary change.

In summary, the theory of the gene and the chromosome theory supplied some of the components at several levels for the new multilevel synthetic theory. Genetics was important at the first level, the level of variation, in both a negative and positive way—it supplied a good theory about the nature of variation that served to eliminate rivals and it provided hypotheses about the behavior of the raw material important at higher levels. Darwin had poor theories about the nature of variation on which selection acted; genetics provided new solutions to that problem crucial to the higher levels. Darwin's theory of natural selection thus received a major new component at the variational level; the synthetic theory can be seen as a refinement of Darwin's theory with new components at the level of variation.

At the populational level, hypotheses about processes producing changes in populations proved crucial to the synthesis. As Darwin had seen, but de Vries did not, the populational level was the site of processes necessary for explaining evolutionary change. Theoretical population genetics and field studies of natural populations contributed to the study of that level. Thus, the synthetic theory represented a revival of this level that had been found in Darwin's theory, but now new components about populations could be added as a result of work applying genetics to populations.

The study of speciation mechanisms during the synthesis relied less on the theory of the gene and the chromosome theory. They played the negative role of eliminating rival theories of speciation based on other theories of variation. The theory of the gene indicated that mechanisms of variation, other than gene mutations and chromosomal aberrations, were unlikely. Furthermore, genetic studies did not indicate that such additional genetic mechanisms to produce species change would be needed. The positive role of genetics at the species level was through its contributions to hypotheses at the populational level, processes that were necessary, but not sufficient, to account for speciation, according to the synthetic theory. The synthetic theory thus added a new level to Darwin's theory to account for speciation.

Once again, the relation of genetics to evolutionary theory shows the importance of finding appropriate levels of organization for solving problems. Strategies for forming interrelations and for changing or adding levels of organization during theory development will be discussed in the final section of this chapter.

Much of the work now seen as part of the synthesis demonstrated that the synthetic theory, once formulated with the new genetics playing important roles in its components, was compatible with evidence in other fields, or, more positively, could contribute to the solution of problems in those fields (Mayr and Provine, 1980). Because genetics played such an important role in the synthetic theory, that work can also be seen as having made genetics, or its implications at the higher levels, compatible with those other fields. Such fields included systematics, paleontology, botany, and morphology. Again, discussing all the kinds of relations among genetics and these fields is outside the scope of this analysis. Suffice it to say that no major incompatibilities arose that necessitated the abandonment of the theory of the gene or the synthetic theory.

Genetics gave more to the synthesis than it received from it. Questions about origins can often be neglected in the study of contemporarily existing mechanisms (so I believe, although this view is controversial; see, e.g., Mayr, 1982). But origin theorists need to know the current state of the mechanisms whose origin they are attempting to explain. Thus, the synthetic theory did not solve problems that arose in Mendelian genetics itself, in the way that molecular biology solved the problems about the nature of gene structure, reproduction, and functioning. The synthetic theory was an attempt to solve problems that arose from questions about evolutionary origins, questions outside the scope of the problems of Mendelian genetics, but questions that Mendelian genetics proved relevant to solving.

13.6 Strategies: Interrelations and Levels of Organization

Previous chapters, especially Chapters 7, 9, and 10, discussed the strategies of **using interrelations** and **moving to a new level of organization**. Both successful and unsuccessful attempts at forming interrelations between genetics and other fields were discussed in previous chapters. These included Bateson's unsuccessful attempts at making connections to physics and Castle's unsuccessful attempts at characterizing genes as small, three-dimensional organic molecules. In contrast, the interrelation between genes and chromosomes was extremely successful. The development of the chromosome theory served in constructing new hypotheses in both fields, resolving anomalies for both, and providing a bridging theory establishing the *kind of relation* that held between genes and chromosomes, namely that genes are *parts of* chromosomes.

This chapter has discussed other historical efforts to relate the theory of the gene, once formed, to the fields of embryology, chemistry, and evolutionary studies. These attempts met with varying degrees of success. The cases in this chapter provide a basis for further development of the strategies of **using interrelations** and **moving to new levels of organization**.

The steps in using an interrelation, developed in previous chapters, may be summarized as follows: (1) find a problem that indicates the need for using other techniques or moving to another level of organization studied by another field, e.g., the "black box" of the internal structure of the genetic "beads"; (2) use methods for determining what other techniques, units, or processes studied by what other fields may be relevant; (3) consider what *kinds of* interrelations might be established, e.g., *identify* the gene with a kind of chemical structure, thereby postulating an interrelation of *identity*; (4) use the postulated kind of interrelation to guide hypothesis formation in one or both fields; and (5) consider the results produced by an interrelation with evidence in its favor, such as the integration of two previously separate bodies of knowledge, or, alternatively, the failure to integrate. The following subsections of this chapter will discuss each of these steps.

13.6.1 Problems Requiring Interrelations

In some cases, the reason for seeking interfield relations is that a problem arises in one field that cannot be solved with the techniques and concepts in that field (Darden and Maull, 1977). The problem can either be a specific anomaly that challenges a component of the theory within the field or it can be a new problem raised by an incompleteness in the field. Linkage was an anomaly for Mendelism that was solved by making the interrelation to cytology. Certain problems get solved at certain levels of organization. The problems of why Mendelian genes segregate and why some show linkage during inheritance were solved at the level of the chromosomes: chromosomes separate during meiosis and genes are linked on chromosomes. (I agree with Kitcher, 1984, that there was and is no need to consider "reducing" segregation and linkage to the molecular level; the problems are solved at the chromosomal level.)

In contrast to linkage, questions about the chemical nature and functioning of genes were not specific anomalies challenging a specific component of the theory of the gene, but were new and unanswered questions. They were neither solved by Mendelian genetics nor were their solutions to be found at the level of chromosomes. The question about the chemical nature of the gene required connecting genetics and biochemistry and could not have been solved using the breeding techniques of Mendelian genetics alone.

New and unsolved problems may be identified by finding gaps in diagrammatic representations, as discussed in Chapters 11 and 12. Blanks or black boxes in the representations indicate unsolved problems that potentially may require other techniques or levels of organization to solve. Sometimes a gap poses a problem that obviously requires moving to other levels of organization to solve. Filling in the gaps between parents and offspring, for example, required moving to progressively lower levels of organization, from germ cells, to chromosomes, to chemical molecules. Similarly, at the beginning of the twentieth century, a representation of the temporal steps involved in natural selection would have shown a blank for the nature of the process producing heritable variation. (See Darden and Cain, 1989, for a representation of the steps in natural selection from a contemporary perspective.) A new theory about heredity and variation would have been expected to be relevant to solving that problem. Thus, the nature of gaps in representations of processes (or depiction of properties of a newly postulated entity) may indicate the level to investigate in order to solve the remaining problems and fill in the gaps.

As a third possibility, an interrelation might be sought, not because a specific problem or an incompleteness is found within the field itself, but because of consideration of a problem outside the field. A complication arises when debate occurs about what problems are inside or outside the field. For Morgan, the problem of the transmission of hereditary characters was the central problem addressed by the theory of the gene. The problem of embryological development was outside the scope of Mendelism, but that problem necessitated the search for some interrelation between genes and stages of development. Other biologists, unlike Morgan, were unwilling to set aside the problem of development in trying to solve problems about heredity. They were still struggling to find one, all-encompassing theory that would be a rival to the theory of the gene and simultaneously solve the other problems (see Saha, 1984; Sapp, 1987). Thus, an unsolved problem may be viewed as outside a field but in need of interrelations to that field in order to be solved; or alternatively, the same problem may be considered to be within a field and serve as an anomaly for an intrafield theory that cannot account for it. The problem of embryological development was viewed in both these ways.

Judgments about the scope of the domain of the theory will determine whether a theory is judged to be successful and then can be related to problems in other fields, or whether the theory may be judged unsuccessful because it does not solve important problems considered to be within its domain. Demands in theory construction and the criteria for theory assessment are intimately related in such a dispute. One approach is to construct a large domain and generate only theories that can account for all its items. Then, when the theory is to be assessed, it already satisfies the criterion of being compatible with phenomena with a wide scope, because it already explains them. In contrast, the approach that was exemplified in the development of the theory of the gene was to narrow the domain, develop a theory with a narrow scope, and then reconsider problems in other areas. Thus, the breadth of the domain of the theory will determine which problems may be considered problems within its own domain and which problems will be in other fields. If narrowing has been used in the early stages of theory construction, then consideration of the relations of the theory to the old unsolved problems may be needed at a later stage. A general strategy for problem finding might be formulated: if a field began by focusing on a narrow problem and by neglecting other problems thought to be closely related to the narrow one, then, once a promising solution to the narrow problem is found, consider again the relations to the neglected ones. Thus, after the theory of the gene was formulated, obvious next steps were to once again consider the problems of embryology and evolution to see how the new theory could be related to them.

If a theory assessment strategy is to assess the relations between a new theory and other accepted theories (or relate it to unsolved problems in neighboring fields), then the problem of the relations between fields calls for analysis and research. If an interrelation to another field (such as the relation between genes and chromosomes) was used in the actual construction of the theory, then the theory will satisfy the assessment criterion of being compatible (in fact, even more closely related than compatible) to the other field. If, however, the problems in the other field were neglected during the development of the new theory, as were problems of embryology and evolution in the formation of the theory of the gene, then re-examination of them becomes necessary at a later stage. Again, we see the close relation between strategies for theory construction and strategies for theory assessment. What constraints were imposed in the generation process will determine what interfield problems remain to be solved by the theory.

Bechtel (1986) suggested additional reasons for "crossing disciplinary boundaries," or, in my terminology, methods for generating problems requiring interfield solutions. Two fields may both attempt to account for the same or partially overlapping domains, but they may not be compatible. Alternatively, a fortuitous link may be found between two fields and then further exploration may prove fruitful for one or both. Finally, developments in one field may look promising for solving problems in another field, even though the original developments were constructed to solve other problems. Thus, a number of different developments may point to the need to consider relations between fields (see Bechtel, 1986, for examples and further discussion).

13.6.2 Methods for Locating Other Techniques and Levels

Some problems will immediately suggest what other techniques or levels will be plausible to pursue. If the question is the chemical nature of the gene, for example, then chemical theories are obviously areas to consider. At other times, however, what specific techniques or levels are needed to solve a problem is itself a problem to be solved. The linkage anomalies did not, by their very nature, point to the level of organization needed to resolve them. As discussed in Chapter 9, Bateson formed hypotheses, first at the level of the factors, and then at the level of entire germ cells. Morgan "discovered" the chromosomal level to be relevant to that problem (see Chapter 9 for more detail). With evolutionary theory, it was unclear whether the level of the production of new variations was sufficient to explain the origin of new species. The synthetic theory showed that other levels, the interbreeding population and geographical isolates, were needed to explain the origin of new species. The colloidal level, in between small molecules and cell organelles, looked promising for solving biological problems, but that promise was not fulfilled. It was not a commonly held view in the 1920s that macromolecules were a promising level for solving problems about the chemical nature of genes. The availability of another level may be determined by the techniques available at the time to study that level. Thus, what other level of organization is relevant to the problem may or may not be easy to find, and may or may not be known in a neighboring field at a given time.

Scientists who explore interrelations to another field to solve their problems may find what they need in the other field, or they may not. A certain amount of caution is in order. The other field may have theories that themselves prove to be wrong, such as the colloidal theory and the protein theory of life. (An older example is that of Darwin's acceptance of Kelvin's age of the earth, an age eventually proved to be much too short.) The inability to connect genetics and development raised the question as to which field of the two was in need of change or further development. As Bechtel remarked, the establishment of successful interdisciplinary relations may require a "critical reconceptualization" in one or both fields (Bechtel, 1986, p. 45). The failure of an interfield attempt at a given time may or may not be fruitful for developing promising lines of research, which when pursued might remove the failure.

Sometimes problems require finding new techniques rather than forming an interrelation to another body of knowledge. Some fields, such as x-ray crystallography, are characterized as "technique" fields rather than subject matter fields. Deciding what new techniques may be relevant to a problem is often a challenging part of developing a new line of research to solve the problem. It was assumed, for example, that a key to answering questions about the nature of genes was to find substances and techniques for causing them to mutate. As it turned out, so many agents are mutagenic that the nature of the mutagens provided less specific information about the nature of the genes than had been anticipated. The cases discussed here, unfortunately, do not provide specific strategies for locating new techniques for solving problems. That would be a profitable question to pursue. One potentially fruitful line of historical research would be to compare the various techniques that were tried in the 1930s and 1940s to investigate the nature of the gene, such as x-ray crystallography, enzyme kinetics, and phage genetics (for further discussion of that history, see Olby, 1974; Judson, 1979).

13.6.3 Kinds of Interrelations

As I have indicated in earlier chapters, I am using the term "interrelation" in my own technical sense to mean a relation between entities or processes that have been studied separately. Thus, the kind of interrelations in which I am interested are ontological ones, such as identity, causality, part-whole, and structure-function. Looser relations may exist between two fields or two bodies of knowledge, such as one supplying an analogy or a technique for the other, but such a relation is not an "interrelation" in my sense. Postulated interrelations constitute actual scientific hypotheses that can be tested empirically. If successful, they become part of the accepted scientific knowledge at the time. Several kinds of interrelations are illustrated in the relations between genetics and other fields, including part-whole, identity, and causal. An example: the chromosome theory established a part-whole interrelation—genes are parts of chromosomes. Chemical theories of the gene postulate an identity relation: genes are identified with a type of chemical molecule.

Part of using the strategy of **forming interrelations** involves forming a hypothesis about the appropriate kind of interrelation. The nature of the problem may indicate what kind of interrelation is to be sought; for example, the problem of the chemical nature of the gene indicated that it was to be *identified* with a chemical structure. The problem of the role of the genes in embryological development indicated that they would have a *causal* role. The issue of the relation between the theory of the gene and evolutionary change, however, gave a less clear indication of what kind of interrelation would be expected; different kinds of relations at different levels of organization were postulated at different times.

A fruitful line of research to pursue is the investigation of what specific kinds of interrelations have been or can be postulated between entities or processes, which are studied with different techniques or which exist at different levels of organization. Such a philosophical research program connects general studies of ontology with specific uses of ontological relations in actual scientific hypotheses. Using the strategy of **forming interrelations** would be easier if analyses existed of the types of relations that can be expected and the features of each type of relation. Philosophers have investigated identity, causal, structure-function, and part-whole relations. A promising line of research would connect those abstract discussions with actual usage of such relations in scientific hypotheses.

13.6.4 Ways Interrelations Guide Hypothesis Formation

The ways that a particular kind of interrelation guides hypothesis formation is an active area of my current research and only tentative conclusions can be suggested here (see Darden and Rada, 1988a, and 1988b for additional discussion). If the relation established is one of identity between two entities, then all properties of one will be expected to be properties of the other. A property in one that is not matched in the other can provide a guide for forming a hypothesis about a corresponding property. If a functional entity is being identified with a structural one, then the structure will be expected to have some means of carrying out that function. The gene, for example, had the functions of producing characters, occasionally mutating, and reproducing itself. Therefore, when genes were identified with colloidal enzymes, they were postulated to have the functions of heterocatalysis and autocatalysis. How gene mutation was to be realized in such a structure was a problem to be solved. Thus, the theories in biochemistry provided the conceptual tools for constructing hypotheses about the structural realizations of gene functions—colloidal particles and enzymatic action were such tools.

The *identification* of gene mutations studied by laboratory techniques with the same gene mutations in wild populations proved critical to interrelating laboratory genetics and field studies of populations, and hence evolutionary change in nature. That identity relation showed that the theory of the gene could serve as one component of a theory of evolution.

In the embryology case, if genes were to be the entities *causing* development, then some means of carrying out the orderly temporal sequence of embryological stages was needed. It seemed unclear how beadlike particles in strings along chromosomes could cause such development. Thus, a causal interrelation between genetics and embryology proved impossible to construct during this period.

In contrast to identity and causal relations, the part-whole relation has received less attention in philosophy (but see, e.g., Wimsatt, 1976; Richardson, 1982; Smith and Mulligan, 1983). What properties of wholes can be expected also to be properties of parts and vice versa is a fascinating and unanswered question. If the whole is known to be composed of certain chemicals, then it can be expected that the parts will contain some of those chemicals. Thus, the chromosomes were known to be composed of proteins and nucleic acids, so hypotheses were formed about the chemical nature of genes on that basis. Maternal and paternal genes were known to randomly assort; therefore, homologous chromosomes were predicted to randomly mix. Further investigation of general aspects of part-whole relations is a fruitful area for further work (see Darden and Rada, 1988a).

This subsection indicates that general strategies for hypothesis formation may be developed by considering both the kinds of interrelations that can be established and the expectations that such kinds of interrelations generate.

13.6.5 Results of Forming Interrelations

The formation of a successful interrelation produces a number of results. Problems in one field may be solved as a result of the interrelation to another. New lines of research connecting the two fields may result. Sometimes an actual interfield theory may be formed, such as the chromosome theory. Such theories serve to integrate separate bodies of knowledge by establishing a specific interrelation between, for example, entities postulated in the two. This kind of integration is not a unification by reduction, but a weaker form of unification in the sense of establishing what the relations are between the two, with each maintaining its integrity. (For further discussion of the contrast between interfield relations and reduction, see Chapter 7; Maull, 1977; Bechtel, 1986, pp. 38-47; and Bechtel, 1988, chs. 5 and 6.)

If a strategy for assessing theories is to consider their **consistency with other accepted theories**, then establishing an interrelation between a theory and the knowledge in other fields should ensure consistency between claims in the two fields. Thus, **using interrelations** in theory construction will "move a constraint into the generator" (as they say in AI). In other words, using the strategy of interrelations in theory construction will introduce the constraint that the new theory be consistent with other accepted theories. Of course, checking consistency with all the claims in another field may require considerable time and effort, so it is a costly constraint to introduce during theory construction. Once again, the tight connection between strategies for theory development and strategies for theory assessment is evident.

If the effort to establish an interrelation is not successful, then the response can be of various kinds. A new intrafield theory may be judged to be inadequate from the perspective of a field to which it cannot be connected; for example, embryologists judged the theory of the gene inadequate. Alternatively, the judgment might have been made that new theories, perhaps at new levels of organization, were needed in embryology in order to make the connection. Which field is in need of further development may not be clear at a particular time.

It is instructive to ask whether, with hindsight about successful interrelations, suggestions can be made for determining that, at a given time, a specific interrelation will be promising. Is there any way, with hindsight, to point to the indications then available, that the colloidal and protein theories of the gene would fail? I would like to have a positive answer to that question, but I do not. Finding criteria for assessing the promise of an interrelation, before tests rule in its favor or against it, is a worthwhile area for further work.

The positive benefits of an interrelation are so great that perhaps pursuing one, even if it may fail, is a good strategy. The continued attempts to form interrelations, and the sensitivity of those in one field to the need to make connections to the other, may in fact drive developments that ultimately will produce a successful interrelation. The successful resolution of the conflict, for example, between Mendelism and Darwinism shows how a conflict can be resolved by new theoretical developments within one of the fields. Further developments in one or both fields, perhaps in response to the past failures, may prove crucial to the establishment of a successful interrelation.

Sometimes successful interrelations produce new interfield theories bridging the two fields. Sometimes the conceptual relation is weaker and is not called a new interfield "theory." Sometimes new disciplines, with all their institutional and social aspects, are produced as a result of successful interrelations. In other cases, however, the original disciplines retain their integrity and a new interdisciplinary area of research grows up between them, with various kinds of institutional concomitants. Bechtel (1986, pp. 32-38) discussed cases that differed in these ways.

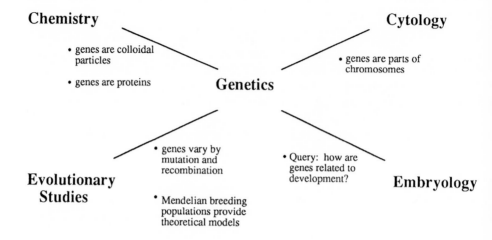

Fig. 13-2. Interrelations between genetics and other fields

He argued that interdisciplinary work will generate a new discipline only if there is a domain (level) in which work is pursued that could not be explored previously (e.g., for lack of appropriate techniques).

Since the object of my work is to find strategies for theory change and not to discuss the institutional aspects of science, I will not pursue here discussion of the disciplinary implications of the formation of interrelations. I have introduced the vague terminology of "interrelations between bodies of knowledge" in order to focus on conceptual issues. I wish to avoid the sociological problem of distinguishing institutionally distinct disciplines from cross-disciplinary research areas.

The strategies of **using interrelations** and **moving to new or different levels of organization** are powerful ones for forming new hypotheses, for testing the results in one field against those in another, and sometimes for resolving specific anomalies. These strategies are important ways in which the search for new knowledge can be guided by old knowledge. Forming interrelations between an extant body of knowledge and an active area of research provides a means of constructing hypotheses that will be integrated with accepted views. Successful interrelations benefit both areas and aid in producing a consistent, growing body of scientific knowledge.

13.7 Conclusion

This chapter has discussed the connections between genetics and the fields of embryology, biochemistry, and evolutionary studies. The strategy of **forming interrelations by moving to other levels of organization** was often a very successful one. Figure 13-2 provides a graphic depiction of the relation between genetics and the other fields discussed in this chapter.

Sturtevant, in reflecting on the history of genetics, concluded:

> The development of genetics is one of the striking examples of the interaction between different disciplines. After 1900, the first such interaction was with cytology, which led

GENETICS AND OTHER FIELDS

to a very rapid development of both subjects. Later interactions were with statistics, practical breeding, evolution theory, immunology, and biochemistry. All of these have led to the utilization of new ideas and new techniques, and to rapid—sometimes spectacular—advances in genetics and in the other fields concerned. (Sturtevant, 1965, p. 135)

Wallace, a student of Dobzhansky, remarked in a similar vein:

The success of modern genetics can be measured by the number of biological disciplines that it has invaded, not in a trivial sense, but, rather, as the hypothesis-generating member of the union: if one of these invaded disciplines is to advance, it must now do so by incorporating modern genetics into its repertoire of concepts and hypotheses.

Thus, from the former pool of physiologists and biochemists—scientists who once scarcely conversed with geneticists—have come modern molecular geneticists. From a second pool, the embryologists, have come today's developmental geneticists. And, from the ecologists of yesterday have come the population biologists and sociobiologists who spend a considerable time predicting the genetic changes that might occur in populations under one or another of a variety of postulated conditions. (Wallace, 1984, pp. 79-80)

Although we have not discussed all these successful interfield interactions in this chapter, the long list shows the fruitfulness of the strategy of **forming interrelations** between a successful theory and problems in neighboring fields.

CHAPTER 14

Summary of Strategies from the Historical Case

This chapter will summarize the major changes in theoretical components of Mendelism between 1900 and 1926 and the major strategies exemplified in those changes. The theoretical components as of 1900-1903 were discussed in Chapter 5 and listed in Table 5.2. Chapters 6 to 11 traced changes to those original components. At the end of this chapter, Table 14.1 provides a summary of all the theoretical components as of 1926.

Each section of this chapter will begin by showing the contrast between the 1900 version and the changed component as of 1926. Then the section will summarize the historical changes, indicate the chapters which discussed them, and summarize my analysis of the strategies exemplified in the changes. The strategies exemplified in the theoretical changes are summarized in Table 14.2. This chapter, thus, is a summary of the major changes and strategies discussed in Chapters 6 to 11.

14.1. Unit-characters to Genes

1900-1903:
C1. Unit-characters
C1.1 An organism is to be viewed as composed of separable unit-characters.

1926:
C1'. Genes and characters
C1.1' Genes cause characters.

Gradual conceptual clarification occurred with the change from the unit-character concept to the postulation of the gene as a cause of characters. No single publication can be pointed to as the one that introduced the concept of an unobservable, material entity postulated as a cause of a visible character. The factor hypothesis developed gradually and debate raged over the possible

226

physical nature of the genetic factors, later called "genes." The development of the claim that a new kind of theoretical entity existed has been traced in a number of ways in the preceding chapters. Chapters 6 to 13 examined the development of the components of the theory in which that claim played a central role, and thus, showed the use to which it was put in explaining domain items. Chapter 11, in particular, explicitly discussed the terminology for referring to the theoretical entity, and the symbols and analog models used to represent it. The changes in claims about genes were driven by growing empirical evidence, concern for clarity of concepts, and the careful attention given to what could and could not be known on the basis of the experimental methods used.

As Chapter 11 discussed, the strategies for conceptual clarification that emerge from the development of claims about the new theoretical entity, the gene, include the following: introducing and manipulating a new symbolic representation; introducing, standardizing, and refining new terminology; explicitly postulating a new theoretical entity and adding to its list of properties.

14.2 Multiple Factors and Multiple Alleles

1900:
C2. Differentiating pairs of characters
    C2.1 In varieties of organisms, the traits by which they differ are
    antagonistic or differentiating pairs of characters.

Between about 1906 and 1910:
    C1.1 (a) One factor produces one visible character.

C2. Differentiating pairs of characters
    C2.1 In varieties of organisms, the factors by which they differ occur
    in allelomorphic pairs.

1926:
C1'. Genes and characters
    C1.2' (a) One gene may cause one character, or
    (b) multiple factors (genes at different loci in linkage groups) may interact in
    the production of one character, or
    (c) one gene may affect many characters.

C2'. Paired genes and multiple allelomorphs
    C2.1' In any one organism, genes occur in pairs (called "allelomorphs," and
    found at the same locus in corresponding linkage groups).
    C2.2' In a population, multiple allelomorphs for a character occasionally
    occur.

The unit-character concept was very important in the earliest work in Mendelism. Although contrary-to-fact claims about history are always dangerous, it may be safe to say that without that concept Mendelism could not have begun. First, seeing the organism as composed of separable unit-characters, rather than in a holistic way, was crucial to the study of both evolution and heredity. Then, making a simplifying assumption of the one-to-one relation between a visible character and the inferred underlying unit was critically important. That assumption allowed the first inferences to be made from data about ratios of characters to the types and behaviors of underlying units. Darwin had made no such assumption about the relation between gemmules and characters; for him, ratios of characters would not have held a clue to inferences about types of underlying units. De Vries and Correns had many crosses with many ratios of characters in their data. Grouping those that showed the 3:1 patterns and then reasoning to the explanation in terms of segregation was a difficult task. The unit-character assumption was important in allowing patterns to be found and inferences to be made.

The introduction of paired symbols (A and a) for the paired unit-characters allowed manipulations of that symbolic notation to produce predictions to test its adequacy. The test cross of a hybrid, for example, with a homozygous recessive $(A + a)$ $(a + a)$ predicted $1Aa:1aa$ ratios for the unit-characters, which could be confirmed by doing such crosses. Testing of the theoretical claims was made possible by the method of first breeding organisms, then counting characters, and then manipulating their symbolic representations. At the beginning, Mendelism explained 3:1 ratios and 9:3:3:1 ratios. The strategy of introducing and manipulating a symbolic notation was discussed in Chapters 3, 11, and 12.

As discussed in Chapters 6 and 11, the need to separate the underlying unit from the visible character emerged gradually. Geneticists gradually realized that the "something" in the germ cells was not the character itself, but the thing that causes, or at least was associated with, the appearance of the character in the developed organism. The need for explicitly postulating a separate "factor" in the germ cells emerged with the attempts to explain dominance in terms of presence and absence, and with the explanation of quantitative variation by the multiple factor hypothesis (see Chapter 6).

Complicating the oversimplified one factor-one character assumption allowed the development of new theoretical components. The new components accounted for items that were originally outside the scope of the domain of Mendelism. The hypotheses of multiple factors and multiple allelomorphs were two such "complications" that extended the domain. The multiple factor hypothesis explained quantitatively varying characters. When more than two alternatives for a given character were found, multiple allelomorphs in a population replaced the general claim that allelomorphs occurred only in pairs. The changes produced refined theoretical components, as my changes to Components 1 and 2 indicate. These theoretical changes resulted in the removal of anomalies and the expansion of the scope of the domain of Mendelism.

The strategies exemplified in these changes are the following: oversimplify and then complicate; overgeneralize and then specialize. If a simplifying assumption is made at the beginning of theoretical development, then it might be expected that complication will be a strategy to employ in the light of anomalies. A question to ask is—what was neglected in forming the simplification? The answer to that question can guide the formation of the next, more complicated theoretical component. In this case, the possibility of several factors interacting to produce one character had been neglected in the one factor-one character simplification. Consequently, postulating multiple factor interactions was an obvious "complication" to make in the light of certain anomalous ratios. Exemplified in the change from the one factor-one character assumption to the multiple factor hypothesis is the following strategy: simplify, work with the simple case until it is well-understood, then gradually complicate, to account for anomalies and expand the domain.

The overgeneralization that all allelomorphs occur in pairs was specialized to two claims: C2.1', that in one organism, at most two different allelomorphs for a character are found, and C2.2', that in a population, more than two allelomorphs for a character may occur. This specialization resulted from new data not known when the overgeneralization was made in the early days of Mendelism, namely, data showing three different allelomorphic characters in a population. The step of specialization involved adding a condition to distinguish whether the alternative characters were "in one organism" or "in a population." Thus, one way of specializing a generalization is by adding a new restrictive condition to indicate the narrow range in which the general claim applies.

In summary, the strategies exemplified in the changes that produced the theoretical components of multiple factors and multiple allelomorphs are the following: simplify, then complicate; generalize, then specialize (see Chapter 6 for more detail on these strategies). The changes served to remove anomalies and expand the scope of the theory.

14.3 Interfield Connection

1900:
C3. Interfield connection
  C3.1 The connections between generations are the germ cells, also called "gametes" (i.e., pollen and egg cells in plants; sperm and eggs in animals).

1926:
C3'. Interfield connection
  C 3.1' Genes are transmitted from parents to offspring in the germ cells.
  (Further development led to the interfield theory, the chromosome theory of Mendelian heredity.)

Mendel formulated his explanation of the 3:1 ratios in terms of types of germ cells; he argued that seemingly identical germ cells in hybrids (such as pollen grains) were, in fact, not identical. Instead, he claimed, they were of one or the other of the pure parental types, with respect to one character. The difference was symbolized by designating the two types as A and a. The "purity of the germ cells," as Bateson called it, was the key new idea as of 1900 that was debated and finally accepted as a theoretical component of Mendelism. The discovery of cells and the cytological work showing that fertilization is the joining of cells from the male and female provided important information for Mendelism. The cell was a necessary level of organization for formulating the key explanatory claim of segregation of different types of germ cells. Thus, the level of the germ cell was crucial for the development of the theoretical components of Mendelism. Nineteenth-century studies based on cell theory had provided that level of organization.

Component 3, as I have stated it, was an implicit assumption in the work of the early Mendelians. (See Chapter 5 for my justification for stating it as a theoretical component of Mendelism.) Its further development led to the chromosome theory of heredity, proposed by de Vries, Sutton, and Boveri (see Chapter 7) and further developed by Morgan and his co-workers (see Chapters 9 and 10). I have not included the chromosome theory of heredity in the revised Component 3' because Morgan did not mention the chromosomes in his statement of the theory of the gene in 1926 (see Chapter 12). He was careful to separate the claims supported by genetic data and the claims associated with the chromosome theory that postulated that genes were parts of chromosomes. Morgan (1926, p. 25) stated the theory of the gene using the terms "genes" and "germ cells," with no mention of chromosomes:

> The theory states that the characters of the individual are referable to paired elements (genes) in the germinal material that are held together in a definite number of linkage groups; it states that the members of each pair of genes separate when the *germ-cells* mature . . . . (Morgan, 1926, p. 25, my emphasis)

One explanation for how germ cells differed was provided by the chromosome theory: germ cells contained chromosomes carrying different genes (see Chapter 10). Other kinds of explanations, using germ cells and not chromosomes, were formulated by other Mendelians. The factor hypothesis claimed that differing germ cells contained different factors, but in Bateson's early formulation no assumptions about the organization of factors within cells were needed to account for segregation. With the discovery of coupled characters, Bateson hypothesized, first, that factors "coupled" or "repulsed," and then, that germ cells selectively "reduplicated" (see Chapter 9 for further discussion). Thus, the nature and behavior of germ cells provided the level of organization for formulating numerous alternative hypotheses to account for breeding data; appeal to the lower level of chromosomes was only one of the alternatives.

The strategy of relating new developments to established knowledge in another field is a fruitful one. The identification of the germ cells as a stage in heredity was important in furnishing a location for the step of segregation, a central component of the theory. The germ cells also played key roles in the alternative hypotheses of reduplication and linkage (as discussed in Chapters 9 and 10). Making available other concepts at other levels of organization is one important role that can be played by the strategy of using interrelations to another body of knowledge. (See Chapters 7, 9, 10, and 13, for more detail on the strategy of using interrelations.)

14.4 Dominance-recessiveness

1900-1903:
C4. Dominance-recessiveness
  C4.1 In a hybrid formed by crossing parents that differ in a single pair of characters, there is some difference such that one character dominates over the other; thus, the character in the hybrid resembles one but not the other of the parents. (Let *A* symbolize the dominant character which appears; *a* the recessive which is not apparent.)

1926:
(No C4': Component on dominance-recessiveness deleted.)

Component 4, as were the other components as of 1900-1903, was stated in terms of characters, not the underlying allelomorphs, factors, or genes. With the clearer separation of characters from the alternative, allelomorphic factors, the dominance component could be stated—one allelomorph dominates over the other in hybrids. By 1926, this claim was not stated as a general claim in the theory of the gene, as discussed in Chapter 6. Dominance was a promissory note, an incompleteness, in the theoretical components that was never cashed. The lack of adequate theoretical components to explain the phenomenon of dominance might have been seen as a major inadequacy for the entire theory of Mendelism; however, such a failure was not viewed that way historically. There are several possible reasons that may help to explain why geneticists adopted this attitude toward the problem of dominance.

First, many exceptions were found to the domain item that one character always dominates over the other. Sometimes blended or new characters were found in hybrids. Thus, the domain item was not a universal, empirical generalization. It was specialized in the light of numerous exceptions. Second, whether the heterozygous hybrid showed dominance, a blend form, or some new form had to be determined empirically for each pair of alternative characters. No theoretical components were successful at predicting whether dominance would occur, and, if so, which character would dominate.

Third, even after the cross was done and the dominant character identified, no theoretical component explained why it dominated. Bateson had attempted to provide the rudiments of an explanation with his presence and absence hypothesis; however, that hypothesis did not survive the discovery of multiple allelomorphs. The explanation of the development of a dominant character was part of the entire problem of embryological development and gene expression, which Mendelism never solved.

Fourth, other theoretical components of Mendelism, as we have seen, had numerous successes. The lack of an explanation for dominance did not hinder development of other parts of the theory. The relation of the promissory dominance component to the other theoretical components was sufficiently modular that the removal of the dominance component did not affect the others (except that lack of dominance would lead to the expectation that 1:2:1 ratios in the F_2 would be found, rather than 3:1 ratios).

Finally, geneticists speculated that dominance was due to some change in chemical molecules; if correct, such speculations would place the problem of dominance (and its solution) at a lower level of organization than either Mendelism or the chromosome theory. The explanation of dominance would thus become part of the domain of some other theory.

All these reasons help to explain the deleting of the promissory dominance component from the set of theoretical components and the demoting of the phenomenon of dominance to a more specialized empirical generalization not explained by the theory.

Deleting dominance from the set of theoretical components allowed Mendelian segregation (i.e., other theoretical components) to be extended to include characters that blended or showed new forms in heterozygous hybrids. Deletion of the empirical generalization about dominance from the domain allowed the scope of the domain to be expanded to include characters that did not show dominance. Thus, deletion of the theoretical component resulted in an expansion of the scope of the domain of the theory and the removal of an uncashed promissory note from the theoretical components.

The strategy that emerges from the elimination of the theoretical component of dominance-recessiveness is the strategy of deleting a problematic theoretical component. This strategy was used when the phenomenon for which the component purported to account had numerous exceptions, when, moreover, the component had not been developed into a satisfactory explanatory or predictive component. This strategy risks the criticism that the lack of explanation of the phenomenon and its exceptions is a critical anomaly for the whole theory. An assessment of that criticism will rest on the successes of other parts of the theory and on judgments about the importance of the phenomenon at issue.

As we saw for Mendelism in its earliest days, issuing promissory notes for explanations is a method of beginning a theory with some incompleteness. In general, it may be difficult to decide at what stage a failure to cash those notes becomes a serious failure. Deleting a component of a theory, and thereby leaving even a more restricted domain item unexplained, might sometimes be viewed as a serious theoretical inadequacy. The dominance case suggests some possible guidelines for using the strategy of deletion of a theoretical component to improve the theory. First, if the theoretical component accounts for a single domain item and if the item is deleted, then delete the theoretical component. Or, if the domain item is an empirical generalization and numerous exceptions to it are found, then consider deleting the theoretical component. Second, if an unsolved problem can be reasonably assumed to require another, as yet unavailable, level of organization to solve it, then shelve the problem at this point. Deleting a theoretical component (or a promissory note for one) and providing nothing to take its place, when a phenomenon (even with numerous exceptions) remains unexplained, sounds like a move that would often result in a worse, rather than better, theory. It may only be justified when other components of the theory are very successful and the deletion can be made without affecting the other parts. Such was the case for the dominance component.

14.5 Segregation

1900-1903:
C5. Segregation
  C5.1 In the formation of germ cells in a hybrid produced by crossing parents
  that differed in a single pair of characters, the parental characters segregate
  or separate, so that the germ cells are of one or other of the pure parental
  types. (Each germ cell has either *A* or *a* but not both; called "purity of the
  gametes.")
  C5.2 The two different types of germ cells form in approximately equal
  numbers. (Equal numbers of germ cells with *A* and *a*.)
  C5.3 When two hybrids are fertilized (or self-fertilization occurs), the differing types of germ cells combine randomly.
  $[(A + a) (A + a) = AA + 2Aa + aa$; appearance: $3A{:}1a]$

1926:
C5'. Law of segregation
  C5.1.1' Parental genes are not modified as a result of being together in a
  hybrid; no new kinds of hybrid genes form.
  C5.1.2' In the formation of germ cells of a hybrid, paired parental genes
  (allelomorphs) segregate so that the germ cells have one or the other of a
  given pair.
  C5.2' The two different types of germ cells are formed in equal numbers.
  C5.3' When two similar hybrids are cross-fertilized (or self-fertilization
  occurs), the differing types of germ cells combine randomly. Symbolized by
  genotype:
  $(A + a)(A + a) = AA + 2Aa + aa$; phenotype: $3A{:}1a.$

Chapter 8 discussed two challenges to segregation: Cuénot's yellow mice case with 2:1 data, and Castle's cases in which pure breeding forms seemed not to re-emerge after a hybrid cross. The theoretical components of segregation survived both challenges and remained as important general claims within Mendelism. The two challenges provided examples that exemplified strategies for localizing and resolving anomalies, as well as strategies for assessing the adequacy of rival hypotheses.

In both the Cuénot and Castle challenges, the analysis of segregation into separable components (C5.1 to C5.3) proved useful for my analysis as to where the alternative hypotheses were localized within the theory. The separate components were seen to be alternative sites for doing hypothesis formation. In the yellow mice case, all the alternative hypotheses to account for the anomaly were localized within the segregation components. Two of the hypotheses, Morgan's alternative to purity and Cuénot's hypothesis of selective fertilization, can be seen as localizing a separable component and asserting its opposite. Morgan's hypothesis focused on C5.1, purity of the gametes, and asserted its opposite, nonpurity. Morgan claimed that the germ cells of hybrids were never pure and that Cuénot's failure to find pure breeding dominant forms (i.e., $0AA: 2Aa: 1aa$) resulted from the appearance of the recessive earlier than is usually the case (see Chapter 8 and Figure 8-1 for more detail). In contrast, Cuénot attempted to explain the $2Aa: 1aa$ anomaly by focusing on the assumption stated in Component 5.3, random fertilization. He hypothesized an alternative—selective fertilization. Something caused the gametes with A not to combine with each other during fertilization, hence no AA forms resulted. Thus, Morgan's and Cuénot's hypotheses can be analyzed as denying theoretical components that had been explicitly stated in the literature. (See Chapter 5 for the historical justification of my analysis of the explicit segregation components.)

Castle and Little also proposed a hypothesis to explain Cuénot's 2:1 ratios; they claimed that the AA combination occurred at fertilization, but the combination proved lethal and the pure breeding yellow mice did not develop. That is, AA died and the result was $2Aa$ and $1aa$. I did not argue that the Castle and Little hypothesis was a denial of an explicitly stated theoretical component. Instead, I interpreted it as having served to uncover an implicit assumption, namely, that all hybrid forms (AA, Aa, aa) were equally viable in normal cases of segregation. Alternative hypotheses to explain an anomaly can thus be localized, not only in explicitly stated theoretical components, but in implicit assumptions not explicitly stated for the normal case. An anomaly may play the important role of a stimulus to uncover implicit components of the theory.

By considering Component 5.2, which was not denied by any of the historically proposed hypotheses (so far as I have found), I was able to suggest another alternative hypothesis to account for the Cuénot anomaly. Component 5.2 claims that equal numbers of types of gametes form in the male and female. I formed an alternative hypothesis by denying equality and suggesting that in either the male or female one type did not form. Thus, the cross would be represented: (No A + a) ($A + a$) = No AA + $2Aa$ + $1aa$. I was able to generate this alternative by systematically considering each of the steps in normal segregation and proposing the opposite of one of the steps. Equal numbers became nonequal numbers in my hypothesis. Systematic consideration of each separable step could have led to its proposal historically if someone had followed the strategy of explicitly delineating each step in the process and considering alternatives to each.

Testing showed Castle and Little's inviability alternative to be correct; dissection of dead embryos eventually confirmed it beyond doubt. No change occurred in the theory to account for the normal segregation process. An anomaly that had challenged key components of the theory was explained as a monstrous case. As other lethal gene combinations were discovered, the malfunction class of lethals became well-known. Deviations from 3:1 ratios could be examined to see if lethals were responsible; they became monster anomalies that were barred from being further problems for the theory. (See Chapter 12 for further discussion of monster anomalies and malfunction classes; Lakatos's strategy of "monster-barring" was mentioned in Chapter 8 and will be discussed in Chapter 15.)

Chapter 8 also discussed another challenge to segregation, namely, Castle's explanation of increased variability in some characters in the F_2 generation in several mammal species. He hypothesized that the allelomorphs for those characters contaminated each other while they were together in the hybrid but before they segregated to form germ cells. Impure genes segregated and resulted in an array of characters in the F_2 generation, each differing slightly from the grandparental characters of the original cross. Castle's hypothesis indicated that Bateson's claim of the purity of the germ cells (Component 5.1) actually contained two separable claims, namely, C5.1.1', the lack of contamination of genes in a hybrid, and C5.1.2', their separation in the formation of germ cells. Thus, the implicit connection between purity and segregation was broken. The two claims were explicitly delineated. Castle denied the first in denying purity and hypothesizing contamination, but left intact the claim about separation of two genes (allelomorphs) in the formation of germ cells in the hybrid. He argued, contrary to the views of numerous other Mendelians, that the claim of pure factors in the germ cells of hybrids was not a necessary component of Mendelism (see Chapter 8 for more detail on Castle's contamination hypothesis).

Castle's hypothesis did not go unchallenged. The *Drosophila* group proposed an alternative explanation for the variability in terms of the multiple factor hypothesis. Modifying factors, they claimed, were reassorted during the hybrid cross and were responsible for the increased variability. The alternative hypothesis of modifying factors was not localized in the segregation components, but in the more general claim about the relation between factors and characters. Multiple factors had already been proposed as an alternative to the one unit-one character assumption in order to account for other cases (see Chapter 6). Muller and Sturtevant argued that this alternative could be used to explain Castle's cases also (see Chapter 8 for more detail).

The arguments and experiments that served in the assessment of the competing contamination and modifying factor hypotheses serve to illustrate various strategies for theory assessment. Castle focused only on his own data to be explained and argued that his hypothesis was simpler and less ad hoc for that narrow domain. Muller argued that numerous other Mendelian crosses indicated purity of germ cells; a hypothesis that left the purity claim intact had greater generality for explaining the entire domain of Mendelism. Tests with a similar anomalous case in *Drosophila* showed modifying genes to be the explanation. Castle, however, was not convinced until he performed a crucial experiment, using his own rats, in which the modification seemed to have taken place. He confirmed his opponents' hypothesis of modifying genes and abandoned his contamination hypothesis. (See Chapter 8 for more detail on the crucial experiment.)

Cuénot's and Castle's challenges to the segregation components served to strengthen the evidence for their generality and to carefully delineate previously implicit, separable steps involved in a typical instance of segregation producing 3:1 ratios.

The challenges to segregation exemplify several strategies for resolving anomalies, as discussed in Chapters 8 and 12. Steps in anomaly resolution may be summarized: (1) reproduce anomalous data; (2) localize the anomaly within a subset of the components of the theory; (3) carefully delineate any implicit theoretical components in the potentially problematic areas; (4) systematically consider a fault in each one as a location of a possible change in the theory; (5) use strategies for producing new ideas to generate alternatives, constrained by the nature of the anomaly and the preservation of nonproblematic components (sometimes, asserting the opposite of each potentially problematic component may serve as a means of generating types of hypotheses to explore); and (6) assess the alternative hypotheses, possibly by devising a crucial experiment (finding a decisive experiment may be the most creative step in this process). Another strategy that was involved in the modifying factor hypothesis is to consider whether a hypothesis, already proposed to resolve another anomaly, can be used to explain the current one. (See Chapter 8 and Chapter 12 for further discussion of the strategy of anomaly resolution exemplified in the challenges to the segregation components.)

14.6 Mendel's Second Law and Linkage

1900-1903:
C6. Explanation of dihybrid crosses [later called "independent assortment"]
   C6.1 In crosses involving parents that differ in two pairs of differentiating characters, the pairs behave independently. Symbolically, $(AB + Ab + aB + ab)(AB + Ab + aB + ab)$ = complicated array that in appearance reduces to 9AB:3Ab:3aB:1ab.
An expanded version of C6 (used for analysis in Section 9.3):
C6. Explanation of dihybrid crosses:
   C6.1 In the formation of germ cells in a hybrid produced by crossing parents that differed in two or more traits, the parental factors segregate or separate, so that the germ cells are of all possible pure parental combinations. Symbolically, $AABB$ x $aabb$ = F_1 hybrid AaBb, which produces germ cells symbolized by AB, Ab, aB, and ab.
   C6.2 The different types of germ cells are formed in equal numbers. Symbolically, equal numbers of germ cells of types AB, Ab, aB, and ab are formed, i.e., gametic proportions are 1:1:1:1.
   C6.3 When two hybrids are fertilized (or self-fertilization occurs), the differing types of germ cells combine randomly. Symbolically, $(AB + Ab + aB + ab)$ $(AB + Ab + aB + ab)$ = complicated array that in appearance (given complete dominance) produces 9:3:3:1 character ratios in the F_2 generation.

1926:
C6'. Assortment, linkage, and crossing-over
   C6.1' Genes are found in linkage groups; groups occur in corresponding pairs.
   C6.2' Genes of different linkage groups assort independently.
   C6.3' Usually genes in the same linkage group are inherited together; at times, however, an orderly interchange, called "crossing-over," takes place between allelomorphs in corresponding linkage groups.
   C6.4' Genes in a linkage group are arranged linearly with respect to each other.
   C6.5' The frequency of crossing-over can be used to calculate the order and relative positions of the respective genes in linkage groups if such disturbing factors as double cross-overs and interference of one cross-over with another are taken into account.

The most extensive revisions to theoretical components of early Mendelism occurred in Component 6, which explained dihybrid and multihybrid crosses, namely, crosses involving more than one differing trait. As discussed in Chapters 5 and 9, no separate, second Mendelian law was formulated during the period between 1900 and 1910. Resolution of the anomalies to 9:3:3:1 ratios produced a number of results, discussed in Chapters 7, 9, and 10: the delineation of Mendel's first law of segregation from what Morgan labeled "Mendel's second law—the independent assortment of genes"; the confirmation of the prediction of the chromosome theory of correlated characters; the new interrelations between sex chromosomes (studied with a microscope) and sex-linked inheritance (studied by breeding techniques); the new theoretical components of crossing-over, linearity, and relative positions of genes along genetic maps; and the discovery of a new level of organization, the "linkage group" between an independent factor and all the genes in an organism.

The resolution of the anomalous ratios for some multihybrid crosses involved the development of an interfield theory, the chromosome theory of Mendelian heredity. As discussed in Chapter 7, the chromosome theory, proposed by Sutton, Boveri, and de Vries, did not fare well in the period before 1910, except for the correlation of sexual differences with specific chromosomes. Few characters confirmed Sutton's prediction that characters would be grouped in inheritance (or "completely linked," to use later terminology), because their allelomorphs were parts of the same chromosome. De Vries's view of pangens being exchanged between homologous chromosomes did not result in such a prediction and thus was more in accord with the data about independent assortment at the time. But the jumping pangens hypothesis was not fruitful in suggesting additional work.

As discussed in Chapter 9, Bateson attempted to explain the "coupling" anomalies that he discovered without appeal to chromosomes. First, he proposed a hypothesis about the coupling and repulsion of factors, and then he developed another hypothesis about selective reduplication of germ cells. The two levels of organization that he used in his hypotheses were thus the level of individual factors and the level of the entire germ cell. Those hypotheses, however, did not fare well on various criteria of theory assessment. Bateson's hypotheses were replaced by the linkage hypothesis, which Morgan and his colleagues developed using interrelations to chromosomes.

Morgan's discovery of a white-eyed mutant in *Drosophila*, which showed sex-linked inheritance, as well as partial coupling of other characters associated with sex, proved to be the anomaly whose resolution produced significant theoretical change. The interrelations between sex chromosomes, sex heredity (characters that were associated with sex in Mendelian crosses), and linked characters were crucial in the early development of new hypotheses. The use of interrelations between cytology and genetics, that is, between chromosomes and genes, produced a useful analog model: the beads on a string model represented genes linearly arranged along the chromosomes. The use of interrelations and the analog model resulted in a burst of creative new hypotheses for Mendelism—partial linkage, crossing-over, double cross-overs, interference, linearity, the new level of organization of the linkage group. Each of these faced challenges from alternative hypotheses, some developed without using the chromosome theory and some developed using different assumptions about the relation of genes to chromosomes, such as three-dimensional rather than linear relations, as discussed in Chapters 9 and 10.

As it turned out, relating the findings produced by one set of techniques to those produced by another proved very fruitful, both for the development of the intrafield theory of the gene and for the interfield chromosome theory of Mendelian heredity. As seen in Chapter 10, Morgan (1926) separated the two theories in his exposition. He stressed that all the hypotheses for Mendelism that were developed using cytological results were also supported by numerical results from Mendelian breeding experiments. In the 1920s, a diversity of opinion existed about the two theories, as discussed in Chapter 13. Some biologists accepted Mendelism, but continued to search for alternative explanations of linkage other than the chromosome theory; others thought chromosomes were implicated in hereditary processes, but did not accept the "discrete unit" view of genes postulated by the theory of the gene. By about 1930, however, the combined theory of the gene and the chromosome theory of Mendelian heredity had much evidence in its favor. As discussed in the conclusion of Chapter 13, the theory of the gene was invading other fields, such as evolutionary biology and biochemistry in the 1920s and 1930s.

A number of strategies were exemplified in the numerous hypotheses formed to account for the anomalies to independent assortment. The separation of the law of segregation from the law of independent assortment occurred in response to the anomalies. That change exemplifies the strategy of delineating separable assumptions and altering one, but not the other. Segregation was delineated and its universality was preserved. However, the second law was specialized: only genes in different linkage groups assort independently. New theoretical components were added to account for linkage. Thus the strategies of **delineate and alter** and **specialize and add** are exemplified in this change. The modularity of the explanations of 3:1 and 9:3:3:1 ratios (which I have labeled Components 5 and 6) thus became apparent—the latter could be altered without changing the former (see Components 5' and 6'). Just as we saw in changes to other components, the linkage anomalies forced implicit assumptions to be made explicit: segregation had implicitly been used to cover both kinds of ratios, but with exceptions to one and not the other, explicit delineation of separable components was needed.

In contrast to the exceptions to 3:1 ratios, the linkage anomalies served to illustrate the difference between what I have called "monster" and "model" anomalies (see Chapters 8 and 12). The anomalies for 3:1 were explained away as monsters; they did not require theory change to account for them. The linkage anomalies, however, were model anomalies because they did require theory change. They represented a new exemplary class of hereditary relationships, namely, relationships between characters that were partially linked in inheritance and thus did not show complete independent assortment. The theory was changed so as to explain linked characters. The scope of the domain was thereby expanded to include a wide range of ratios beyond the 9:3:3:1 ratios in dihybrid crosses. The exemplars for sex linkage and autosomal linkage entered the explanatory repertoire of the Mendelian geneticist.

As discussed in Chapters 9 and 10, the formation of the new hypotheses that resolved the linkage anomalies exemplify the strategies of **using interrelations, postulating a new level of organization, introducing and manipulating an analog model**, and **using an analog model to develop new theoretical components with new theoretical terminology**. The interrelations were between genetics and cytology. The new level of organization was the linkage group; it constituted a level between that of the individual genes and of all the genes in the entire germ cell. Two analog models played roles: Castle's rat-trap, three-dimensional model and the *Drosophila* group's linear model of beads along a string.

As discussed in Chapter 10, the bead model served to provide what Boyd called a "theory-constitutive metaphor," namely, a metaphor that becomes at least for a time an "irreplaceable part of the linguistic machinery of a scientific theory.... Such metaphors are *constitutive* of the theories they express" (Boyd, 1979, p. 360). The terminology of "crossing-over," "double cross-overs," "interference," and "linear genetic maps" all became a part of the theory itself. The terms gained meaning from the visual analog model of beads along a string. Furthermore, although historical evidence is lacking, the model may have furnished the ideas used in formulating the hypotheses of double cross-overs and interference to resolve various anomalous ratios for the linear linkage hypothesis.

14.7 The New Component of Mutation

1926:

C7'. Mutation

    C7.1' Genes occasionally mutate and then cause a different character.

    C7.2' Such point mutation does not alter their linear relation to other genes in the linkage group.

The components of Mendelism in the period 1900-1903 (see Chapter 5) did not include a theoretical component for mutation. The alternative characters that were used in the early Mendelian crosses, such as yellow and green peas, occurred naturally. As discussed in Chapter 10, de Vries discovered dramatically differing forms of *Oenothera*, evening primroses, living side-by-side in the wild and brought them back to his experimental garden. Subsequent generations yielded dramatic mutations. *Oenothera* became the model organism on which he developed his mutation theory. Some mutations were "digressive," according to de Vries and showed segregation. Others, however, were "progressive" and produced large-scale changes that bred true. The latter, he claimed, were species-forming mutations. Had de Vries been correct, then genetics would have had two parts—studies of Mendelian genes and studies of species-forming genes. As discussed in Chapter 13, no separate field of population genetics would have been needed to study speciation.

As others did breeding experiments to search for species-forming mutations, numerous smaller scale mutations were found that did segregate. When Morgan discovered a white-eyed male in a culture of red-eyed *Drosophila*, he called it a "mutant." Intensive study of *Oenothera*, as well as increasing knowledge about other variations, allowed Mendelians to explain de Vries's numerous progressive mutations as due to smaller scale, segregating mutations (or lethal gene combinations or abnormal chromosome behavior), as discussed in Chapter 10. Successive editions of books by the *Drosophila* group showed the introduction and expansion of sections on mutation (see Chapter 10 for details). Muller argued that the study of mutation would be fruitful for a new research program to tackle questions about the problems of the size and physical nature of the gene. It was hoped that by understanding what caused a gene to change, its physical nature could be determined. Experimental studies of mutagenic agents, including chemicals, x-rays, and light of other wave-lengths, were lively areas of research in the 1930s and 1940s.

The strategy exemplified in the introduction of the component of mutation into the theory of the gene is the following: take an old idea outside the theory and quantitatively change it to produce a modified idea that can be built into a new theoretical component. Showing that supposedly "large scale" mutations were within the scope of Mendelian experimental studies served to expand the domain of the theory of the gene. It also served to establish a new research program with new experimental techniques. Thus, adding a new component to a theory may result in substantial changes to the field and may be fruitful for future work.

14.8 Additional Strategies from the Case

Some strategies discussed in Chapters 5 to 13 have not yet been summarized in this chapter. They are strategies that are reflected less directly in changes to specific theoretical components than the strategies discussed in Sections 14.1 to 14.7. Especially important are the strategies associated with the diagrammatic representation of the theory, discussed in Chapters 11 and 12, and the strategies of theory assessment discussed throughout previous chapters.

Problem-finding strategies are suggested by the diagrammatic representations of the theory discussed in Chapters 11 and 12. Chapter 11 discussed a (framelike) representation of a list of general properties and of the specific ones that were filled in as claims about genes, the new theoretical entities, changed. Chapter 12 represented the theory of the gene as filling in blanks in pedigree diagrams for the step between parents and offspring. When a diagrammatic representation can be made and blanks exist, they are instances of "blackbox incompleteness" in our knowledge that need to be filled. Incomplete information about the step from parents to offspring was such an incompleteness and it was successively filled in at lower levels of

---

C1' Genes and characters
  C1.1' Genes cause characters.
  C1.2' (a) One gene may cause one character, or (b) multiple factors (genes at different loci in linkage groups) may interact in the production of one character, or (c) one gene may affect many characters.

C2' Paired genes and multiple allelomorphs
  C2.1' In any one organism, genes occur in pairs (called "allelomorphs" and found at the same locus in corresponding linkage groups).
  C2.2' In a population, multiple allelomorphs for a character occasionally occur.

C3' Interfield connection
  C3.1' Genes are transmitted from parents to offspring in the germ cells. (Further development led to the interfield theory, the chromosome theory of Mendelian heredity.)

(No C4': Component on dominance-recessiveness deleted)

C5' Law of segregation
  C5.1.1' Parental genes are not modified as a result of being together in a hybrid; no new kinds of hybrid genes form.
  C5.1.2' In the formation of germ cells of a hybrid, paired parental genes (allelomorphs) segregate so that the germ cells have one or the other of a given pair.
  C5.2' The two different types of germ cells form in equal numbers.
  C5.3' When two similar hybrids are cross-fertilized (or self-fertilization occurs), the differing types of germ cells combine randomly. Symbolized by genotype: $(A + a)(A + a) = AA + 2Aa + aa$; phenotype: $3A:1a$.

C6' Assortment, linkage and crossing-over
  C6.1' Genes are found in linkage groups; groups occur in corresponding pairs.
  C6.2' Genes of different linkage groups assort independently.
  C6.3' Usually genes in the same linkage group are inherited together; at times, however, an orderly interchange, called "crossing-over," takes place between allelomorphs in corresponding linkage groups.
  C6.4' Genes in a linkage group are arranged linearly with respect to each other.
  C6.5' The frequency of crossing-over can be used to calculate the order and relative positions of the respective genes in linkage groups if such disturbing factors as double cross-overs and interference of one cross-over with another are taken into account.

C7' Mutation
  C7.1' Genes occasionally mutate and then cause a different character.
  C7.2' Such point mutation does not alter their linear relation to other genes in the linkage group.

Table 14-1. Components of the theory of the gene, 1926

organization. Theories about progressively lower levels of organization began with sexual fluids, then moved to germ cells, then chromosomes, then genes, and even later, DNA. Problems of incompleteness are to be distinguished from anomalies that result from a failed prediction and, thus, indicate some sort of incorrectness.

When blanks in the diagrammatic representation of parents to offspring were filled in, new steps for normal processes of heredity could be represented. Specific crosses, such as those of yellow and green peas, provided exemplary problem solutions. Abstracting the details, such as the specific color of peas, from such a problem provided an abstract problem-solving schema for application to similar cases. The steps of segregation of genes furnished by the exemplar thus became part of the abstract, diagrammatic representation of the steps in normal hereditary processes. Additional exemplars for blending inheritance, independent assortment, sex linkage, and non-sex linkage were added to the explanatory repertoire as Mendelism developed. The typical steps for each such process also could be diagrammatically represented and applied to other instances of the same kind.

The strategy of representing exemplary problem solutions in an abstract way and using them to solve other similar problems is a powerful one. As discussed in Chapter 12, this method combines (1) Kuhn's (1970) insight that science students often learn by doing exemplary problems and learning to apply them appropriately, and (2) Kitcher's (1981) view of theories as explaining domain items by instantiating an abstract schema. Thus, the development of the theory of the gene, on this analysis, is viewed as the filling in of blanks in the steps between parents and offspring and as the development of a growing set of exemplary cases for different kinds of hereditary processes. This analysis of theory development is similar to that of Thagard's (1988) in that a scientific theory may be represented by abstract problem-solving schemas that generate explanations.

As discussed in Chapter 12, the diagrammatic representation of the steps in normal segregation or normal independent assortment cases allows anomalies for predicted 3:1 or 9:3:3:1 ratios to be plausibly localized in the steps normally producing those ratios. Then the question can be raised as to whether the anomaly is a monster or a model. It may be a monster anomaly that does not require a change in theory but is only a quirky case or a class of malfunctions, such as lethal gene combinations (discussed in Chapter 8). Alternatively, it may be a model anomaly requiring theory change. Model anomalies indicate that the purported schema for normal cases is inadequate. The normal diagram may need to be altered to correctly represent the normal case. Or, perhaps, additional schemas need to be added to expand the explanatory repertoire. They can be produced by abstracting from the exemplary model case, such as the linkage anomalies discussed in Chapters 9 and 10. (For more discussion of the process of abstraction, see Darden, 1987; Darden and Cain, 1989.)

Discussions of alternative hypotheses throughout Chapters 6 to 13 often revealed criteria for assessing the competitors. In some cases, scientists explicitly argued for or against a hypothesis on specific grounds; for example, Chapter 8 discussed strategies for hypothesis assessment that included generality, simplicity, lack of ad hocness, empirical adequacy, predictive adequacy, and the number of additional problems raised. My analysis of criteria of theory assessment in Chapter 8 is thus an example of the use of history of science that Losee (1988) suggested, namely, the analysis of historical cases to try to determine the actual criteria used by scientists in judging theories. In other cases, the criteria were less explicitly stated in the published literature, but the eventual fate of the hypotheses could, nonetheless, be traced. In those cases, I used assessments by other historians or I suggested myself what criteria the competing hypotheses satisfied. In

S1. Strategy: conceptual clarification
   Change: unit-character—>genes cause characters
   Result: postulation of theoretical entity

S2. Strategy: complicate, specialize, add
   Change: one-one—>one-many, many-one, many-many
   Results: expanded scope of domain; new theoretical components

S3. Strategy: postulate interfield or interlevel relations
   Change: explicit identification of germ cells
   Result: other concepts and levels available for use in other components

S4. Strategy: delete
   Change: eliminate dominance component with many exceptions
   Results: overgeneralization removed; expansion of scope of domain

S5. Strategy: make explicit previously implicit separable assumptions; (possibly) deny and propose opposite
   Change: purity and segregation: together—>separate
   Result: newly delineated components and (possibly) alternative hypotheses ready for testing

S6. Strategies: delineate and alter; specialize and add; use interrelations and analog model to generate new ideas
   Changes: a. the law of independent assortment separated from the law of segregation;
       b. the law of independent assortment specialized to apply only to independent assortment of genes in different linkage groups;
       c. addition of new components of linkage, crossing-over, linearity
   Results: one law separated into two and significant new components added to theory limiting the generality of the second law and resolving model anomalies

S7. Strategy: add new component by quantitatively altering old idea
   Change: de Vriesian mutation—>smaller scale mutations
   Results: addition of new component to theory (and new research program in studying mutation); expansion of the scope of the domain

Table 14-2. Major strategies for changing the theory of the gene

Chapter 9, for example, I discussed Cock's (1983) assessment of the inadequacy of Bateson's reduplication hypothesis on the grounds of its inability to explain the facts and its invocation of an ad hoc and unexplained phenomenon (namely, selective reduplication of germ cells).

Chapter 13 indicated how the strategy of using interrelations can function in hypothesis assessment, as well as in hypothesis formation. If an interrelation to another body of knowledge was used in forming the hypothesis, then the hypothesis will be integrated with that body of knowledge. If the strategy used at the beginning of theory formation was to narrow the domain, then a strategy for assessing the theory at a later stage will be to consider its interrelations to bodies of knowledge neglected at the outset. The theory of the gene was assessed differently by biologists in different fields, depending on its promise for solving problems in those fields. In the 1920s, it was promising for evolutionary problems but not for embryological ones. Thus, the strategy of using interrelations to other fields in theory assessment may produce different results, depending on which other field is considered.

In summary, the diagrammatic representation of the theory and the consideration of criteria for the adequacy of hypotheses both produced additional strategies for theory change, discussed especially in Chapters 7, 8, 9, 12, and 13.

14.9 Conclusion

Table 14-1 summarizes the components of the theory of the gene as of 1926. This chapter has compared those components with the theoretical components of Mendelism as of 1900-1903 (summarized in Table 5-2) and summarized the strategies exemplified in the changes that have been the subject of Chapters 6 to 13. Some strategies, numbered to correspond to the components in Table 14-1, are summarized in Table 14-2. Chapter 15 will place the strategies exemplified in the gene case within the context of a more systematic summary of strategies for theory change.

CHAPTER 15

General Strategies for Theory Change

The strategies discussed in previous chapters are ones that were exemplified in changes in the theory of the gene and the chromosome theory. The method of analysis in previous chapters might be called a "data-driven" or "inductive" analysis: strategies have emerged from the case, albeit because I was looking for them there. An alternative method to devise strategies is to consider work by philosophers of science and others who have studied scientific reasoning. This "theory-driven" method, to be used in this chapter, serves to put the case-related strategies into a broader context and to relate the analysis of theory change in this case to more general work.

Examination of this broader context is instructive for exploring the "space of possible strategies" in theory change. Some of the strategies exemplified in the gene case are strategies that have been previously discussed by philosophers; other strategies I devised in the light of the changes in the case. On the other hand, some strategies that have been discussed in other work have not yet been mentioned here because they were not exemplified in the gene case. So, this chapter aims at a more systematic discussion of the various types of strategies. Of course, even this more systematic list is not a complete list. Probably there is no complete list; surely new strategies arise. This list is extensive, however, and many of these strategies are important ones.

Consider once again the idea that there are stages in the development of a theory and strategies for making theoretical changes. Figure 15-1 is the diagram introduced in Chapter 2 and augmented in subsequent chapters as additional strategies were discussed. The types of strategies include: (1) strategies for producing new ideas, (2) strategies for theory assessment, and (3) strategies for anomaly resolution and change of scope. Each group (1-3) of strategies will be discussed in this chapter.

15.1 Strategies for Producing New Ideas

Table 15-1 provides a list of strategies for producing new ideas. The strategies are not meant to be mutually exclusive. Several might be used simultaneously. For example, postulating a new interrelation between two bodies of knowledge in two different fields (**using interrelations**) may simultaneously involve a **move to a different level of organization** and may be accompanied by the **introduction of a new symbolic representation**. Such a case is illustrated by the postu-

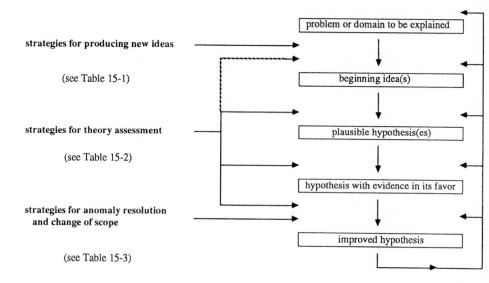

Fig. 15-1. Refined stages and strategies, IX.

lation that genes were parts of chromosomes that could be represented by the model of beads on a string. Because theory development may be, in general, an incremental process (as it was in the gene case), the strategies for producing new ideas do not have to produce a full-blown, complete theory at the outset. They need only contribute to finding a seminal (perhaps vague) idea, which can then be refined.

Figure 15-1 shows the process of theory construction beginning with a problem to be solved. The problem arises in some already established context, based on past scientific work and expectations. Previous chapters have provided little explicit discussion of strategies for finding problems. One can imagine an abstract possibility that could generate a new problem: a problem is posed when a new regularity, a new pattern is observed; then an explanation of the regularity is sought. New hypotheses are generated to account for the regularity. Alternatively, within the context of an already proposed hypothesis, criteria for theory assessment can pose problems for a hypothesis that fails to satisfy a given criterion. Thus, the problem that begins an episode of hypothesis development may be a problem for an already existing hypothesis. Problems faced by a hypothesis fall into at least two categories: incompleteness or incorrectness. When a hypothesis contains black boxes, promissory notes, gaps in proposed steps, it is incomplete and in need of further specification. Chapters 11 and 12 discussed strategies for finding "blanks" that needed to be "filled in," that is, for determining what additional information is needed to remove incompleteness. In contrast, anomalies arise when the proposed hypothesis is judged to be incorrect. Empirical anomalies result from a failed prediction or an anomalous domain item that the hypothesis cannot explain. More specific strategies for empirical anomaly resolution are discussed in Section 15.3.

15.1.1 Use Analogy

Analogical reasoning has the potential of being a useful strategy for the construction of new hypotheses. One of the general strategies that has been explored in this book is that new ideas can be constructed from old ones by various processes of transformation and combination.

1. Use analogy
2. Invoke a theory type
3. Use interrelations
4. Move to another level of organization
5. Introduce and manipulate a symbolic representation
6. Introduce a simplifying assumption, then complicate
7. Begin with vague idea and successively refine

Table 15-1. Strategies for producing new ideas.

Analogical reasoning is a good example of such a process. Prior knowledge, in the form of an analog, can be used to form new hypotheses.

Analogical reasoning in hypothesis construction may be divided into four processes: retrieval, elaboration, mapping, and justification (Kedar-Cabelli, 1988). In order to form a hypothesis in a subject area (also called the "target"), an appropriate analog (also called the "source" or "base") must be found. Finding an appropriate analog is called "retrieval." One important aspect of analogical reasoning is how retrieval is guided by the nature of the problem to be solved. After an analog is retrieved, it may have to be elaborated to make explicit features that are relevant to the problem situation. Then, there is a mapping of features from the analog to the subject area. Features of the analog usually must be transformed in various ways during the mapping in order to make them applicable to the subject area. Mapping is the process of putting into correspondence the features of the subject and the analog.

After mapping, the new hypothesis (or hypotheses) formed about the subject area must be justified. Strategies for theory assessment (to be discussed later) can be used in the justification step. To the four processes of retrieval, elaboration, mapping, and justification may be added an optional fifth: consolidation (Hall, 1989). Consolidation involves, first, analyzing the results of a particular episode of analogical reasoning and, second, compiling hindsight for use in the future about the successes and failures of that particular episode. All of these processes have been the subject of recent study by philosophers, cognitive scientists, and AI researchers (e.g., Clement, 1988; Helman, ed., 1988; Prieditis, ed., 1988; Hall, 1989; Holyoak and Thagard, 1989). They will be discussed only briefly here.

A rough distinction can be made between what I call "close" and "distant" analogy. In retrieval, if an analog is found within the same field as the problem, then this may be called reasoning by "close analogy." For example, if one has an exemplary problem solution, then other similar problems may be solved in the same way. In Mendelian genetics, once one knows how to explain 3:1 ratios for a cross between yellow and green peas, that can serve as an exemplar (a close analog) for explaining a cross between tall and short peas or between red-eyed and white-eyed fruit flies (Chapter 12 discussed such exemplars). On the other hand, retrieval produces a "distant analogy" if fewer features are shared between analog and subject, or if the analog comes from a different field (e.g., an analog from acoustics for an optics problem).

The distinction between close and distant is fuzzy. One might imagine a continuum from identical (a thing is only identical to itself) to inductively similar (one can class the two things together and form a scientific law about them) to merely analogous (the two things are clearly from different fields but have some similarities) to completely different (the two things have no similarities). It may not be clear exactly where inductive similarity stops (e.g., both these two bodies have mass, so a general law can be formed encompassing them) and analogy begins (e.g., both sound and light are waves and are thus analogous). Inductive similarity and distant analogy may or may not be sharply separable. Analogy researchers have disagreed about this; I do not have a firm opinion on the topic. On the one hand, the term "analogy" has been used to designate any similarity relation, whether inductive similarity or cross-field similarity; alternatively, it has been used to distinguish cross-field similarities, which are designated analogies, from inductive similarities, which are not called analogies. Hesse (1966, pp. 72-77) discusses differences between inductive similarity and analogy. Gentner (1983) argues for a similar distinction that she labels "literal similarity" versus "analogy."

Another aspect of retrieval, in addition to the amount of similarity between the analog and subject, is how an appropriate analog is found during a hypothesis formation episode. Retrieval has been studied both in humans and in the context of building computational models of analogy. In computational approaches, some sort of indexing scheme for analogs is often used; of necessity, such a scheme limits the kind of retrievals that the AI system can make. Hall (1989, p. 98) summarized the factors affecting retrieval: (1) what the reasoner attends to in the problem situation; (2) what is available in the store of potential analogs; and (3) the degree to which the reasoning context during retrieval matches the indexing scheme for the stored analog. The latter condition may be recast in terms of the amount of similarity between the subject and the potential analogs.

If retrieval depends on similarity between the subject and the analog, then the question arises as to how similarity is recognized and measured. Assume an ontology consisting of objects, properties of those objects, and relations among objects. Similarity might be based on having the same number of objects in the subject and analog. Or it might be based on the number of properties (or kinds of properties) shared by the objects in the analog and subject. Or it might be based on identical kinds of relations among the objects in the analog and subject.

Attempts have been made to develop metrics for measuring the number of shared properties, but such a method obviously depends critically on how the properties are differentiated and enumerated. Also, in any given use of an analog, some properties may be more important than others; thus, a measure of total properties in common may not be a good measure for evaluating a good analogy. Gentner (1983) argued that good analogies should be based, not on shared properties, but on shared relations. Of particular importance, she argued, are systematic causal relations within the analog which can be mapped to the subject. Gentner (1983) discussed the systematic causal relations exhibited by the solar system—a central object with other objects moving in closed paths around the central one. Such a causal system serves as an analog for a model of the atom. If one can specify the causal system independently of the details of the analog, then analogical reasoning becomes what is discussed later as reasoning by **instantiating an abstract schema**. The abstract schema is a representation of the relations shared by the analog and the subject. When an analogy is used to construct a complex theory, a whole system of causal relations may be provided by the analog for use in the subject field. Such a method of invoking an abstract theory type will be discussed in the next section. Gentner's claim that good analogies are based on shared relations, not properties, has been criticized by Turner (1988, p.19), who enumerated instances of good analogies based on shared properties. In sum, analyses of how relevant features (either properties or relations) aid retrieval is still an active area of research and disagreement.

After an analog is retrieved, then a mapping must be constructed between the objects, properties, and relations (or some subset thereof) of the analog and the subject. Hesse (1966), for example, discussed the relations between the wave theory of sound and that of light. The causally related properties of sound are expected to map to similar properties of light. Unexplored areas of potential similarity (called by Hesse "neutral analogy") provide areas for further development of the theory, such as whether light moves in a medium. Selecting causally related properties within the analog as candidates for mapping is a way of attempting to ensure that a good mapping will be constructed. A question that always arises for analogical reasoning is the following: if the analogy and subject share some features but not others, what evidence is there that any given additional feature will be a shared one? A good mapping is one that provides some evidence that the mapped (hypothesized) property or relation should be expected in the subject; in other words, a good mapping provides plausible hypotheses about properties or relations in the subject area. If a set of causal relations exists in the analog, then a similar set may be expected in the subject. Additional methods, besides restricting mapping to causally related properties, to constrain and focus mapping have been explored in computational studies (see, e.g., Holyoak and Thagard, 1989). The various methods aim at answering one of the key questions in analogical reasoning: how can useful transfers be made between the analog and subject?

After an analog is retrieved, in order to find a useful mapping, some elaboration of the analog may be necessary. Consider the case of the analogy between beads on a string and genes along the chromosome (discussed in Chapter 10). An idea about the analog is elaborated—a knot in one part of the string interferes with the formation of another knot nearby. This idea is transformed into the hypothesis that one cross-over interferes with the formation of another cross-over nearby on the chromosome. The hypothesis of interference, thus, could have been formed by elaboration, mapping, and transformation.

After a hypothesis is formed using an analogy, it is in need of justification. Philosophers have disagreed as to whether using analogy provides any justification for a new hypothesis thus formed (Schaffner, 1974a). My view is that analogy provides a weak form of plausibility for the new hypothesis (Darden, 1976). If a hypothesis is constructed by analogy to an already known (kind of) entity or process, then the hypothesis is more plausible than a hypothesis that postulates the existence of something entirely new. Such plausibility does not, of course, guarantee correctness. Theory assessment within the domain of the hypothesis must still be made. To extend Hesse's example with sound and light waves, it was initially more plausible to suggest that light traveled in a medium than to propose that it consisted of mediumless waves. Only with repeated failure to justify the hypothesis of the purported medium, called the "ether," did nineteenth-century physicists abandon this plausible hypothesis and move to postulate a new kind of phenomenon, electromagnetic waves (Oppenheimer, 1956). Elsewhere, I called this strategy **reasoning by a failed part of an analogy**. Such a strategy is a very creative way to use an analogy, but it produces a hypothesis that is less plausible initially than one formed by analogy to something that is familiar. Thus, hypotheses formed by a nonfailing analogy may gain a weak measure of plausibility from that method of construction, but they must also be evaluated using the usual strategies of theory assessment, to be discussed in more detail later. If such assessment fails, then the initially plausible hypothesis will have to be modified or abandoned. One modification strategy is to use only a part of a failed analogy, if some parts may be plausibly severed from others.

One final method of hypothesis construction by analogy will be discussed. If a single analog does not suffice, two or more may be used in a piecemeal way. I call this strategy **reasoning piecemeal from more than one analog** (Darden, 1982a). Retrieving, elaborating, and mapping from more than one analog introduces additional complications in the theory construction process. Especially important is the need to ensure that the various components of a theory, constructed by using different analogs, are consistent with each other; also, they must function together to produce a systematic theory. Burstein (1986; 1988) has discussed the use of two or more analogies and some of the problems of integration that such use entails. He proposes that one analogy may be used to "debug" faulty hypotheses generated by using another analogy. Such use of multiple analogies in anomaly resolution is a relatively unexplored area of research and may be a fruitful area for further work in philosophy of science.

The strategy of **using an analogy** did not figure prominently in the analysis of strategies from the gene case. De Vries (1900) did draw explicit analogies between his pangens and chemical molecules, but this analogy played little role in the specific development of components of the theory of the gene. Although no single analog provided a set of causal relations that was imported into genetics, several analog models of genes and their relations were proposed by geneticists, such as beads on a string (see Chapter 10). The beads on a string model provided a graphic representation of the interrelation between genes and chromosomes. That model shows a relationship between two strategies: the strategy of **using interrelations** and **using an analog model** to depict the nature of the interrelation.

The analog model of intertwining threads provided a "theory-constitutive" metaphor, in Boyd's (1979) sense, as discussed in Chapter 10. The theoretical terms of "crossing-over," "double crossing-over," and "interference" became constituent parts of the theory of the gene. Those new terms gained meaning from the visual depictions of those processes in the analog model: strings crossed-over each other and one knot interfered with the formation of another. Thus, even when an analogy does not provide a whole system of causal relations to structure the entire theory, some components of the theory may be produced by **using an analog model**.

In summary, the strategy of using an analogy can be a powerful one for constructing a new theory, if one or more good analogs are available to be used in elaboration, mapping, and transformation. The new hypotheses thus formed must be assessed in the subject domain and may have to be improved in the light of anomalies. If no analogy provides the overall structure for the entire theory, parts of analogs may be used in constructing individual components of the theory, either during original theory construction or during anomaly resolution.

15.1.2 Invoke a Theory Type

Invoking a theory type is closely connected to **reasoning by analogy**. Consider the idea that two analogs share a common abstract structure (Genesereth, 1980; Gentner, 1983; Darden, 1982a; 1987). The common abstract structure shared by the solar system and a similar model of the atom, for example, is that of a central object with other objects moving in closed paths (circles or ellipses) around the central one. In reasoning directly from an analog, the abstraction might never be constructed. In other words, someone reasoning about the structure of the atom, based on the solar system, might reason that the nucleus is like the sun, without ever explicitly constructing an abstract pattern with a "central object" labeled. If an abstract structure is constructed to show what two analogs share, however, then an explicit abstraction has been extracted from the detailed instances. Abstractions can be at various levels of detail; for example, an abstraction of the structure of the solar system might specify that the objects "move in elliptical paths," or more abstractly, "move in closed-curve paths." Once such an abstract structure of a type of theory has been formed, it can then play a role in additional instances of theory construction without invoking the detailed analogs (Darden, 1982a, 1987; Darden and Cain, 1989).

Invoking an abstraction of a theory type is a means of getting an overall structure for constructing a new theory. One of the best examples in biology of a recurring type of theory is a selection theory. Once Darwin had constructed his theory of natural selection, it became a theory available for abstraction and use in other instances of theory construction. Another selection theory is the clonal selection theory of antibody formation (Burnet, 1957; Darden and Cain, 1989). Instantiating a type of theory in a new domain is potentially a powerful method for new theory construction; the abstract structure of the entire theory is supplied at once. A realistic interpretation of types of theories is that they abstractly represent types of relations or processes common in the natural world. Selection is such a process.

Shapere (1974b) discussed two types of theories—compositional and evolutionary. Compositional theories explain a theoretical problem posed by a domain by postulating relations among constituent parts of the individuals making up the domain. Evolutionary theories solve a theoretical problem by postulating historical changes in individuals making up the domain. Examples of compositional theories include Daltonian atomism and the periodic table of chemical elements. Examples of evolutionary theories are theories of stellar evolution and theories of the evolution of the chemical elements (Shapere, 1974b, p. 534).

Shapere analyzed the way patterns of relations among domain items may suggest pursuing a given theory. Discrete, periodic order among domain items (such as the periodic table shows) suggests searching for a compositional theory. Order in the domain showing a factor that decreases or increases (especially if interpreted as having a temporal direction) may direct the search for an evolutionary explanation. He focused more on how the nature of the data could point to the plausibility of pursuing a type of theory than on detailing the abstract structure of compositional or evolutionary theories.

Shapere extended his analysis of compositional theories to argue that the theory of the gene was an example of a discrete compositional theory (Shapere, 1974a, p. 197). Shapere correctly claimed that some similarity exists between atomism and Mendelism. Both had laws with discrete numerical values that were eventually explained by postulating discrete particles. Genes were postulated as discrete, hidden, causal factors to explain the ratios of characters in Mendel's laws. Yet, there are also differences between the cases. Geneticists explained the Mendelian ratios, at the outset, not by appeal to material particles but by postulating different types of germ cells. The hypothesis that genes were discrete material particles emerged as the field developed, as discussed in Chapter 11. It was a controversial claim. To explain Mendel's laws, some difference between types of discrete *germ cells* was needed, but other hypotheses were considered that did not include different *material particles* within different germ cells. Thus, the generally accepted theoretical components of Mendelism in its early days did not include a hypothesis about discrete material hereditary particles.

An even more important disanalogy exists between atoms, which *compose* samples of elements, and genes, which *cause* characters. Atoms do not *cause* elements; atoms are *constituent parts* of samples of elements. In contrast, genes do not *compose* characters; they *cause* characters. Shapere did not adequately distinguish the relations of "part of" and "cause." Either may be the relation between postulated hidden entities and the individuals whose behavior they explain. (See Darden and Rada, 1988a for more discussion of these different types of relations). The theory of the gene is a causal theory, but it is not a compositional theory. The theory of the gene postulated hidden causal factors, not hidden components of characters. Thus, I am arguing that the theory of the gene is not a compositional theory and its construction does not follow the pattern of compositional reasoning discussed by Shapere for the atomic case (Shapere, 1974b, p. 541).

(As an aside, it is worth noting that, in contrast to the theory of the gene, the chromosome theory of Mendelian heredity is a part-whole theory. Genes are parts of chromosomes. The formation of the chromosome theory, however, was not the subject of Shapere's (1974a) discussion. Also, laws about chromosomes with integral values did not point to the need for discrete underlying components of chromosomes, an important step in the pattern of compositional reasoning in Shapere's analysis. The postulation of a part-whole relation resulted from similarity of properties of chromosomes and Mendelian factors, as discussed in Chapters 7, 9, and 10. Therefore, although the chromosome theory may be considered a compositional theory, it does not follow the pattern of reasoning that Shapere discussed for the atomic case, in which laws about elements with integral combining ratios made plausible the postulation of underlying discrete atoms composing the elements.)

Therefore, neither the theory of the gene nor the chromosome theory seems to be a candidate for theories that could have been constructed by **invoking a theory type**, in the sense of having a single abstract structure representing a typical process. Segregation, independent assortment, linkage, and crossing-over are not clear examples of components of some sort of type of abstract structure. Nor does it seem promising to try to form such an abstract structure based on the theory of the gene in order to make it available for future instances of theory construction; those processes seem unlikely to be types that recur throughout nature.

However, the idea of a type of hypothesis, construed somewhat differently, was exemplified in several instances of anomaly resolution discussed in previous chapters. The sense of "type of hypothesis" is not that of an overall theory type; instead, it is an idea that some vague type of hypothesis might serve to resolve an anomaly. For example, when hypotheses to account for the 2:1 anomaly were proposed (see Chapters 8 and 12), the anomaly was localized in one of several steps in the normal process. Next, to account for the fault, the opposite of the normal step could be suggested as a hypothesis of a given type. One step was the random fertilization of germ cells; thus some hypothesis of a "nonrandom type" was plausible. Cuénot's hypothesis of selective fertilization (germ cells with the dominant genes were postulated not to combine) was an instance of such a nonrandom type. This sense of a type of hypothesis does not provide an elaborate structure for all the components of a new theory, but it indicates that an abstract, or vague, new idea can function in the formation of a new hypothesis during anomaly resolution. The hypothesis, once constructed, becomes one of several components within a larger theory. Thus, one way of resolving an anomaly is to localize one or more components that may be failing and consider what type of component could replace it to resolve the anomaly. The generation of the type is guided by the nature of the problematic component and by the nature of the anomaly, as discussed in Chapters 8 and 12.

It would probably be preferable to have two different terms to distinguish the ideas of, first, a structure of a "global type" of theory such as selection theories, and, second, the much more local sense of a vague idea about a "type of hypothesis" that may figure in a step of anomaly resolution to change a component of a "larger" theory. (Perhaps the latter could be called a type of "component modifying hypothesis.") The larger theory may or may not be an instance of a global type; I argued earlier that the theory of the gene is not.

The strategy of **invoking an abstract type of theory** is a powerful method for theory construction, when a structure of an appropriate type is available. The entire structure of the whole theory is provided by the abstraction, and the task becomes to fill in the variables with specific details from the case. Hanson (1961) discussed reasoning to a type of hypothesis in his discussion of retroductive or abductive reasoning; he suggested that inverse square hypotheses were one example of a type. The idea of instantiating an abstract schema is a powerful one in computational approaches to theory formation because the overall structure of the theory is supplied by the schema. (For more discussion of theories as schemas, see Kitcher, 1981; Thagard, 1988.) The history of science can provide evidence about types of theories; a promising research program is to extract the abstract structures from those theories to make them available for future instances of theory formation. A catalogue of abstract theory types used, historically, in science would constitute what I have called "compiled hindsight" (Darden, 1987; Darden and Cain, 1989). Holton (1973) suggested various "themes" that have recurred in the development of theories, such as constancy and change, discontinuity and continuity, mechanism versus teleology. It would be interesting to see how such recurring themes relate to abstract types of theories.

Such a research program raises numerous questions: How many types exist? Can they be classified? What kinds of problems point to what types (as, for example, the problem of explaining adaptation points to selection theories)? At what level of abstraction should the types be represented? Those questions cannot be pursued here. Although it was not a method exemplified in the gene case, **invoking a theory type** is a powerful strategy when it can be used, because it provides an overall structure for a theory.

15.1.3 Use Interrelations

Another powerful strategy for theory construction is **postulating specific interrelations** between two bodies of knowledge produced by different techniques (discussed in Chapters 7, 9, and 10). Postulating an interrelation, in the technical way I am using the term here, is not to postulate some sort of similarity between two bodies of knowledge, such as an analogical relation, nor is it to borrow a technique from one field for use in another. Instead, it is to postulate a much more specific kind of interrelation that may be characterized as an "ontological" relation between entities or processes investigated by two fields or by different techniques. Examples of such interrelations are identity, part-whole, causal, and structure-function. One might call these "physical relations" between entities or processes, although when relations are postulated between cognitive and neurobiological entities that term may be less apt (see Savitt, 1979; Bechtel, 1988). Another possible label for such relations is "world structural relations," to indicate that they refer to relations in the world (Jerrold Levinson, personal communication). Analogical relations are not "physical" or "in the world." The term "interfield theory" may be applied to a postulated interrelation if the two bodies of knowledge can be clearly distinguished as being part of two separate scientific fields and the interrelation is sufficiently developed to be called a "theory" (Darden and Maull, 1977). An interfield theory, or more generally, the claim that a specific interrelation obtains, is itself a scientific hypothesis, which evidence can serve to support or falsify.

The key idea in using the strategy of interrelations in theory construction is to play one idea off against another, using each to inform the other. For example, examining information about the purported structures that carry out some function may be a way of forming new hypotheses about constraints on functions; alternatively, such a structure-function interrelation may direct the search for particular kinds of structures.

Hypotheses formed by **using interrelations** between accepted knowledge in two fields are more plausible at the outset than those generated by **reasoning by analogy** (Darden, 1980). When, for example, genes were postulated to be parts of chromosomes, then the relation of part-whole served in the formation of a number of hypotheses about both genes and chromosomes, as discussed in Chapters 7, 9, and 10. With an analogy, some properties of an analog map to the subject area and some do not. For any hypothesis formed using an analogy, uncertainty exists as to whether it was formed using a property that should be mapped. With interrelations such as part-whole, however, one can form more plausible hypotheses about which aspects of the part will have some realization in the whole. If some of the parts in a set of wholes are assorting randomly, for example, then the wholes may also be randomly assorting (given that the parts do not separate from their wholes). This principle was applied in the prediction that maternal and paternal chromosome sets randomly assorted, given random assortment of Mendelian factors. It would be interesting to know what general mappings would be expected, given a type of interrelation. Are there general expectations, for example, about what will map between structurally characterized entities and their functions?

The extent to which interrelations are useful in the development of hypotheses in each field depends on the state of development of knowledge in the two fields at the time the interrelation is made. If neither field is well-developed, then the interrelation may not be very fruitful in hypothesis formation. If one field has a theory that has been extensively developed and is well-supported by evidence, then it may provide most of the information to be transferred. This kind of case is similar to analogical reasoning when the analog is well understood and does not change as a result of being used in the analogy. Yet, Black's (1962) interaction view of analogy suggests that analogs are changed too, by being put into the analogical relation.

Another possibility, the one illustrated by the chromosome theory of Mendelian heredity, is that both fields (genetics and cytology) had developed well-supported bodies of knowledge when the interrelation was made. In the chromosome theory case, mapping went both ways, that is, hypotheses were formed in each field, based on information from the other, as discussed in Chapter 7.

When an interrelation to another field is used in hypothesis formation, one of two possibilities may obtain. First, a known phenomenon in one field may be used as the basis for a hypothesis in the other. As discussed in Chapter 9, Morgan might have used visual intertwinings between homologous chromosomes in his formation of the hypothesis of crossing-over (although breaking and rejoining had not yet been observed). Second, an unknown phenomenon in one field may be predicted to exist, based on the presumed interrelation. As discussed in Chapter 9, Bateson predicted selective reduplication of germ cells, a phenomenon with no cytological basis, to account for the coupling anomalies. At the outset, Morgan's hypothesis was more plausible than Bateson's because Morgan's was based on an observed phenomenon in cells, not on an as yet unknown phenomenon. Had the unexpected phenomenon of selective reduplication of germ cells been discovered, however, it would have provided strong evidence for Bateson's reduplication hypothesis. Thus, there is a tension between the strategy of **using interrelations to known phenomena** to produce a more plausible hypothesis and a strategy of theory assessment that gives preference to theories that **make novel predictions**.

Postulating an interrelation between two well-supported bodies of scientific knowledge not only aids hypothesis formation, but, if the interfield theory is supported by evidence, it serves to unify the two fields. If the two retain their own integrity, such as genetics and cell studies did, then perhaps the term "unify" is too strong. They were not united in the sense of completely merging; nonetheless, how the two were related was specified. Perhaps a better term for the product of successful interrelations is "integration" (Bechtel, personal communication). Thus, successful interfield theories play a role in the integration of scientific knowledge, in the sense of specifying kinds of interrelations between entities and processes postulated by the two fields. (For additional discussion of the advantages of forming interfield theories, see Darden and Maull, 1977; Bechtel, 1974; 1988. For an argument that interfield theories, when available, are better sources of hypotheses than analogies, see Darden, 1980; Staats, 1983. For discussion of interfield theories as an alternative to reduction for unifying science, see Maull, 1977; Bechtel, 1986; 1988.)

In the gene case, we have seen that the strategy of **using interrelations** provided major components of the theory of the gene (the components of linkage and crossing-over), established a new interfield theory (the chromosome theory), aided in the development of the analog model of beads on a string, and served to integrate genetics and cytology. Attempts to relate genetics to fields other than cytology were sometimes less successful in the period before 1930, as discussed in Chapter 13. A promising topic for further work is to explore the contrasts between successful and unsuccessful interfield relations. Can any general conditions for expecting success at a given historical period be found? Does a failed interrelation, such as that between genetics and colloidal biochemistry, point to the need for a reconceptualization of one or the other field? (See Bechtel, 1984, for more on reconceptualizations in interfield relations.)

15.1.4 Move to Another Level of Organization

If phenomena to be explained can be put into a hierarchy, a way of producing new ideas (in order to explain the phenomena) is to form hypotheses about the behavior of entities and processes at a different level of organization. If other fields have studied that level, then the interlevel relation may also be an interfield relation. Thus, the **using interrelations** strategy discussed in the previous section includes **using an interlevel relation** when a body of knowledge exists at the appropriate other level. If no other appropriate level is known to exist, however, then the strategy **move to a new level** is less like the interrelations strategy; the latter postulates a relation between *known* information. Phenomena may point to the existence of an as yet unexplored level, often at a lower level of organization.

Exactly how to characterize a "level" may be problematic. For structural levels, size may be a good way to distinguish and order different levels; smaller sized structures are at lower levels. How to decompose the higher level structure into its component parts may not be a straightforward procedure, however. Even more difficult is the problem of finding criteria to clearly identify hierarchies of functional levels (for more discussion, see Wimsatt, 1976; Kaufmann, 1976; Richardson, 1982; Grene, 1987). Burian (1987) suggested that the level to which an entity or process belongs depends on its place in the causal order—what sorts of entities it interacts with, what sorts of forces it is subject to, and what sorts of perturbations it can withstand. The perturbation question is especially important in distinguishing levels: is there a major relevant class of perturbations that disaggregate or disrupt one set of entities but not another?

The relations between genes, characters, and chromosomes are complex, when viewed from the issue of what levels of organization are being postulated. For genetics, the formation of the chromosome theory might be viewed as moving the level upward, from the small unobservable genes to larger entities of which they are a part, namely, observable chromosomes. Or, the chromosome theory might be claimed to have moved the level downward, from observable characters to microscopic chromosomes. As I argued in Chapter 7, the chromosome theory is not obviously an "interlevel" theory in the sense of connecting two bodies of knowledge about two different structural levels of organization.

It is interesting to ask what kinds of problems suggest the use of the strategy **move to another level**. The phenomenon of reversion to grandparental characters suggested to both Darwin and Mendel that something, below the level of observation, must be postulated as present in the parental generation. Darwin proposed unobservable, dormant gemmules; Mendel proposed something that was "recessive" and that determined a germ cell of a given type. The appearance, disappearance, and reappearance of a character suggested a hidden causal factor. Filling in a causal sketch of a process may necessitate moving to another level of organization to find entities that can play the necessary causal roles.

In previous chapters, I proposed the (metascientific) hypothesis that certain problems get solved at certain levels of organization and could not have been solved before that level was discovered. The problem of heredity (to explain the cause of the resemblance of offspring to parents) could not have been solved prior to the discovery of cells. As previously discussed, Mendel's law of segregation was formulated, at the outset, in terms of types of germ cells. The problem of what characterized the difference between different types of germ cells necessitated going to a level below that of the cell, to something that accounted for the differences between the types of germ cells. Thus, the general problem of heredity required for its solution, first, the level of germ cells, and, then, the level of the macromolecule. Similarly, the problem of why genes segregate was solved at the level of the chromosomes; homologous maternal and paternal chromosomes separate during germ cell formation, carrying the genes with them. No lower level explanation was needed to explain the segregation of genes. Of course, cytologists could still inquire about the mechanisms of pairing and separation, and molecular biologists are exploring the molecular mechanisms of breaking and rejoining in chromosomal recombination. One could also ask for an evolutionary explanation of the origin of segregation. But those were additional questions beyond that of asking for a mechanistic explanation of why genes segregate. It is plausible to argue that the chromosome theory provided an adequate mechanistic explanation of Mendelian segregation.

In contrast to segregation, the problem of the physical nature of the genes was not solved at the chromosomal level, nor was the colloidal level the correct chemical level for solving the problem, as discussed in Chapter 13. Solving the problem of the physical nature of the gene required the level of the macromolecule, which was not identified until the 1930s (Olby, 1979; Kay, 1986). An issue in need of more analysis in philosophy of science is exactly what accounts for the judgment that a problem, such as explaining why genes segregate, has been solved at a certain level of organization.

Often problems are solved by the postulation of entities at smaller levels of organization. Examples include the postulation of enzymes to explain cell-free fermentation, the postulation of the viruses to explain diseases caused by agents that could pass through filters that trapped bacteria, and the postulation of macromolecules to explain the molecular weight values produced by the ultracentrifuge (Darden, 1986). However, sometimes the new levels are higher, such as Burnet's (1957) claim that cells form clones during the antibody response, rather than the antibody molecules themselves reproducing (as Jerne, 1955, had proposed; discussed in Darden and Cain, 1989). Another example of an upward move is Stanley's (1975) controversial claim that species selection, at a higher level of organization than the natural selection of adapted individual organisms, is needed to explain certain trends in the paleontological record.

Sometimes a new level is sandwiched in between two previously known levels. The linkage group introduced an intermediate level that grouped genes that were part of the same chromosome. It was a level of organization in between independent genes and all the genes in an organism. As we have seen, "linkage group" was one of the theoretical terms used by Morgan in stating the theory of the gene in 1926 (Morgan, 1926, p. 25; discussed in Chapters 4, 9, and 10).

Part of the task of theory formation in science is to learn to "cut nature at its joints" (Boyd, 1979; Plato, *Phaedrus*, 265e, p. 511). Finding new levels of organization, such as the linkage group, is an important part of that task. **Moving to another level of organization** can be a powerful strategy for theory formation. Using information from one level to construct hypotheses about another is potentially a fruitful way of using old ideas to guide the formation of new ideas.

15.1.5 Introdce and Manipulate a Symbolic Representation

The strategy of **introduce and manipulate a symbolic representation** actually consists of numerous different strategies. I introduced this vaguely named strategy in Chapter 3 in the context of trying to imagine a hypothetical path to the discovery of segregation. A key step in explaining the 3:1 ratios was the step of introducing the algebraic, symbolic representation in terms of $(A + a)(A + a) = 1AA + 2Aa + 1aa$. Once that move is made, then the symbols A and a can be manipulated. The question can be asked—what might they represent?

One might designate the use of mathematical symbolism as the use of a formal analogy, but such a designation does not distinguish between **using an analogy** and the more general strategy of **manipulating a symbolic representation**. Any use of a model falls under this general strategy, either mental models, diagrammatic representations, scale models, computer simulations, or formal systems of equations. The important feature uniting these is that they all stand in a relation of representation to the natural system being investigated. Furthermore, the activity of manipulation is a part of using this strategy. The symbolic representation serves as a substitute for the natural system. Numerous kinds of manipulations of the representation are possible: it can be changed in one way and then changed in another; it can be put into states or contexts that may be impossible to do experimentally with the natural system; it can serve as an aid to visualization; it can be played with to see what it will do. Additional components of the representation may be introduced because of some internal requirement of the representation itself, and, then, the question can be asked—does this new component actually represent something in the natural system? Eventually Mendel's symbols came to represent segregating and recombining genes. Another example was provided by Watson (1968) in *The Double Helix*. He recounts his manipulation of cut-out models of molecules in the discovery of the base-pairing in DNA. Manipulation and visual inspection of the geometric models was an essential step in the discovery. Thus, **introducing and manipulating a symbolic representation** is a strategy that can lead to new theoretical ideas.

In Mendelian genetics, manipulating the symbols, often represented in diagrams or tables (such as Punnett's square), became the way to make predictions. The physical nature of the entities that the symbols represented could be left as a black box, something to be filled in later, even while the symbols aided in making successful predictions for ratios of characters. Not only can the symbols be manipulated to yield predictions, they can function in hypothesis formation. I manipulated the symbolic notation in generating alternative hypotheses in the face of anomalies, as discussed in Chapters 8 and 9. The numbers of germ cells of A or a type could be adjusted to produce the anomalous ratios. Thus, **manipulating a symbol system representing hypothetical entities** is a way of forming new hypotheses about numbers and behaviors of those entities in the light of anomalies.

Numerous questions about the use of this strategy can be posed: How are appropriate symbols chosen? How does the choice of one symbol system rather than another affect hypothesis formation? Does manipulation of symbols, freed from representing specific data points, aid in the move from data to theory? The issues raised by this strategy are too numerous for this brief discussion to do them justice. The literature on mental models, simulations, scale models, and other forms of representation is a rich source for ideas about this strategy for hypothesis construction.

15.1.6 Introduce a Simplifying Assumption, Then Complicate

Using the strategy of **introducing a simplifying assumption** is a way to begin theory construction. The example discussed in the gene case was the introduction of the assumption that one hereditary unit caused one character, which I called the "unit-character" concept. A similar simplifying assumption was introduced in chemistry. Dalton (1808) assumed a simple relation between atoms and elements: in compounds, assume one atom for each combining element. Thus, water molecules were considered to consist of one atom of oxygen and one atom of hydrogen. In both cases, no access to the theoretical entities was available, other than by inferences from the data. The data in the two cases consisted of data about ratios of characters in genetic crosses and data about combining proportions of elements in chemical reactions. In both cases, it was a plausible assumption that underlying units were present and that their numbers and behavior could be inferred from the data. The simplest hypothesis was proposed in each case: a one-to-one relationship between a theoretical entity (gene, atom) and a differentiable, observable entity (an element, a hereditary character).

As geneticists and chemists gained more facility in doing experiments and making inferences about the underlying theoretical entities, the means for complicating the oversimplification emerged. In genetics, the multiple factor hypothesis, as well as genetic mapping (discussed in Chapters 6 and 10), provided the means of moving beyond the oversimplification of one unit-one character. In chemistry, investigations of combining volumes (along with some changes in theoretical assumptions about the nature of gases) provided new methods for determining the proportions of atoms in molecules. Then, data were collected relevant to determining the numbers of atoms of each element present: water molecules have two hydrogen atoms and one of oxygen (Greenaway, 1966). Using the strategy of **introducing a simplifying assumption** to begin theory construction thus leads to the expectation that complications will later have to be introduced, either to resolve anomalies or to extend the scope of the domain.

Chapter 6 also discussed **beginning with an overgeneralization and specializing**; that strategy is similar to the one discussed earlier, **oversimplify, then complicate**. Thus, the strategy of **overgeneralize, then specialize**, is not discussed separately here, but will be discussed later as a strategy for anomaly resolution. (For further discussion of simplifications and idealizations, see Shapere, 1984, pp. 318-356; pp. 356-368.)

15.1.7 Begin with a Vague Idea and Successively Refine

A number of recent historians and philosophers of science have suggested that theories may begin as vague ideas that are developed in stages. Discovery is not necessarily an instantaneous process, producing a complete new theory in a blinding flash of insight. Instead, theories may be constructed incrementally, often beginning with vague ideas that are successively refined and modified (Shapere, 1974b; Monk, 1977; Boyd, 1979; Gutting, 1980; Nickles, 1987d; Kohn, 1980; Holmes, 1985). This book has extended the view of incremental theory development, from that of a single individual incrementally developing a theory, to development of a theory by groups of geneticists over a thirty-year period.

As discussed earlier, the early stages of the formation of the theoretical components of Mendelism were marked with oversimplifications, promissory notes, incompleteness, and

1. Internally consistent and nontautologous
2. Systematicity vs. modularity
3. Clarity
4. Explanatory adequacy
5. Predictive adequacy
6. Scope and generality
7. Lack of ad hocness
8. Extendability and fruitfulness
9. Relations with other accepted theories
10. Metaphysical and methodological constraints
11. Relation to rivals

Table 15-2. Strategies for theory assessment.

applicability to a small domain. An area of concrete achievement (explaining 3:1 ratios) provided the basis for further developments. Vagueness in the details about the theoretical entities did not hamper the early development of the theory, in part because precise predictions could be made about observable characters.

If the vague idea involves the postulation of a new kind of theoretical entity, then there are different ways it can be vague. As discussed in Chapter 11, a frame-structured representation of the properties of the entity can be constructed with specific kinds of properties not yet filled in. For example, what kind of entity the gene was—a material particle or something else—was not known for many years. (AI researchers would call this: not knowing in what "a kind of" hierarchy—or, ako hierarchy—to place the new entity.) Postulating a hidden causal factor, whose nature is as yet unknown, is a frequent beginning point for theory construction. Another form of vagueness is found when the kind of thing (e.g., a material particle) is postulated, but very few properties are ascribed to it at a given stage. Having few specific values for types of properties is a form of vagueness that differs from the vagueness of not knowing what kind of entity the new theoretical entity is.

The use of the strategies of **beginning with vague ideas and promissory notes** in constructing a theory is not problematic, if methods for making and testing precise predictions are available. Anomalies are likely to be found when the predictions are tested. Then, refinements to the theoretical components can be made. In assessing a theory at a given time, judgments have to be made about which problems of incompleteness are serious failings of the theory and which, on the other hand, can be successfully shelved pending further work (possibly at another level of organization). As a strategy for producing new ideas, however, it may be a good strategy to **begin with a vague idea and successively refine it.** Then, strategies for anomaly resolution and expansion of scope become particularly important in constructing an adequate theory, as will be discussed later.

15.2 Strategies for Theory Assessment

In this book, I have explored the view that theory construction and theory assessment are intimately tied. I have been especially concerned with the role that assessment strategies can play in constraining or directing the search for new ideas. In addition, I have focused on the role that assessments can play in generating anomalies, whose resolution leads to theory development. Intimate relations exist between strategies in theory formation and strategies in theory assessment. My list of assessment strategies in Table 15-2 has been constructed with these concerns in mind. It is similar to the list quoted from Newton-Smith (1981) in Chapter 2. Our task now is to discuss these strategies of theory assessment, especially the roles they can play in theory construction.

Just as it is difficult to find verbal statements of theories by scientists, it is often difficult to find them appealing to specific criteria of theory assessment, with the notable exceptions of explanatory and predictive adequacy. Sometimes appeals to generality, lack of ad hocness, or simplicity are present. At other times, however, the criteria being used in an assessment must be inferred from the specific arguments given in favor of or against a hypothesis. (Additionally, it is possible to assess a historical theory in the light of considerations relevant at its time, but using criteria that no historical person seems to have used.) Several previous chapters discussed plausible inferences about the assessment criteria used in episodes of hypothesis evaluation (or that could have been used). Chapter 14 summarized those assessment strategies. This chapter now provides a more systematic discussion of criteria of theory assessment.

The criteria for assessing a theory, listed in Table 15-2, are arranged in an order. The list begins with criteria that can be applied before or independently of empirical testing; next are issues about empirical adequacy, followed by questions about the theory's future prospects. The list concludes with issues about metaphysical and methodological constraints, and with issues about relations to other theories (both rivals and those with which it does not compete but must be compatible). This order is not necessarily the order in which criteria are or should be applied to theories, especially if the criteria play roles in strategies used during hypothesis construction; it is merely a convenient one for grouping related types of criteria.

These criteria should not be viewed as specifying necessary and sufficient conditions for being a good scientific theory. My task here is not to develop such a list or to argue for its adequacy. My task is to develop a list of some criteria that can function in theory change, both as constraints in theory construction and as evaluative criteria. Thus, these "criteria" for theory assessment I also call "strategies" for theory assessment, because how they can be and should be used in any given problem-solving episode is an open question. The outcome of applying one criterion may conflict with the outcome of applying another. It is interesting to consider how decisions to weight one criterion more than another would affect both theory construction and theory choice.

15.2.1 Internally Consistent and Nontautologous

An infrequently questioned criterion is that theories be **self-consistent** (i.e., contain no internal contradictions among their components). Further, a rather minimal demand is that the theory **not be a tautology** (i.e, a trivially true statement with no empirical consequences). These criteria have functioned in some historical cases. Consistency, for example, arose as an issue for the phlogiston theory (Conant, ed., 1948), which may have been inconsistent in asserting that phlogiston sometimes had positive weight and at other times had negative weight (but weight was not an important property within the framework of the phlogiston theory, so whether this was an *internal* inconsistency of the theory is not unproblematic). During theory construction, the requirement of internal consistency may not be constantly at the forefront of attention. A scientist may entertain inconsistent hypotheses because insufficient attention has been focused on all components of a complex theory (Holmes, 1985). Working out the logical relations between components may require some period of time. And it may even be useful to consider generating hypotheses inconsistent with some other component; maybe the other component is the problematic one. Even though this criterion may be relaxed at some stages of theory construction, internal consistency seems likely to be a requirement of a good theory.

That a good scientific theory be nontautologous, that is, that it have empirical consequences, is likely to be demanded of good theories. It is a criterion that has figured in some scientific debates. The charge of tautologousness is still raised against Darwin's theory of natural selection when it is stated using the oversimplified phrase "survival of the fittest." If no definition of fitness can be given, other than as those that survive, then the phrase becomes the tautology "those that survive are those that survive," which has no empirical content (see Sober, 1984, pp. 63-85). Various satisfactory answers have been given to this charge, including the reply that Darwin's theory is much more than just "survival of the fittest" and the reply that "fittest" is to be characterized in terms of "engineering fitness," that is, how good an organism's traits are for living in a given environment (Gould, 1977).

It is likely that, in constructing a theory, **internal consistency** and **lack of tautologousness** will be constraints to introduce early and maintain except in very occasional circumstances.

15.2.2 Systematicity versus Modularity

The systematic interconnections of the various components of the theory are related to the criterion (to be discussed later) of **lack of ad hocness** of a given component. If a component is "tacked on" to the theory, without having systematic connections to other components, then the theory as a whole is less systematic. For example, a theory formulated in terms of hidden genetic factors would be less systematic if another component was added that did not make use of such factors to explain some aspect of heredity. In such a case, satisfying the systematicity requirement would require the construction of a "larger" theory that somehow encompassed both the factor components and the nonfactor component.

Nevertheless, there is a tension between simultaneously insisting that all the components be systematically interconnected and the demand that the components be modular to facilitate anomaly resolution. **Modularity** aids in the process of localizing an anomaly and resolving the anomaly by modifying a single component. **Systematicity** may make such localization difficult because of the systematic interconnections among components: changing one may require changes in another. More will be said about systematicity and modularity later in discussions of other criteria.

Both **systematicity** and **modularity** were found in the theoretical components of Mendelism, as I have stated them for 1900-1903 (in Chapter 5) and 1926 (in Chapter 14). "Genes cause characters" was a component that played a role in all the other components. Because the theory of the gene made use of the same theoretical factors in all its components, it showed systematic connections among its components. Yet, the theory also exhibited modularity in that some components could be removed without affecting the other components. Dominance, for example, was a modular component that was removed as a general component when numerous exceptions were found, with no effect on the other theoretical components (see Chapter 6). The theory of the gene thus showed systematic connections among its components, but some components were modules that could be changed without affecting the others.

15.2.3 Clarity

Theoretical claims should be stated **clearly**, and the nature of theoretical entities and processes specified in detail. Debate occurred regarding the clarity of the claims about Mendelian factors. The factors were postulated entities associated with characters, but few properties were attributed to them in the early stages of Mendelism. Nonetheless, even though only some of its properties were accessible to experimental investigation, the claims about genetic factors were sufficiently clear to allow theory development: conceptual problems emerged, conceptual clarifications were made, and the theory was improved (see Chapter 11).

Different ways concepts and components can be vague were discussed above in Section 15.1.7, in analyzing the strategy of **beginning with a vague idea**. A decision to use the strategy of beginning with a vague idea, thus, would dictate that the criterion of **clarity** should not be imposed too early in the stages of theory development. This intimate relationship between these two strategies is one of the numerous examples of the close relationship between strategies for producing new ideas and strategies for assessing them.

15.2.4 Explanatory Adequacy

A theory is expected to **explain a domain of phenomena**. The philosophy of science literature contains a number of analyses of explanation: explanation is providing a derivation (Hempel, 1965), explanation involves fitting something into a schema or pattern (Schank, 1986; Kitcher, 1989), explanation consists of providing a mechanistic cause. The topic of explanation is too complex to discuss in detail here; but some means of determining when theoretical components adequately account for domain items is needed in order to apply the constraint of **explanatory adequacy** in theory construction and evaluation.

Consideration of explanation in the context of theory construction, rather than merely in theory evaluation, raises important issues. Previous chapters have viewed genetic theory, as well as the domain it explains, in a piecemeal way: somewhat modular theoretical components function in explaining modular domain items. Which items are included in the scope of the domain plays a role in determining which components are needed in a theory to explain them. Exploring the relations between the inclusion or exclusion of domain items and the structure of the resulting theory is a fruitful area for further research.

As we have seen, which phenomena are inside and which are outside the domain to be explained may be a matter of debate. Thus, the strategy of evaluating a theory for its **explanatory adequacy** is closely tied to the strategy of assessing the **scope** of the domain of the theory (to be discussed later). Numerous questions were raised about the scope of the domain of Mendelism. It can be argued that the domain of Mendelism was discovered as the theory changed and expanded. Certainly prior to 1900, there was no clearly delineated domain of phenomena, consisting of ratios in the transmission of characters in hybrids, for which a general theory of heredity was actively sought. The domain for general theories of heredity in the late nineteenth century included embryological development, and sometimes normal growth of organisms, as well as the origin of new species. Mendelism began as an explanation of a very small domain of phenomena, consisting of ratios in hybrid crosses for a few varieties of plants. Its domain of applicability was determined, as additional characters in additional species were tested, and as quantitative, blending, and linked characters were explained by expansions in the theory. Had all those types of phenomena been used to evaluate Mendelism in 1900, it would certainly have failed to be an adequate explanation. In addition, had Mendelian theory been assessed for its promise of being a very general theory for all hereditary phenomena at that stage, it would not have been judged to be a promising theory. As Chapter 13 discussed, some biologists were never willing to accept the theory of the gene because of its lack of promise in explaining embryological development. Thus, the scope of the domain for which the theory is expected to account may continue to be a matter of debate, even after the theory has been successful within a narrow domain.

In Chapter 12, I argued for the view that Mendelian genetics can be viewed as providing exemplary schemas for kinds of Mendelian crosses. Explaining a data point involves fitting it into the schema. The schemas in genetics can be viewed as providing abstract characterizations of genetic mechanisms; when the mechanism is operating properly, the expected output is produced, such as 3:1 ratios. On this view, a set of exemplary schemas constitutes the theory and determines what kinds of data the theory can explain, namely, those data points that fit into a schema. This mechanistic view of explanation as fitting the item to be explained into an abstract schema proves useful in doing anomaly resolution, as Chapter 12 argued. Furthermore, Chapter 12 discussed the way "model" anomalies can be transformed into additional schemas that explain what was previously an anomalous result; this view lends itself to a view of incremental theory development, driven by the need to explain empirical anomalies.

In sum, **explanatory adequacy** provides an important constraint in theory construction: a theory must explain the items in its domain. This strategy is closely tied to the strategy to be discussed later of determining the **scope** of the domain of the theory.

15.2.5 Predictive Adequacy

The ability to make predictions can function as a strong constraint in theory construction, while the ability to make **successful** predictions can function as a strong constraint in theory assessment. The ability of Mendelism to provide precise, quantitative predictions was a major argument in its favor. The confirmation of many predictions figured largely in its success. Furthermore, the resolution of anomalies that conflicted with predictions drove theory change, as seen in previous chapters.

Issues about predictive adequacy have been much discussed in philosophy of science and a lengthy reprise of those debates cannot be given here. **Predictive adequacy** is closely connected to other strategies for theory assessment, especially **explanatory adequacy** and **relation to rivals.** A lively issue of current debate is the relative importance of a theory's explaining an already known fact versus predicting a novel fact. The logical relations between a hypothesis and a fact are the same, regardless of the time when the fact was discovered (i.e., whether it was discovered before or after the hypothesis was constructed). On the other hand, the ability to predict novel facts seems, somehow, to indicate that one theory is better than another theory that can merely explain already known facts. In order to apply the criterion of predicting novel facts, one would have to specify which facts were used in theory construction, and which were only found after the theory was constructed (or were not used by the discover during the theory construction episode). Such a temporal order may or may not be determinable from historical evidence.

Giere (1984) has argued that a good prediction should be derivable from the theory (plus initial conditions), verifiable with current techniques, and improbable in the light of background knowledge. The improbability criterion is another way of trying to capture the intuition that a good theory predicts novel phenomena. As discussed in the section on **using interrelations** in theory construction, however, there is a tension between the strategy of **using interrelations to known phenomena** to construct a more plausible hypothesis and a strategy of theory assessment that gives preference to theories that **make novel or improbable predictions.** Another tension has been pointed out by Brush (1989): the ability of a new theory to explain a recalcitrant anomaly for an old theory may count more in favor of the new theory than the new theory's ability to predict an entirely new kind of phenomenon.

These various tensions once again show the close connection between processes of theory construction and theory evaluation. It would be interesting to construct an experimental system in which different weights could be given to the different criteria so that the outcome of employing one strategy versus another could be measured.

Geneticists spent considerable amounts of time designing experiments to test the predictions of their hypotheses. Crucial experiments were sometimes decisive in choosing among alternative hypotheses (e.g., see Chapter 10). The strategies discussed in this book have been focused on hypothesis formation and revision activities. This case might also be used to investigate **strategies for designing experiments** to discriminate among competing hypotheses, which is an important component of scientific reasoning. Although discussion of experimental strategies is beyond the scope of this book, it should be realized that using the assessment strategy of **predictive adequacy** requires using strategies for designing experiments and interpreting the results. (Cognitive psychologists and AI researchers have investigated reasoning in experimental design; see, e.g., discussion of the MOLGEN system by Friedland and Kedes, 1985.)

An area in need of additional work is exploring disconfirming versus confirming strategies during prediction testing (Gorman and Gorman, 1984). Such strategies can aid in determining the scope of the domain. Furthermore, the heuristic of looking at extreme cases (Lenat, 1976) is another method for determining the theory's scope.

15.2.6 Scope and Generality

Two criteria of theory assessment are closely related: the **scope of the domain** explained by the theory and the **generality of the theory**. Both are closely connected to the criteria of **explanatory** and **predictive adequacy**. The scope of the domain refers to the number of domain items the theory explains (or number of kinds of domain items, if "kinds of" items can be determined; see Thagard, 1988, on "consilience" as a method for measuring breadth of the domain based on the kinds of items explained). It is often argued that theories of a larger scope are to be preferred; in other words, a theory that adequately explains a large domain is to be preferred over one that explains only a subset of that domain. The theory of the gene, for example, has a larger scope if it explains the inheritance of all hereditary characteristics than if it only explains the inheritance of qualitatively different characters (with some other theory required to explain quantitatively varying characters). Generality refers to the claims made by the theoretical components; for example, a theory that claims that all genes independently assort is more general than a theory that claims that some are independent while others are linked.

Although scope and generality are related, they are not the same criterion. This point is best illustrated with a hypothetical example. Suppose two different theories are rivals for explaining the same large domain of data. Theory One contains one theoretical component for every data point in the domain; such a theory has a domain of large scope, but its individual theoretical components are not general components. Theory Two contains only a few theoretical components, each of which is involved in explaining many domain items (and many of the domain items are themselves empirical generalizations covering many data points). The theoretical components in Theory Two are more general than those of Theory One. Theory Two is the more general theory, even though the scope of its entire domain is the same as that of Theory One. (Alternatively, someone might want to argue that the difference between the two should be characterized in terms of **simplicity** rather than generality; Theory Two is simpler because it contains fewer theoretical components. In addition, **modularity versus systematicity** is at issue here. Because of the numerous components it contains, Theory One wins on modularity, but why those independent modules are even considered to be a single scientific theory may be questioned. Theory One seems merely to be a set of ad hoc assumptions. Theory Two clearly wins on systematicity because its theoretical components are more systematically connected with each other.) Theory Two would be judged to be the better theory, in the sense of having theoretical components with greater generality.

Nevertheless, striving for ever more general theoretical components may not always be the best strategy to follow. A different scenario illustrates this point. A theory explains a narrow domain and has a very general component that applies to that entire, but small, domain. By making the general component more specialized and adding an additional component, the theory may be able to explain a larger domain.

The genetics case presented actual examples. The theoretical component that each gene causes one character was a universal generalization; it applied to the domain of all qualitatively different characters (such as yellow and green pea color) that showed Mendelian segregation. But explanation of the inheritance of quantitative characters (such as height in humans) was outside the scope of the domain. By specializing the one gene-one character component and adding the multiple factor component (many genes-one character), the domain of the entire Mendelian theory was expanded to include quantitative characters. That expansion of the scope was gained at the cost of specializing an overly general component and adding another one, as Chapter 6 discussed. Another such example was the component that all genes assort independently (Mendel's second law); it, too, was specialized and another component was added. The new components were (1) all genes in different linkage groups assort independently and (2) genes within a linkage group are linked and sometimes cross over, as discussed in Chapters 9 and 10.

Specialization and addition in these two genetic examples resolved anomalies; the resulting, altered theory explained a larger domain. Thus, the scope of the entire theory can sometimes be increased by making a theoretical component less general (using the strategy of **specializing an overgeneralization**) and adding an additional component. These cases show that there may be a trade-off between, on the one hand, the generality of the individual components of a theory and, on the other hand, the overall generality of the entire theory. The degenerate case of the use of the strategy of **specialize and add**, however, is the case of Theory One, in which every component has been specialized to account for only one data point. Consequently, decisions may have to be made about how special or how general to make components and about the scope of a domain that an entire theory can be expected to cover.

It might be suggested that the scope of a domain may be determined by what a theory's rivals can explain. In other words, if a rival theory can explain an item, then that item must also be included in the domain of the theory in question (Laudan, 1977). Such a relation may apply in some cases, but (as will be discussed in Section 15.2.11) the theory of the gene, as a whole, had no single serious rival. Bateson never succeeded in formulating a theory that challenged the theory of the gene. Various German biologists developed theories of the role of the cytoplasm in inheritance and development (Sapp, 1987), but those theories did not have the predictive success of the theory of the gene. No other theory of transmission genetics ever emerged. Thus, the explanatory scope of the theory of the gene was determined by what its components could account for, but rivals to it did not determine that scope.

15.2.7 Lack of Ad Hocness

If a theoretical component is ad hoc, it plays a role only in accounting for one, or a few, otherwise anomalous domain items; it lacks systematic interconnections to other theoretical components;

and it does not aid in successfully predicting new domain items (Leplin, 1975). If theoretical components can be clearly differentiated, then theories with fewer components are simpler. **Simplicity in the sense of lack of numerous ad hoc hypotheses** is often claimed to be a mark of good theories (see, e.g., Thagard, 1988, pp. 82-86, for discussion of this close relation between **simplicity** and **lack of ad hocness**). Moreover, theories that are easily **extendable** do not require ad hoc additions.

The issue about the ad hoc addition of a new component was explicitly raised in several debates in the gene case. It arose, for example, in the debate between Castle and the *Drosophila* group about the hypotheses of contaminated genes versus modifying genes, as discussed in Chapter 8. Each charged the other with making unnecessary ad hoc additions to Mendelism. At a given stage in theory development, it may not be easy to find good methods for distinguishing between illegitimate ad hoc additions to the theory and good, newly added theoretical components. It may take some time to explore the consequences of adding a new theoretical component, to see if it aids in making new predictions, and to see if it can be systematically connected to other theoretical components. Thus, judgment about ad hocness may require hindsight analysis.

There is a tension between the anomaly resolution strategy of **specialize one component and add another** and the assessment strategies of **avoiding ad hocness** and of **preferring simple theories with few theoretical components**. Trade-offs have to be made among these demands in theory construction.

15.2.8 Extendability and Fruitfulness

Some criteria of theory assessment function in strategies for assessing a theory's *past* successes and can best be applied with hindsight. One such criterion is **explanatory adequacy**, that is, how well the theory has accounted for the items in its domain. Other criteria are more forward-looking, and useful in strategies for assessing the promise of a theory for future work. Nickles (1988) refers to such forward-looking assessments as the "heuristic appraisal" of theories. He is correct that philosophers of science have paid too little attention to this aspect of scientific inquiry. **Extendability** and **fruitfulness** are two criteria that can figure in forward-looking appraisals (although one can also look back and ask how extendable or fruitful a theory has been).

If a theory is **easily extendable**, then it can easily accommodate changes and extensions to explain new domain items. Newton-Smith (1981) called this criterion "smoothness." As discussed in Chapter 9, for example, Bateson's reduplication hypothesis was not easily extendable, whereas the linkage hypothesis was. As soon as it was proposed, one could see that the linkage hypothesis could easily accommodate three-factor crosses, whereas such crosses required very complicated adjustments to the reduplication hypothesis. Thus, linkage was more extendable than was reduplication, and this feature contributed much to its success.

Fruitfulness is a measure of the theory's fertility in suggesting new experiments or new ideas for its further development (Newton-Smith, 1981; McMullin, 1982; 1984). At the beginnings of Mendelism, some biologists quickly saw its fertility for suggesting new interpretations of results from old hybridization experiments and for suggesting new experiments. Bateson immediately recognized it as fruitful. Similarly, the chromosome theory in the hands of the *Drosophila* group after 1910 was very fruitful in making predictions, suggesting the new methodology of coordinating genetic and cytological findings, and suggesting new hypotheses, such as linkage. By the 1930s, the theory of the gene was not especially fruitful. Many of the Mendelian phenomena were by then well understood. Answers to new questions required new research techniques. Distinguishing new, fertile areas from those where much of the work has been done is an important process in choosing research programs to pursue.

As discussed in Chapters 11 and 12, if diagrammatic representations can be made and blanks are indicated, then those are problem areas in need of solutions. The theory (or theories) figuring in such representations may be a fertile source of suggestions for filling in the blanks. Once the blanks have been filled in, the theory is less fruitful in suggesting new problems to be solved.

Further research by philosophers of science is needed to develop good appraisal strategies.

15.2.9 Relations to Other Accepted Theories

The relations between a new theory and other accepted, nonrival theories can be of numerous kinds. Minimally, the new theory does not contradict any claim in any other theory, that is, the new theory is consistent with other accepted theories. Or, a new theory may be faced with an anomaly because it is inconsistent with some other claim. When a successful theory is developed in one field and then proves to be inconsistent with a successful theory in another field, deciding which theory is at fault may not be easy. An example is the conflict between genetic evidence that DNA played a role in heredity and the biochemical view that DNA was a simple repeating polymer, without sufficient complexity to play the role of the genetic material. Eventually new biochemical analyses of DNA, using new techniques, determined that the biochemical hypothesis about its simple structure was incorrect (Olby, 1974; Judson, 1977).

Introducing the constraint that a new theory must be **consistent with other accepted theories** is a strong constraint in theory construction and it is a conservative methodology. A less conservative strategy in theory construction is to allow **explanatory** and/or **predictive adequacy** for one domain to take precedence; only at a later stage introduce the constraint of consistency with other accepted theories. The less conservative strategy would allow inconsistences to be found and the previously accepted theories to be questioned because of interfield conflicts.

Consistency is a weak relation between two bodies of knowledge. Stronger kinds of relations between a new theory and other accepted theories may exist. Rather than as a constraint in theory assessment, another body of knowledge may provide information of use in strategies for producing new ideas. For example, **reasoning by analogy** or **invoking a theory of a given type** in theory construction may result in two bodies of knowledge with similar types of theoretical entities or processes. Insisting on analogies to other theories as a constraint in theory construction is too conservative a strategy, but using it can result in a more plausible theory at the outset, as I argued in the section on reasoning by analogy in theory construction.

Stronger relations can exist between a new theory and other accepted theories than mere consistency or analogy. "Interrelations," in the technical sense in which I am using that term in this book, entail the claim (on a realistic interpretation of theories) that the newly postulated entities or processes are physically related to other entities or processes. For example, the chromosome theory was based on the part-whole interrelation between genes and chromosomes. Consequently, genetics was much more tightly related to cytology than by mere consistency. The chromosome theory served to integrate genetics and cytology by postulating the specific physical relations between entities investigated in the two fields. Unification of science, in the sense of knowledge about the interrelations among entities and processes, is often claimed to be a goal of science. Thus, a theory that is constructed **using interrelations** to another, accepted body of knowledge, will contribute to satisfying this goal. Using the strategy of **interrelations** in theory construction may result in a theory that satisfies the criterion of being **integrated with the other theories** that were used in the interrelation. (Alternatively, postulating an interrelation may necessitate consistency checks between other components in the two bodies of knowledge to ensure no problems with integrating the two.)

Thus, relations to other accepted theories can range from mere consistency, to analogy, to a tight interrelation. A new theory may be constructed with one of these relations as a constraint; if so, then it may already satisfy the criterion that **new theories be compatible with other accepted theories.** Alternatively, a new theory may be constructed without such a constraint and then the strategy of **assessing a new theory in relation to other accepted theories** will be needed.

15.2.10 Metaphysical and Methodological Constraints

The category of metaphysical and methodological constraints is a broad category encompassing many kinds of issues in theory assessment. How prior assumptions function in theory assessment, or in strategies for producing new ideas, is not yet well understood by philosophers of science, but is an active area of research (see, e.g., Brown, 1977; Laudan et al., 1986).

One example of a methodological constraint is the demand that a theory be **experimentally testable.** It became important in early twentieth-century biology (Allen, 1969b). The theory of the gene was a paradigm example of a theory satisfying the constraint of experimental testability. It also satisfied the methodological dictum that theories should **make quantitative predictions,** an argument in its favor stated explicitly by Morgan (1926, p. 25).

The demand that a theory be **simple** (according to some criterion of simplicity) could be either a methodological or a metaphysical constraint in theory construction or assessment. One might advocate searching for simpler theories on the grounds that they will be more easily testable, in which case simplicity is a methodological constraint. On the other hand, one might have a metaphysical assumption that nature is simple and that simple theories are more likely to be true. As discussed earlier, the lack of numerous ad hoc components in a theory may be taken to be a measure of the simplicity of that theory, if components can be clearly individuated. Alternatively, the number of (kinds of) theoretical entities or processes postulated by the theory may also be a measure of its simplicity. Those criteria did figure in the debate between the contamination hypothesis and the modifying factor hypothesis, as discussed in Chapter 8.

Metaphysical or methodological constraints can range from very general considerations to much more specific ones. A general concern might be whether a theory is compatible with a "world view." One prominent example was the mechanistic world view in which theories were sought that postulated mechanisms; the mechanisms were often characterized in terms of the behavior of material particles. Such a general world view can function as a constraint in producing new ideas—for instance, only mechanistic ones will be proposed. Or, strategies for producing new ideas could be less constrained, and mechanism could be imposed as a criterion for choice among the new ideas produced by a less constrained process.

This category of metaphysical and methodological criteria also includes general assumptions that may influence data collection. Philosophers of science have been concerned with the extent to which commitments to prior theories can affect observations (Hanson, 1961; Kuhn, 1962). Such an effect has been called the "theory-ladenness" of observations. The gene case illustrated one way in which a prior assumption affected the kind of data collected by Mendelians. The collection of Mendelian data was based on the unit-character concept, the assumption that organisms could be viewed as composed of independently variable, hereditary characters. The data were certainly "laden" with that concept, as discussed in Chapters 5, 6, and 11. But the use of that concept, or category, in data collection, did not determine what the characters were, nor what their numerical ratios were. As Scheffler argued, categories reflect an advance resolution to individuate, group, and separate data along certain lines. They do not, however, determine any particular assignment of particular data points to those categories (Scheffler, 1967, p. 40). For those who preferred to view the organism as a whole, not individuatable into independent characters, the theory of the gene was unacceptable. The data based on the categorization into unit-characters were irrelevant to the holistic problems posed from that different way of viewing the organism. The holistic challenges to genetics, such as that by Russell (1930) discussed in Chapter 13, were challenges that resulted from a broader perspective in which theories were sought that accounted for the organization of the organism as a whole. The theory of the gene failed to be a holistic theory and thus failed on this (metaphysical?) criterion. The material particles postulated by the theory of the gene to explain the production of independently variable characters did not seem promising as explanatory factors for the organization of the whole organism.

Another metaphysical presupposition is that science should be unified because it is a body of knowledge about a single natural world. Such a unity-of-science assumption may be closely connected to the strategy of **using interrelations** in theory construction and the strategy of assessing the adequacy of theories in **relation to other, accepted theories**. Such a metaphysical assumption provides the grounds for adopting these strategies.

How scientists formulate and choose among very general metaphysical and methodological perspectives and goals is an important area of research in contemporary philosophy of science concerned with understanding scientific change (see, e.g., Shapere, 1984).

15.2.11 Relation to Rivals

The previously mentioned criteria for theory assessment can all be applied to a single theory, without consideration of possible competitors to it; however, an important issue in assessing a theory is whether it has rivals and whether it is better than they are. Also relevant may be whether rivals have been sought and how systematic the search for them has been. One view of what counts as a rival theory is that a rival purports to account for all or some of the same domain; thus, the formation of rivals can aid in determining the scope of the domain for which a single theory should be sought.

The use of "abductive" reasoning is critically dependent on the search for rival hypotheses. Hanson's schema for abductive reasoning was discussed in Chapter 2: abductive reasoning begins with puzzling phenomena; next, the claim is made that the phenomena could be explained by a hypothesis (or a type of hypothesis); the conclusion is that the hypothesis is plausible. In order for the judgment of the plausibility of a hypothesis to be a good one, as Achinstein (1970) argued, alternatives to it, which would also explain the phenomena, must be considered. Many "wild" hypotheses might explain the phenomena.

Abductive reasoning has received attention in AI and computational philosophy of science (see Thagard, 1988, ch. 4, for kinds of abductive reasoning in discovery). Josephson (et al., 1987) equates abductive reasoning and inference to the best explanation. He characterized this form of reasoning as follows:

D is a collection of data (facts, observations, givens),
Hypothesis H explains D (would, if true, explain D),
No other available hypothesis explains D as well as H does.

Therefore, H is probably correct.

The confidence of the conclusion, Josephson continued, depends on a number of factors including: (1) how good H is by itself, independent of considering the alternatives; (2) how decisively it surpasses the alternatives; (3) how thorough the search was for alternative explanations; and (4) pragmatic considerations, including (a) how strong the need is to come to a conclusion at all, especially considering the possibility of seeking further evidence before deciding, and (b) the costs of being wrong and the benefits of being right (Josephson et al., 1987).

This is a good list of factors relevant to assessing rival hypotheses. Factor (1), how good the hypothesis is by itself, may be assessed by using the criteria discussed in the strategies for assessment. Factor (2), how decisively H surpasses its rivals, will depend on judgments about the relative weighting of the various criteria for assessment. Factor (3), the thoroughness of the search for alternatives, is intimately related to the systematic generation of alternative hypotheses. Concern with such an issue in theory evaluation has not traditionally been discussed in philosophy of science, probably because the focus has been on assessing the adequacy of whatever hypotheses were available, with no concern for the reasoning in generating the hypotheses. In addition, philosophers have often claimed that an infinite number of hypotheses could explain any given fact; searching an infinite space seemed futile. Constrained, heuristic search, however, is an alternative (Buchanan, 1982; 1985). When types of hypotheses can be ruled out or ruled in, all the (possibly infinite number of) instantiations of those types need not be considered.

The approach taken both in computational philosophy of science and in this book is that strategies for constraining hypothesis generation are intimately related to strategies for hypothesis assessment (and anomaly resolution). Therefore, in assessing the adequacy of a given hypothesis, the adequacy of the search for alternatives becomes a relevant concern, as well as evaluating whatever rivals have been generated. The methods for hypothesis formation (in the context of resolving anomalies), discussed in Chapters 8, 9, 10, and 12, explicitly addressed the issue of systematically generating alternative hypotheses. I was able to show how those methods could have been used to generate the hypotheses that were proposed historically. Also, I used them to generate plausible alternatives not actually proposed historically (although none of mine were as good as the confirmed ones turned out to be). As the gene case illustrated, a single scientist usually proposed one alternative and began testing predictions from it; other scientists did likewise. Yet, by using more thorough methods of hypothesis generation, one scientist might have systematically generated the alternatives that were proposed by others, as well as the ones I proposed. The evidence from the gene case suggests that an alternative strategy for scientists to consider is to use methods for more systematic generation of rival hypotheses, rather than quickly jumping to one hypothesis and beginning to test it. (A similar point is made in Platt, 1964.)

As already discussed, the entire theory of the gene did not have a well-developed rival for explaining transmission genetics, so this case has not been a good one for illustrating competition between the theory of the gene and any equally good rivals. Numerous hypotheses, however, were proposed as alternative modifications to *components* of that theory. The discussion of those competing hypotheses has illustrated the issues of systematic search for competitors and criteria for choosing among alternative hypotheses.

This discussion has assumed that what counts as a rival to a hypothesis can be unproblematically determined. Sometimes that is the case; other times it is not. Whether two hypotheses are mutually exclusive rivals or whether both (or parts of both) could be correct may not be obvious at a given time. The history of science shows numerous disputes in which two rivals later came to be encompassed in a single theory. The history of evolutionary theory has several examples in which rival hypotheses were later incorporated as components in a single theory. Rival explanations of fossil forms were Lamarck's species transmutation view and Cuvier's hypothesis of extinction. Darwin showed that evolution (transmutation) and extinction could be explained by a single theory (Bowler, 1989). In the late nineteenth century, natural selection and isolation were viewed as rival explanations for the origin of new species; the evolutionary synthesis showed that the two processes were compatible and could be combined to explain the splitting of one gene pool into two (Bowler, 1989).

Two issues that are in need of further work by philosophers of science are (1) the development of methods to generate rival, alternative hypotheses, and (2) the problems of determining whether two purported rivals can be somehow made compatible.

15.3 Strategies for Anomaly Resolution and Change of Scope

Anomaly resolution involves, first, the localization of the problem within one or more components of the theory; then, the generation of one or more new hypotheses to account for the

---

1. Confirm that an anomaly exists
    a. Reproduce anomalous data
    b. Reanalyze problem
2. Localize problem
    a. Outside theory
        i. Monster-barring
        ii. Outside scope of theory's domain
    b. Inside theory
3. Alternative ways of changing the theory
    a. Alter a component
        i. Delete
        ii. Generalize
        iii. Specialize
        iv. Complicate
        v. Simplify
        vi. Delineate and change one but not other
        vii. "Tweak"—alter slightly
        viii. Propose opposite
    b. Add a new component
        Use strategies for producing new ideas with added constraints
    c. For (a) and (b), design an altered or new theoretical omponent in the light of
        the following constraints:
        i. The nature of the anomaly—the new theoretical component must account for it
        ii. Maintain a systematic connection with other, non-problematic theoretical components, that is,
            avoid ad hocness
        iii. Other criteria of theory assessment may be
            introduced as constraints, such as extendability and fruitfulness
4. If (a) or (b) resulted in a change in the theoretical components, assess the new version of the theory by
    applying criteria of assessment to the new component(s).
5. If the above steps fail to resolve the anomaly, consider whether the anomaly is sufficiently
    serious to require abandoning the entire theory, or whether work with the theory can continue despite
    the unresolved anomaly.

Table 15.3 Strategies for anomaly resolution and change of scope

anomaly; and finally, an evaluation of those hypotheses to determine which one best satisfies criteria of theory assessment. Those issues will be the subject of following subsections of this chapter.

Strategies for anomaly resolution are closely related to strategies for change of scope. Change of scope can be (1) an expansion of the theory to explain items originally outside the domain, (2) expansion to explain new items just discovered, or (3) specialization to exclude anomalies and preserve adequacy within a narrower domain. The latter possibility is one way change of scope is used in anomaly resolution.

As discussed in the earlier sections on explanatory adequacy and scope, what is inside the domain and what is outside the domain of a theory at any given time may be a matter of debate. Thus, a given item, such as an experimental result, might be considered an anomaly for the theory because the item is inside the domain to be explained by that theory. Alternatively, the same item, it might be argued, is not an anomaly for the theory because that item is outside the scope of the domain of the theory; some other theory is expected to account for it. If theory construction begins with a large domain to be explained and then anomalies arise, **narrowing the scope of the domain** may aid in anomaly resolution. If a proposed hypothesis is very general and an anomaly arises for it, then one strategy is to **specialize the overgeneralization** and exclude the anomalous item from the domain. On the other hand, if theory construction begins with a hypothesis that applies to a very narrow domain, as did Mendelism, then theory change occurs with **expansion of the scope of the domain,** by including items originally not part of it. Because of this close relationship between the scope of the domain and the identification of an anomaly, strategies for anomaly resolution and change of scope are closely related and are considered together here.

The term "anomaly" usually refers to a problem posed by data that the theory cannot explain. If the anomaly results from a failed prediction, then it indicates some kind of incorrectness in the theory; however, theories may face other kinds of problems besides those posed by empirical anomalies. As discussed in Chapter 12, a theory may be incomplete, even though no empirical anomaly indicates that it is incorrect. (For more on incompleteness, see Shapere, 1974; Leplin, 1975.) In addition to problems of incorrectness or incompleteness, a theory may face conceptual problems of various kinds, which were discussed in Chapter 11. Discussion of general strategies for resolving all the kinds of problems that a theory may face would make this chapter even longer than it is already. Hence, the focus here will be on empirical anomalies that result from a failed prediction.

The lists of strategies for producing new ideas and strategies for theory assessment in Tables 15-1 and 15-2 were not presented as sequential strategies. In contrast, anomaly resolution entails four sequential steps. Table 15-3 shows four steps: (1) Confirm that an anomaly exists, (2) Localize the problem, (3) Change the theory, and (4) Assess the results. However, the subcomponents of Steps 2 and 3, representing ways of resolving the anomaly, are alternative, not sequential steps. Alternative strategies for resolving an anomaly include: **monster-barring, altering an existing component,** and **adding a new component.** These are presented in an order—from the strategy that produces the least change in the theory, namely, **barring the anomaly** as a monstrous case not requiring theory change, to the strategy that produces the greatest change, namely, **adding new theoretical components.** If the four stages of anomaly resolution fail, then Step 5 indicates the choices: either abandon the entire theory or shelve the anomaly for the present and continue work on the theory anyway. The gene case has proved to be an excellent case for illustrating strategies for anomaly resolution and change of scope. Previous chapters (especially 8 and 12) have discussed episodes that exemplified many of these strategies. The following subsections will summarize the stages and strategies.

15.3.1 Confirm Anomalous Data or Problem

Before efforts are made to resolve an anomaly, the correctness of the anomalous data needs to be confirmed. Repeating experiments is one way to verify the existence of an anomaly. There is no need to pursue subsequent steps if the purported anomaly can be shown to be due to incorrect assumptions about initial conditions or to be due to faulty experimental equipment or faulty procedures. The gene case provided few instances of a purported anomaly being explained away as based on faulty data. Presumably many such instances were never published in the scientific literature; scientists corrected them before publishing.

15.3.2 Localize the Anomaly

After an anomalous result has been confirmed to exist, then the next step is to localize the anomaly. If the anomaly can be viewed as an unusual case, then it may be localized as outside the normal domain of the theory. In other words, it may be barred as a monster (**monster-barring** was discussed in Chapters 2 and 12). In the case of **monster-barring**, localization serves to remove the anomaly from being a problem for the theory. Lethal gene combinations, for example, were hypothesized to be the cause of anomalous ratios; the explanation of the anomaly thus required no change in the theoretical components that explained normal gene ratios. (Ideally, the theory can be used to explain how the normal process has failed in the monstrous case.) As discussed in Chapter 12, two kinds of monsters are possible, unique ones and ones that belong to classes. Both kinds of monsters—unique ones and malfunction classes—are barred from necessitating a change in the theory. They are explained, or better, explained away.

Another way to localize the anomaly outside the theory is to change the scope of the domain. If it can be argued that the anomaly is outside the scope of a (perhaps narrowed) domain, then the anomaly is removed from being a problem for the theory; some other theory must account for it. If the anomaly cannot be barred as a monster or removed by a change of scope, then its localization within the theoretical components needs to be considered (labeled Step 2b in Table 15-3).

Two different strategies for localization have been discussed in previous chapters (Chapters 6, 8, and 12). One depends on representing the theoretical components in sentences; the other depends on representing the theory diagrammatically, in terms of a sequence of steps in a process. A sentential representation allows explanations to be construed as arguments; the theoretical components involved in explaining a particular domain item can be identified because of the roles they play in the arguments. Some components more "directly" account for some items, such that, if the domain item is removed, the directly connected theoretical component becomes unnecessary. When an anomaly arises for that domain item, the theoretical component(s) directly accounting for it is a likely site for modification; however, other, more general theoretical components may also be involved in accounting for a given domain item and thus they too become potential sites for modification. Assigning credit to these potential sites involves determining exactly which theoretical components are used in the explanation of the anomalous item. One further complication is introduced if implicit assumptions are also involved in the explanations; localization may then necessitate uncovering implicit assumptions as the site of failure. Thus, using a sentential representation in anomaly resolution necessitates investigation of the network of connections in the explanatory argument patterns. Localization of potential sites of failure involves tracing the paths in such networks between particular domain items and particular theoretical components. This view was developed by Glymour, who suggested illustrating the skeleton of the argument for explaining a particular domain item. Then, the theoretical components used in the explanation can be seen and localized (Glymour, 1980, p. 197).

The gene case provided examples of direct localization, localization in a more general component, and localization in previously implicit assumptions. The direct localization strategy worked for the resolution of anomalies to dominance. Exceptions to dominance were localized in the dominance component of the theory (see Chapter 6). Furthermore, some exceptions to 3:1 ratios were plausibly localized by Castle in the segregation components (see Chapter 8). On the other hand, Muller explained Castle's anomalous ratios by localizing the problem in a more general component, the claim that one character is caused by one gene. Using the already available new component of multiple factors, Muller explained the anomaly with modifying factors (see Chapter 8). In another anomaly resolution episode, lethal genes required uncovering the implicit assumption that all gene combinations were viable (see Chapter 8).

Anomalies may require further articulation of the theory before they can be localized. When exceptions to 9:3:3:1 ratios were found, those anomalies served to show that separate theoretical components for segregation and for independent assortment were needed, as Chapters 9 and 10 discussed. The anomaly necessitated the delineation of the theoretical components for independent assortment from the segregation components. The case illustrates the strategy of **delineate one component into two and alter one but not the other**. The components in which the anomaly is to be localized may not have been clearly stated as separate from other components until the anomaly showed that delineations were needed. Localization of anomalies thus may drive the explicit articulation of the theory.

Another method for localization was discussed in Chapter 12: represent the theory as typical steps in typical hereditary processes. Detect which step(s) failed; for instance, no germ cells produced, or fewer embryos produced than expected. Localize the anomaly in one or more steps of the process by using additional information about which steps in the process occurred and which did not. Such additional information may have to be generated by additional experiments to detect which steps were reached and which were not. This kind of localization is analogous to a diagnostic reasoning task, as discussed in Chapter 12. It is like localizing a faulty module in a device by seeing, for example, which module has electricity coming in but puts none out. Queries about inputs and outputs to the steps of the process aid in localization. (See Sembugamoorthy and Chandrasekaran, 1986, for a discussion of this kind of diagnostic reasoning about devices.)

Localization requires modular theoretical components. The gene case illustrated such modularity. Localization in a few theoretical components proved possible for the anomalies that the theory of the gene encountered. Philosophers' pessimism about localization of anomalies within a theory is not justified by the gene case (see Chapter 8 for more detail).

15.3.3 Alternative Ways of Removing An Anomaly

Changing the components is like a redesign task. Reasoning in design involves designing something new to fulfill a certain function, in the light of certain constraints. Redesigning the theory involves constructing a component that will account for the anomaly, with the constraints of preserving the unproblematic components of the theory and producing a theory that satisfies criteria of theory assessment. Just as AI work on diagnostic reasoning is suggestive for localization strategies, AI work on redesign in the light of constraints can be suggestive for strategies for changing the theory. Not surprisingly, AI research has shown that implementing methods for redesigning faulty modules is more difficult than the first step of localizing the problem. Similarly, creatively constructing new hypotheses is, in general, a more difficult task than discovering the need for a new hypothesis.

To have a term to contrast to "monster" anomalies, in Chapter 12, I introduced the term "model" anomalies, by analogy with "model" organisms. Model anomalies require a change in the theory. The model anomalies serve as a model for what is normal; in other words, model anomalies are not actually anomalous at all. They serve as models of normal types of processes that are commonly found. Model anomalies are resolved, either by changing an existing theoretical component or by adding a new one or both (e.g., one component may be specialized and an additional new component added).

Table 15-3 divides strategies for changing a theoretical component into two categories: (3a) alter a component and (3b) add a new component. A number of different strategies exist for altering an existing component. They are listed in Table 15-3, Step 3a, and will be discussed later.

15.3.3.1 Alter a Component of the Theory

Deleting a component is obviously an easy kind of change to make. Its successful use depends on the modularity of the deleted component, such that the removal does not affect the rest of the theory. As discussed in Chapter 6, after the discovery of numerous exceptions to complete dominance and the inclusion of blending within the scope of Mendelism, dominance as a general component of the theory of the gene was eliminated. If the deleted component explains other items in the domain in addition to the anomalous ones leading to its deletion, then some other component(s) will have to be modified or added to account for those items. The gene case did not provide an example, but I discussed such a case in the change from Darwin's 1868 hypothesis of pangenesis to de Vries's 1889 modified version in his intracellular pangenesis (Darden, 1976). One component of Darwin's hypothesis was the claim that hereditary units circulated throughout the body. The circulation component was used to explain inheritance of acquired characters (gemmules from the longer neck of the giraffe circulate to the reproductive areas and are passed on to the baby giraffe). Additionally, circulation explained how a new individual can bud off from the body part of its parent, as is common in plants (circulating gemmules congregated in the area of budding and were passed on to offspring). By 1889, de Vries argued that doubt had been cast on the claim that acquired characters were inherited; thus, the circulation of hereditary units throughout the body was no longer needed in a hereditary theory; however, de Vries still had to explain the domain item of budding. He used a new component in his "intracellular" pangenesis to explain budding, namely, the assumption that all the hereditary units were contained in the nucleus of most cells. The pangens in the nuclei of, for example, a leaf cell could produce a new plant by budding. When a theoretical component was deleted because of the deletion of only one of the domain items for which the component accounted, then a new component was needed to account for the remaining items.

Another method for slightly changing a theoretical component in the light of an anomaly is to **generalize** or **specialize** the component. Generalization and specialization are methods of modifying a working hypothesis that have been extensively used in studies of induction and concept learning in AI (Mitchell, 1982; Dietterich et al., 1982). **Generalization** expands the scope of a hypothesis; **specialization** narrows the scope. If generalization occurs by dropping a condition, then it is like **simplification** (Darden, 1987). Specialization by adding a condition is one form of **complication**. The strategies of **generalization-specialization** and **simplification-**

complication were discussed in Chapter 6. Chapter 6 also discussed the strategy of **specialize and add**: in the light of an anomaly, one component is specialized to apply to fewer domain items and another component is added to cover other cases that were previously anomalous.

If bold, general, simplifying assumptions marked the beginning stages of theory construction, then **specialization** and **complication** will be likely strategies to use as anomalies arise. If the theory was constructed originally in a very conservative way, carefully specialized to apply to a narrow domain, then **generalization** will be a way of expanding the scope of its domain, even if no specific anomaly is at issue.

A more systematic strategy for anomaly resolution is, at the outset, to consider explaining the problematic data in both the most general and the most specialized ways consistent with the data. Then, alternative hypotheses—varying along a spectrum from general to specific—become candidates for future development. This method of systematically considering a range of hypotheses, from the most general to the most specific, is called the "version space" method of hypothesis formation in AI (Mitchell, 1982). The version space is the space of all hypotheses between the most general and the most specific that account for a given set of data. Then refinements are made in the light of new data points, which may be considered anomalies in our current terminology. Exceptions to generalizations are resolved by adding conditions to make a general hypothesis more specific. A new instance not covered by a specific hypothesis drives the formation of a more general hypothesis by dropping conditions from the specific one. I know of no instances in the gene case of such systematic consideration of the space of hypotheses consistent with a data set. I doubt that scientists typically engage in such systematic generation of alternative hypotheses.

"**Tweaking**" is a term for the strategy of **changing a component slightly** to account for an anomaly or a new instance. Schank (1986, p. 81) used the term similarly, when he suggested explaining anomalies by invoking past explanation patterns and changing them slightly to apply to the new situation. He was not, however, discussing explanations in science. I am using the term as an eclectic class of strategies for ways of changing a component. "Tweaking" strategies are strategies for making slight changes in the theory that do not fit into any of the more specific ways of making slight changes discussed above. An example is slightly changing the parameters in a quantitative model to account for a quantitative anomaly that is just a little off from the predicted value.

Yet another strategy for altering a component is to **propose the opposite** of the component localized as failing. Numerous instances of proposing the opposite of a component have been discussed in previous chapters. Purity was changed to nonpurity; random fertilization was changed to selective fertilization; equal numbers of types of gametes was changed to unequal numbers; independent assortment was changed to linkage. Each of these alternatives raised new issues: how was nonpurity produced? Why did some gametes selectively fertilize others? How were unequal numbers produced? What accounted for linkage? Thus, proposing the opposite was only a part of producing a satisfactory alternative component. As discussed earlier, it is a strategy for producing the new idea that a certain *type* of hypothesis would account for the anomaly. Additional strategies for producing new ideas were needed to instantiate the type of hypothesis that the strategy of **propose the opposite** suggested.

15.3.3.2 Add Something New

The strategies already discussed for altering a component produce slightly new hypotheses; however, those strategies may all prove to be inadequate and a new component of the theory may be needed. In such a case, strategies for producing new ideas (discussed in Section 15.1) will need to be invoked. If some components of the theory are not candidates for modification, preserving consistency with them will be an added constraint that is not present in the original construction of an entirely new theory. The most important additions to Mendelism were the components of linkage and crossing-over. As discussed in Chapters 7, 10, and 11, their construction exemplified the strategies for producing new ideas of **using interrelations** and **move to another level of organization**. Linkage and crossing-over were consistent with the segregation components.

Thus, resolving anomalies by adding new components to a theory requires use of the strategies for producing new ideas, constrained by preserving consistency with the unproblematic components of the theory.

15.3.4 Assessing the Hypotheses to Resolve the Anomaly

Hypotheses proposed as modified theoretical components have to be evaluated, using the criteria of theory assessment. The stages diagram in Figure 15-1 can be used to represent anomaly resolution. The anomaly is the problem to be solved. The strategies for altering a component or the strategies for producing new ideas provide one or more hypotheses as candidates for the modified theory components. The systematic generation of alternative localizations and alternative hypotheses at those locations is an important part of the task of anomaly resolution. Then the criteria for theory assessment are used to evaluate the hypotheses, with the added constraint that the new components must be consistent with the unmodified components; also, the criterion of a systematic relation among all the components of the theory becomes important. Furthermore, important criteria were the criteria of explanatory adequacy, predictive adequacy, and the lack of ad hocness. What counts as a legitimate addition to the theory and what is an illegitimate ad hoc change may be a matter of debate.

15.3.5 Unresolved Anomalies

All of the strategies for resolving an anomaly may fail. In such a case, scientists working on the theory will have to decide whether the anomaly is sufficiently serious to require abandoning the entire theory or whether it can be shelved as a problem requiring resolution, while work on other parts of the theory continues (Step 4 in Table 15-3). Again, no decisive criteria may be present to choose among these alternatives. Even if the entire theory is to be abandoned, the anomaly may well provide a pointer to the components of the theory most at fault and provide hints as to what a new theory should contain in order to avoid having the same anomaly. The gene case did not include the abandonment of the entire theory of the gene. Instead, in the gene case, hypothesis testing provided guidance in subsequent stages of modification: anomalies caused the abandonment of proposed changes to theoretical *components*. Geneticists did sometimes abandon hypotheses on the basis of falsifying evidence.

15.4 Conclusion

We have now discussed all the parts of the diagram in Figure 15-1 of the stages and strategies of theory development. The gene case has provided a rich source of examples of some of the strategies, especially those for anomaly resolution. As with any part of an ongoing research program, I hope that this discussion will provide a springboard for future work to evaluate it and refine it in the light of any (metalevel) anomalies. Other cases and other analyses will also, I hope, aid in refinements of, and additions to, the lists of types of strategies.

CHAPTER 16

Implications for Further Work

This book has discussed strategies for theory change, including strategies for producing new ideas, strategies for theory assessment, and strategies for anomaly resolution and change of scope. Chapters 6 to 13 analyzed the strategies exemplified in the development of the theory of the gene between 1900 and 1926. Chapter 14 summarized those strategies, and Chapter 15 placed those strategies into a more systematic list of strategies. Thus, Chapters 14 and 15 serve as summaries. The results of this analysis are working hypotheses about theories and theory change. Because they are based on only one case, they are in need of further exploration in order to test their scope, to refine them in the light of anomalies, and to determine their implications for other work. This chapter will discuss directions for future work and implications of this analysis for the fields of history and philosophy of science, science, science education, cognitive science, and AI.

The working hypotheses fall into three categories: representation of theories, general aspects of theory change, and an analysis of strategies for theory change. This study has explored two different methods for representing theories—sentential and diagrammatical. They each had advantages and disadvantages. The sentential representation served to clarify the claims in the scientific literature and it aided in articulating the claims on which geneticists agreed and disagreed. It also provided a way of showing which components of the theory were used in explanations of which domain items. Stating the theory in separate but related sentences aided in detecting both the systematic connections among the components and their modularity. That analysis could be pushed further in the direction of an explicit axiomatization so that explanatory links would be even more clear. Such a move, however, would take the representation of the theory even further from the way it was presented in the scientific literature. Biologists rarely stated theories in sentences in their published work; they certainly did not provide axiomatizations. I found that it was a difficult task to devise an adequate list of sentences, ones that captured the theoretical content at a particular time. It is likely that other historians could give good arguments for alternative lists and that other philosophers could produce axiomatizations, perhaps even ones that are not logically equivalent. A sentential representation is underconstrained by the historical record because of the record's vagueness and imprecision and lack of explicitly stated claims. Interestingly, such vagueness did not prevent biologists from understanding, using, and developing Mendelism. Thus, it is a challenge to philosophers and cognitive scientists to understand how scientists represent, understand, and use scientific theories in ways that are less rigorous than formal logic demands.

The alternative representation of the theory that I devised—a diagrammatic representation based on abstraction from exemplars—was more faithful to (at least some of) the scientific literature. The diagrams depict typical steps in typical hereditary processes. The theory can be represented by a set of abstract diagrams, constituting the set of typical processes whose details are filled in by the theory. That set is the explanatory repertoire of the theory. Explanation, once the theory has some successful patterns, becomes the following task: find an appropriate exemplar (or pattern abstracted from that exemplar), and fit the data to be explained into that exemplary pattern. Such a diagrammatic representation of steps in a typical process proves useful in localizing anomalies, that is, the failure of one or more typical steps.

As Chapter 12 discussed, this diagrammatic and exemplar approach draws on other work in philosophy of science. Kuhn (1970) provided suggestions about the importance of exemplars that have proved fruitful in this analysis and may be useful in others. Kitcher's (1981) and Thagard's (1988) views that theories provide abstract schemas has been extended here. The extent to which this view of theories is applicable to other cases will be a fertile area of exploration of philosophers of science. It will also be fruitful to explore how new schemas can be constructed to add to the explanatory repertoire. Because AI knowledge representation techniques can easily be used to represent sequential steps with inputs and outputs (Moberg and Josephson, 1990), a schema based on this form of representation may be especially promising within computational philosophy of science.

In addition to exploring methods for representing a theory, this analysis has also illuminated aspects of the development of a successful scientific theory. Whether the development of the theory of the gene has features that will generalize to other cases remains to be seen. A brief summary of those features here will aid others who wish to do comparisons with other cases. Representation of Mendelian theory as a *modular system* (either modular sentences or modular abstract patterns) proved useful in tracing changes in modules as the theory developed. The development of the theory of the gene was a temporally extended process, as modules were changed piecemeal over time. That process was marked by a number of features: a narrow scope to a larger domain; oversimplifications to more complicated theoretical components; overgeneralizations to more specialized claims; implicit ideas uncovered in order to produce explicit components; the deletion of a component; the delineation of one component into two separate components; and the addition of new components. Change was driven by resolving empirical anomalies, expanding the scope of the theory, and solving conceptual problems. In the face of anomalies, specific modular theoretical components were localized as the potential sites of failure. Alternatives to possibly failing components were proposed. Diverse assessments were made. Controversies raged. Consensus sometimes emerged and a successful theory was built in a piecemeal way.

No one scientist systematically generated a large set of competing hypotheses to solve a particular problem. Furthermore, the alternative hypotheses that were proposed in a given historical episode sometimes did not exhaust the space of plausible hypotheses. These features of theory development suggest that scientists might find it useful to attend, explicitly, to their own strategies for hypothesis generation and also to employ strategies for more exhaustive hypothesis generation. (This point is also made in Platt, 1964.)

In addition to the study of the representation and the development of scientific knowledge, this analysis has undertaken the task of devising strategies for producing that knowledge. The particular way that I used a historical case in order to construct strategies for theory change is a method of my own devising. The historical case provided evidence that a change from Component X to Component X' occurred; I then considered what strategy of reasoning *could have produced* such a change. No claim is made that any historical person consciously employed the strategy. They may have done so, but the historical evidence does not provide evidence about their actual reasoning processes. In constructing the strategies, I have drawn on the study of reasoning by researchers in a number of fields, including philosophy of science, AI, cognitive science, science teaching (e.g., Clement, 1988), and the study of creativity (e.g., Perkins, 1981). In instances of theory change where I was unable to find an appropriate strategy that had already been analyzed in the literature, I was forced to devise and name a new one. The strategy of **delineate and alter**, for example, involved making two components from one, by explicitly separating two claims that had previously been lumped together. I know of nothing in previous literature about the reasoning involved in making implicit assumptions explicit, and then independently manipulating the newly uncovered assumptions. The nature of the case itself drove my construction of this strategy; it is certainly not a strategy that I would have conceived in the abstract, without the impetus from the historical case. Thus, this work has developed a new methodology for devising strategies. First, changes in a historical case are traced and represented explicitly. Next, reasoning strategies, which could have produced the historical changes, are devised. Then, the strategies are named and stated in a general way. This is a method that can be applied to other cases.

A number of questions about strategies are raised, but not answered, by the analysis in this book. Is the history of science, indeed, a rich source of specific changes that exemplify general strategies? To what extent are the strategies that are discussed here applicable in other cases? To what extent are any of the specific strategies discussed here actually used by scientists (given evidence about their reasoning)? To what extent are the strategies discussed here used outside of science? Historians of science and cognitive scientists studying reasoning may be able to use the strategies discussed in this book as hypotheses about the reasoning strategies that scientists actually use and provide evidence to test those hypotheses.

Philosophical questions can also be raised about the strategies themselves. What kind of more systematic analyses of strategies would be worthwhile? How can strategies be codified so that one would know what strategy to use for what problem? What are the relations between the strategies; for example, are some more general and others more specific? How can strategies for assessment be used in hypothesis construction? Should some strategies be used prior to others or in preference to others? How are strategies to be evaluated, that is, how are good strategies to be separated from bad ones? To what extent is the later success or failure of a hypothesis related to the strategies used in constructing it? What kinds of biases are introduced by adopting a given (type of?) strategy? (For a discussion of strategies and their biases, see Wimsatt, 1980.) To answer these questions, much work remains to be done by philosophers of science interested in reasoning strategies.

Scientists may also find this discussion of strategies of interest and of use. Some scientists are undoubtedly better than others at doing their scientific work. If good strategies can be devised and articulated, then some scientists may be able to improve their reasoning methods by adopting appropriate ones. Moreover, even the best scientists may not reflect on their own strategies or be effective in articulating what strategies they use. Explicit articulation of their own, good strategies might be of use to their colleagues and their students (for a similar argument, see Bartholomew, 1982). Perhaps the strategies discussed in this book will be useful to scientists engaged in the task of forming, assessing, and refining hypotheses. The strategies may be viewed as distilled patterns of reasoning that could have been used in concrete past scientific achievements. If similar situations again arise, then a similar pattern of reasoning may yield productive results. I have suggested that a strategy not often followed in this case was the systematic consideration of plausible alternatives. Perhaps scientists should be wary of a tendency to find and to promote only one hypothesis, thereby failing to consider alternatives.

Of course, it may be difficult (or even detrimental) to reflect on what one is doing in the midst of doing it, especially if one is doing it effectively and rapidly. I am sympathetic to scientists who say that they had rather do science than consider what philosophy of science they employ. (I too had rather do history and philosophy of science than write about its methodology.) Perhaps metalevel analyses of science should be done by those of us in science studies. In that case, better methods need to be developed for collecting data about scientists' reasoning, other than the usual laboratory and thinking notebooks, which are woefully incomplete. The use of thinking-aloud protocols during routine problem-solving has proved fruitful, but additional methods are needed to capture the reasoning in the development of a new theory over a period of years.

The results of the study of reasoning strategies may be of particular benefit, not to scientific experts, but to science students, who have not yet gone through a long apprenticeship. Science teachers, at both the undergraduate and graduate levels, might explicitly teach reasoning strategies. Such instruction might have the advantage of reducing the time that students spend as apprentices, time spent implicitly absorbing methods of reasoning (while explicitly studying knowledge and techniques). One of the reasoning strategies with special implications for science teaching is the strategy of abstracting a problem-solving schema from an exemplary, concrete problem solution (as discussed in Chapter 12). All teachers know the importance of providing good examples. The view of theories as abstract schemas shows that theories may be taught by, first, providing a concrete usage and then explicitly abstracting the pattern from the exemplar. That pattern is what is used (or sometimes slightly modified) to explain additional, similar instances. Mendelian genetics is usually taught by using the theoretical assumptions to explain the results of specific hybrid crosses. The diagrams provide the depiction of the abstract patterns to be filled in to provide similar explanations. Experts come to know that to explain, for example, a 3:1 ratio, the segregation schema is to be used. More explicit awareness of the role of abstraction formation might aid the process of teaching the patterns, and, thereby, of converting novices into experts.

More generally, the teaching of scientific reasoning could impart an alternative view of science than the one often conveyed by science classes; that is, it could promote the view of science as an exciting intellectual enterprise, rather than as a static body of knowledge to be memorized. Furthermore, hypothesis formation, testing, and refinement is a general pattern of reasoning in numerous nonscientific tasks, such as diagnostic reasoning in situations where devices malfunction. Teaching such strategies both to science and nonscience majors can thus be justified as useful to them outside the classroom.

I am experimenting with methods for teaching strategies in my introductory philosophy of science courses. Class discussions of general strategies are illustrated with historical scientific examples, or with what could have happened in a historical situation. Students then use computer simulations as environments in which to practice hypothesis formation, testing, and refinement. They are asked to report on the reasoning strategies that they used. Guidance on the use of certain strategies could be added to simulations, and "what if" situations could be posed to allow exploration and use of alternative strategies. The strategies for hypothesis formation and refinement discussed here could be coupled with teaching strategies for experimentation (e.g., a confirmation versus a disconfirmation strategy; see Gorman et al., 1984). Discussion of the relations of strategies for theory change and strategies for experimentation are outside the scope of this book, but that is a fruitful topic for further work. The potential exists for developing course software that teaches both reasoning and experimental strategies and also is fun to use.

One of the strategies is a particularly fruitful one for further work in a number of directions—the strategy of **forming interrelations between two bodies of knowledge**. Researchers in a number of fields (e.g., Staats, 1983; Gökalp, 1987; Bechtel, 1988) have found earlier discussions of interfield theories (Darden and Maull, 1977; Maull, 1977; Darden, 1980) useful in analyzing relations between scientific fields. Issues about the unity of science are complex; the older view that science becomes more unified by reducing one theory to another has been shown to have limited applicability (e.g., Schaffner, 1974b; Bechtel, 1988). The idea, developed here, of finding interrelations that bridge two different bodies of knowledge is a promising one; it shows one way that areas of science become more integrated. A focus in this discussion has been the *kind of interrelation* postulated between two bodies of knowledge: identity, part-whole, structure-function, and causal. A potentially fruitful area for further work is the exploration of the ways such interrelations can guide hypothesis formation. If a part-whole interrelation is postulated, for example, what hypotheses about wholes can be formed based on information about the parts, and vice versa? Work in the formal ontology of parts and wholes (e.g., Smith and Mulligan, 1983) may aid in producing general strategies for using part-whole relations and contribute to the work in AI on the representation of knowledge in part-whole hierarchies.

AI studies the representation of knowledge and computational methods of reasoning. I have found work in AI very suggestive as I have puzzled over ways to represent a scientific theory and ways to devise strategies for theory change. Perhaps some of the work here will be suggestive for further AI work. Chapter 6 discussed generalizing and specializing claims; generalization and specialization have been extensively investigated as methods in machine learning (e.g., Mitchell, 1982). The gene case introduced more complicated patterns of generalizing and specializing than those studied in the machine learning work on learning from examples. Further comparisons between the strategies in Chapter 6 and AI work might yield insights.

AI researchers have explored the idea of discovery as search through a space of possibilities (e.g., Buchanan, 1982). That method is suggestive of the strategy of **systematically generating alternative hypotheses in the face of an anomaly**. How to characterize a search space in a given problem-solving episode is a fruitful area for further work. In addition, Chapter 11 explored the advantages and disadvantages of using a frame-structured representation to represent historical changes in claims about genes. More extensive knowledge representation of this case, such as putting "gene" into "kind of" and "part of" hierarchies, raises interesting questions about knowledge representation (see Darden and Rada, 1988a, for more discussion). In Chapter 12, I suggested an analogy between anomaly resolution and diagnostic reasoning, a pattern of reasoning studied extensively in AI. Using AI methods for debugging theories is a fruitful area for further work. Additional analysis of the gene case to introduce more precision is required to build an AI system to simulate anomaly resolution; it would diagnose the site(s) of failure and redesign the failing components, using strategies for altering and adding components (for a beginning on this work, see Moberg and Josephson, 1990).

There are a number of reasons to build AI systems in order to simulate past scientific achievements. An adequate simulation demonstrates the internal consistency and sufficiency of the knowledge and the reasoning strategies that the model encodes (Thagard, 1988). Further-more, implementations hold the promise of testing the sufficiency of particular strategies for producing particular results. In given episodes, I would like to have been able to demonstrate that the strategy I proposed was sufficient for producing the given change. A simulation would allow a proof of sufficiency. Furthermore, such a simulation would make experiments possible. The analysis in this book might be analogized to doing observational science. I have observed the scientific changes in past scientific episodes. AI simulations hold the promise of transforming history and philosophy of science from an observational into an experimental field. For example, one could explore the consequences of using alternative strategies, and, also, one could evaluate the efficacy and efficiency of given strategies for solving types of problems (for similar work, see Langley et al., 1987; Thagard, 1988). Success in building such simulations will provide further evidence for my proposal that (1) the history of science is a rich source of hindsight about how to do good science, and (2) such hindsight can be compiled into reasoning patterns and implemented in computer programs to model good scientific reasoning (Darden, 1987). The kinds of strategies captured by modeling past scientific achievements will, it is hoped, continue to be of use in relevantly similar problem-solving episodes in present science (for arguments for this methodology, see Langley et al., 1987; Thagard, 1988).

Finding good methods for doing computational science will contribute to the current project on the matrix of biological knowledge (Morowitz and Smith, 1987). That project's goal is to find computational methods for representing biological knowledge and for making new discoveries, using data stored in computers. Huge amounts of data are now available in on-line databases. The time is now ripe for automating scientific reasoning, first, to form empirical generalizations about patterns in the data, then, to construct new explanatory theories, and finally, to improve them over time in the light of anomalies. Thus, the development of computational models for doing science holds promise for areas where large amounts of data overwhelm human cognitive capacities (Schaffner, 1986b; Morowitz and Smith, 1987). The goal is to devise good methods for doing science, whether or not the methods are ones actually used by humans. The goal is not the simulation of scientists, but the making of discoveries about the natural world, using methods that extend human cognitive capacities. In sum, computational models allow exploration of methods for representing scientific theories, as well as methods for reasoning in theory formation, testing, and improvement. My hope is that the strategies discussed in this book will be suggestive for that work.

It is an exciting time to work at the interface of historical, philosophical, and computational approaches to science in order to explore methods for developing new ideas. Future prospects beckon.

Bibliography

[Note: When a reprint edition is listed, page references are to the reprint.]

Achinstein, Peter (1970), "Inference to Scientific Laws," in Roger Stuewer (ed.), 1970, *Historical and Philosophical Perspectives of Science, Minnesota Studies in the Philosophy of Science*, V. 5. Minneapolis: University of Minnesota Press, pp. 87-111.

Achinstein, Peter (1987), "Scientific Discovery and Maxwell's Kinetic Theory," *Philosophy of Science* 54:409-434.

Adams, Mark (1980), "Sergei Chetverikov, the Kol'tsov Institute, and the Evolutionary Synthesis," in E. Mayr and W. Provine (eds.), 1980, *The Evolutionary Synthesis*. Cambridge, Massachusetts: Harvard University Press, pp. 242-278.

Alexander, Jerome, and Calvin B. Bridges (1928), "Some Physico-Chemical Aspects of Life, Mutation, and Evolution," in J. Alexander (ed.), *Colloid Chemistry*, V. 2, New York: The Chemical Catalog Co., pp. 9-58.

Alexander, Jerome, and Calvin B. Bridges (1929), "Some Physico-Chemical Aspects of Life, Mutation and Evolution," *Science* 70:508-510.

Allen, Garland (1969a), "Hugo de Vries and the Reception of the 'Mutation Theory'," *Journal of the History of Biology* 2:55-87.

Allen, Garland (1969b), "T. H. Morgan and the Emergence of a New American Biology," *The Quarterly Review of Biology* 44:168-188.

Allen, Garland (1975), *Life Science in the Twentieth Century*. New York: Wiley.

Allen, Garland (1978), *Thomas Hunt Morgan*. Princeton, New Jersey: Princeton University Press.

Allen, Garland (1985), "Heredity under an Embryological Paradigm: The Case of Genetics and Embryology," *Biological Bulletin* 168 (suppl.):107-121.

Avery, O. T.; C. M. MacLeod; and Maclyn McCarty (1944), "Studies on the Chemical Nature of the Substance Inducing Transformation of Pneumococcal Types," *Journal of Experimental Medicine* 79:137-158. Reprinted in J. A. Peters (ed.), 1959, *Classic Papers in Genetics*. Englewood Cliffs, New Jersey: Prentice Hall, pp. 173-192.

Baltzer, Fritz (1967), *Theodor Boveri: Life and Work of a Great Biologist, 1862-1915*. Translated by Dorothea Rudnick. Berkeley: University of California Press.

Barr, Avron, and Edward Feigenbaum (eds.) (1981), *The Handbook of Artificial Intelligence*, V. 1. Los Altos, California: William Kaufmann.

Bartholomew, G. A. (1982), "Scientific Innovation and Creativity: A Zoologist's Point of View," *American Zoology* 22:227-235.

Bateson, William (1894), *Materials for the Study of Variation*. New York: Macmillan.

Bateson, William (1898), "Progress in the Study of Variation, II," *Science Progress* 2. Reprinted in R. C. Punnett (ed.), 1928, *Scientific Papers of William Bateson*, V. 1, pp. 357-370.

Bateson, William (1900), "Problems of Heredity as a Subject for Horticultural Investigation," *Journal of the Royal Horticultural Society* 25. Reprinted in W. Bateson, 1928, *William Bateson, F.R.S., His Essays and Addresses, with a Memoir by Beatrice Bateson*. Cambridge, England: Cambridge University Press, pp. 171-180.

Bateson, William (1902), *Mendel's Principles of Heredity—A Defense*. Cambridge, England: University Press.

Bateson, William (1903), "On Mendelian Heredity of Three Characters Allelomorphic to Each Other," *Proceedings of the Cambridge Philosophical Society*, II. Reprinted in R. C. Punnett (ed.), 1928, *Scientific Papers of William Bateson*, V. 2. Cambridge, England: Cambridge University Press, pp. 74-75.

Bateson, William (1905), "Letter to Adam Sedgwick," in W. Bateson, 1928, *William Bateson, F.R.S., His Essays and Addresses, with a Memoir by Beatrice Bateson*. Cambridge, England: Cambridge University Press, p. 93.

Bateson, William (1907), "Facts Limiting the Theory of Heredity," *Science* 26. Reprinted in R. C. Punnett (ed.), 1928, *Scientific Papers of William Bateson*, V. 2. Cambridge, England: Cambridge University Press, pp. 162-177.

Bateson, William (1909), *Mendel's Principles of Heredity*. Cambridge, England: Cambridge University Press.

Bateson, William (1913), *Problems of Genetics*. New Haven, Connecticut: Yale University Press.

Bateson, William (1916), "Review of *The Mechanism of Mendelian Heredity*," *Science* 44. Reprinted in R. C. Punnett (ed.), 1928, *Scientific Papers of William Bateson*, V. 2. Cambridge, England: Cambridge University Press, pp. 452-463.

Bateson, William (1922), "Evolutionary Faith and Modern Doubts," *Science* 55. Reprinted in W. Bateson, 1928, *William Bateson, F.R.S., His Essays and Addresses, with a Memoir by Beatrice Bateson*. Cambridge, England: Cambridge University Press, pp. 387-398.

Bateson, William (1928), *William Bateson, F.R.S., His Essays and Addresses, with a Memoir by Beatrice Bateson*. Cambridge, England: Cambridge University Press. Reprinted by Garland, New York, 1984.

Bateson, William (unpublished), *Bateson Papers*, microfilm copy, American Philosophical Society, Philadelphia, Pennsylvania.

Bateson, William, and R. C. Punnett (1905), "A Suggestion as to the Nature of the 'Walnut' Comb in Fowls," *Proceedings of the Cambridge Philosophical Society* 13. Reprinted in R. C. Punnett (ed.), 1928, *Scientific Papers of William Bateson*, V. 2. Cambridge, England: Cambridge University Press, pp. 135-138.

Bateson, William, and R. C. Punnett (1908), "The Heredity of Sex," *Science* 27. Reprinted in R. C. Punnett (ed.), 1928, *Scientific Papers of William Bateson*, V. 2. Cambridge, England: Cambridge University Press, pp. 179-182.

Bateson, William, and R. C. Punnett (1911a), "On Gametic Series Involving Reduplication of

Certain Terms," *Journal of Genetics* 1:239-302. Reprinted in R. C. Punnett (ed.), 1928, *Scientific Papers of William Bateson*, V. 2. Cambridge, England: Cambridge University Press, pp. 206-215.

Bateson, William, and R. C. Punnett (1911b), "On the Interrelations of Genetic Factors," *Proceedings of the Royal Society*, B, 84. Reprinted in R. C. Punnett (ed.), 1928, *Scientific Papers of William Bateson*, V. 2. Cambridge, England: Cambridge University Press, pp. 216-220.

Bateson, William, and R. C. Punnett (1911c), "The Inheritance of the Peculiar Pigmentation of the Silky Fowl," *Journal of Genetics* 1. Reprinted in R. C. Punnett (ed.), 1928, *Scientific Papers of William Bateson*, V. 2. Cambridge, England: Cambridge University Press, pp. 188-205.

Bateson, William, and E. R. Saunders (1902), "The Facts of Heredity in the Light of Mendel's Discovery," *Reports to the Evolution Committee of the Royal Society*, Report I. London: Harrison and Sons. Reprinted in R. C. Punnett (ed.), 1928, *Scientific Papers of William Bateson*, V. 2. Cambridge, England: Cambridge University Press, pp. 29-73.

Bateson, William; E. R. Saunders; and R. C. Punnett (1905), "Further Experiments on Inheritance in Sweet Peas and Stocks: Preliminary Account," *Proceedings of the Royal Society*, B, 77. Reprinted in R. C. Punnett (ed.), 1928, *Scientific Papers of William Bateson*, V. 2. Cambridge, England: Cambridge University Press, pp. 139-141.

Bateson, William; E. R. Saunders; and R. C. Punnett (1906), "Experimental Studies in the Physiology of Heredity," *Reports to the Evolution Committee of the Royal Society III*. Reprinted in R. C. Punnett (ed.), 1928, *Scientific Papers of William Bateson*, V. 2. Cambridge, England: Cambridge University Press, pp. 152-161.

Baxter, Alice Levine (1974), "Edmund Beecher Wilson and the Problem of Development," Ph.D. dissertation, Yale University, New Haven, Connecticut.

Baxter, Alice Levine (1976), "Edmund Wilson as a Preformationist: Some Reasons for his Acceptance of the Chromosome Theory," *Journal of the History of Biology* 9:29-57.

Baxter, Alice Levine, and John Farley (1979), "Mendel and Meiosis," *Journal of the History of Biology* 12:137-174.

Bayliss, William M. (1923), *The Colloidal State in Its Medical and Physiological Aspects*. London: Henry Frowde and Hodder and Stoughton.

Bechtel, William (1984), "Reconceptualizations and Interfield Connections: The Discovery of the Link Between Vitamins and Coenzymes," *Philosophy of Science* 51:265-292.

Bechtel, William (1986), "Introduction: The Nature of Scientific Integration," in W. Bechtel (ed.), *Integrating Scientific Disciplines*. Dordrecht: Nijhoff, pp. 3-52.

Bechtel, William (1988), *Philosophy of Science: An Overview for Cognitive Science*. Hillsdale, New Jersey: Erlbaum.

Black, Max (1962), *Models and Metaphors*. Ithaca, New York: Cornell University Press.

Boveri, Theodor (1902), "On Multipolar Mitosis as a Means of Analysis of the Cell Nucleus," in B. H. Willier and J. Oppenheimer (eds.), 1964, *Foundations of Experimental Embryology*. Englewood Cliffs, New Jersey: Prentice Hall, pp. 75-97.

Boveri, Theodor (1903), "Über die Konstitution der chromatischen Kernsubstanz," *Verhandlungen der deutschen zoologischen Gesellschaft zu Würzburg* 13:10-33.

Boveri, Theodor (1904), *Ergebnisse über die Konstitution der chromatischen Substanz des Zellkerns*. Jena: Gustav Fischer.

Bowler, Peter J. (1983), *The Eclipse of Darwinism*. Baltimore: Johns Hopkins University Press.

Bowler, Peter J. (1984), *Evolution: The History of An Idea*. Berkeley: University of California Press.

Bowler, Peter J. (1989), *Evolution: The History of An Idea*. Revised edition. Berkeley: University of California Press.

Boyd, Richard (1979), "Metaphor and Theory Change: What is 'Metaphor' a Metaphor For?" in A. Ortony (ed.), *Metaphor and Thought*. Cambridge, England: Cambridge University Press, pp. 356-408.

Brachman, Ronald J., and Hector L. Levesque (eds.) (1985), *Readings in Knowledge Representation*. Los Altos, California: Morgan Kaufmann.

Bridges, C. B. (1913), "Non-disjunction of the Sex Chromosomes of Drosophila," *Journal of Experimental Zoology* 15:587-606.

Bridges, C. B. (1914), "Direct Proof through Non-disjunction that the Sex-linked Genes of Drosophila are Borne by the X-chromosome," *Science* 40:107-109.

Bridges, C. B. (1916), "Non-disjunction as Proof of the Chromosome Theory of Heredity," *Genetics* 1:1-52, 107-163.

Brown, Harold I. (1977), *Perception, Theory, and Commitment: The New Philosophy of Science*. Chicago: University of Chicago Press.

Brush, Stephen (1978), "Nettie M. Stevens and the Discovery of Sex Determination by Chromosomes," *Isis* 69:163-172.

Brush, Stephen (1989), "Prediction and Theory Evaluation: The Case of Light Bending," *Science* 246:1124-1129.

Buchanan, Bruce (1982), "Mechanizing the Search for Explanatory Hypotheses," in Peter Asquith and Thomas Nickles (eds.), *PSA 1982*, V. 2. East Lansing, Michigan: Philosophy of Science Association, pp. 129-146.

Buchanan, Bruce (1985), "Steps Toward Mechanizing Discovery," in K. Schaffner (ed.), *Logic of Discovery and Diagnosis in Medicine*. Berkeley: University of California Press, pp. 94-114.

Buchanan, Bruce, and Edward Shortliffe (eds.) (1984), *Rule-Based Expert Systems: The MYCIN Experiments of the Stanford Heuristic Programming Project*. Reading, Massachusetts: Addison-Wesley.

Burian, Richard M. (1987), "Ontological Progress in Science," unpublished manuscript. Department of Philosophy, Virginia Polytechnic Institute and State University, Blacksburg, Virginia.

Burian, Richard M. (1987), "Challenges to the Evolutionary Synthesis," *Evolutionary Biology* 43:247-269.

Burian, Richard M.; Jean Gayon; and Doris Zallen (1988), "The Singular Fate of Genetics in the History of French Biology, 1900-1940," *Journal of the History of Biology* 21:357-402.

Burnet, F. M. (1957), "A Modification of Jerne's Theory of Antibody Production Using the Concept of Clonal Selection," *The Australian Journal of Science* 20:67-69.

Burstein, Mark (1986), "Concept Formation by Incremental Analogical Reasoning and Debugging," in R. S. Michalski, J. Carbonell, and T. Mitchell (eds.), *Machine Learning*, V. 2. Los Altos, California: Morgan Kaufmann, pp. 351-369.

Burstein, Mark (1988), "Combining Analogies in Mental Models," in David Helman (ed.), *Analogical Reasoning*. Dordrecht: Reidel, pp. 179-203.

Cannon, William A. (1902), "A Cytological Basis for the Mendelian Laws," *Bulletin of the Torrey Botanical Club* 29:657-661.

Carlson, Elof A. (1966), *The Gene: A Critical History*. Philadelphia: Saunders.

Carlson, Elof A. (1981), *Genes, Radiation and Society: The Life and Work of H. J. Muller*. Ithaca, New York: Cornell University Press.

Carothers, Eleanor E. (1913), "The Mendelian Ratio in Relation to Certain Orthopteran Chromosomes," *The Journal of Morphology* 24:487-509. Reprinted in B. Voeller (ed.), 1968, *The Chromosome Theory of Inheritance: Classic Papers in Development and Heredity*. New York: Appleton-Century-Crofts, pp. 175-177.

Castle, W. E. (1903), "Mendel's Law of Heredity," *Science* 18:396-406.

Castle, W. E. (1906), "Yellow Mice and Gametic Purity," *Science* n.s. 24:275-281.

Castle, W. E. (1907), "Experimental Evolution in Lawrence Hall," *The Harvard Graduates' Magazine* 16:244-248.

Castle, W. E. (1909), "The Behavior of Unit Characters in Heredity," in *Fifty Years of Darwinism*. New York: Henry Holt, pp. 143-159.

Castle, W. E. (1911), *Heredity in Relation to Evolution and Animal Breeding*. New York: D. Appleton.

Castle, W. E. (1912), "The Inconstancy of Unit-Characters," *American Naturalist* 46:352-362.

Castle, W. E. (1914), "Mr. Muller on the Constancy of Mendelian Factors," *American Naturalist* 49:37-42.

Castle, W. E. (1916a), *Genetics and Eugenics*. Cambridge, Massachusetts: Harvard University Press.

Castle, W. E. (1916b), "Pure Lines and Selection," *Journal of Heredity* 5:93-97.

Castle, W. E. (1919a), "Are Genes Linear or Non-Linear in Arrangement?" *Proceedings of the National Academy of Sciences* 5:500-506.

Castle, W. E. (1919b), "Is the Arrangement of the Genes in the Chromosome Linear?" *Proceedings of the National Academy of Sciences* 5:25-32.

Castle, W. E. (1919c), "Piebald Rats and Selection, A Correction," *American Naturalist* 53:370-376.

Castle, W. E. (1919d), "Piebald Rats and the Theory of Genes," *Proceedings of the National Academy of Sciences* 5:126-130.

Castle, W. E. (1920), *Genetics and Eugenics*. 2nd edition. Cambridge, Massachusetts: Harvard University Press.

Castle, W. E. (1924), *Genetics and Eugenics*. 3rd edition. Cambridge, Massachusetts: Harvard University Press.

Castle, W. E., and Glover M. Allen (1903a), "Mendel's Law and the Heredity of Albinism," in *The Mark Anniversary Volume*, Article 19, pp. 379-398.

Castle, W. E., and Glover M. Allen (1903b), "The Heredity of Albinism," *Proceedings of the American Academy of Arts and Sciences* 38:603-622.

Castle, W. E., and Alexander Forbes (1906), "Heredity of Hair-Length in Guinea Pigs and Its Bearing on the Theory of Pure Gametes," *Carnegie Institution of Washington*, Publication No. 49, pp. 5-14.

Castle, W. E., and C. C. Little (1910), "On a Modified Mendelian Ratio among Yellow Mice," *Science* 32:868-870.

Castle, W. E., and John C. Phillips (1914), "Piebald Rats and Selection: An Experimental Test of the Effectiveness of Selection and of the Theory of Gametic Purity in Mendelian Crosses," *Carnegie Institution of Washington*, Publication No. 195, pp. 1-56.

Chandrasekaran, B., and Sanjay Mittal (1983), "Deep versus Compiled Knowledge Approaches to Diagnostic Problem-solving," *International Journal of Man-Machine Studies* 19:425-436.

Chargaff, E. (1950), "Chemical Specificity of Nucleic Acids and Mechanism of Their Enzymatic Degradation," *Experientia* 6:201-209. Reprinted in J. Herbert Taylor (ed.), 1965, *Selected Papers on Molecular Genetics*. New York: Academic Press, pp. 245-253.

Charniak, Eugene, and Drew McDermott (1985), *Introduction to Artificial Intelligence*. Reading, Massachusetts: Addison-Wesley.

Churchill, Frederick B. (1970), "Hertwig, Weismann, and the Meaning of Reduction Division circa 1890," *Isis* 61:429-457.

Churchill, Frederick B. (1974), "William Johannsen and the Genotype Concept," *Journal of the History of Biology* 7:5-30.

Churchill, Frederick B. (1987), "From Heredity Theory to *Vererbung*, The Transmission Problem, 1850-1915," *Isis* 78:337-364.

Clement, John (1988), "Observed Methods for Generating Analogies in Scientific Problem Solving," *Cognitive Science* 12:563-586.

Cock, A. G. (1983), "William Bateson's Rejection and Eventual Acceptance of Chromosome Theory," *Annals of Science* 40:19-60.

Coleman, William (1965), "Cell, Nucleus, and Inheritance: A Historical Study," *Proceedings of the American Philosophical Society* 109:124-158.

Coleman, William (1970), "Bateson and Chromosomes: Conservative Thought in Science," *Centaurus* 15:228-314.

Conant, James B. (ed.) (1948), "The Overthrow of the Phlogiston Theory: The Chemical Revolution 1775-1789," *Harvard Case Histories in Experimental Science*, V. 1. Cambridge, Massachusetts: Harvard University Press, pp. 67-115.

Correns, Carl (1900), "G. Mendel's Regel über das Verhalten der Nachkommenshaft der Rassenbastarde," *Berichte der deutschen botanischen Gesellschaft* 18:158-168. English translation: "G. Mendel's Law Concerning the Behavior of Progeny of Varietal Hybrids," *Genetics* 35 (1950):33-41. Reprinted in C. Stern and E. Sherwood (eds.), 1966, *The Origin of Genetics, A Mendel Source Book*. Translated by Leonie Kellen Piternick. San Francisco: W. H. Freeman, pp. 119-132.

Crane, Diana (1972), *Invisible Colleges*. Chicago: University of Chicago Press.

Creighton, Harriet B., and Barbara McClintock (1931), "A Correlation of Cytological and Genetical Crossing-Over in *Zea mays*," *Proceedings of the National Academy of Sciences* 17:492-497. Reprinted in J. A. Peters (ed.), 1959, *Classic Papers in Genetics*. Englewood Cliffs, New Jersey: Prentice-Hall, pp. 155-160.

Cuénot, L. (1902), "La Loi De Mendel Et L'Hérédité De La Pigmentation Chez Les Souris," *Archives de Zoologie Expérimentale et Générale* 3 series, notes, pp. 27-30.

Cuénot, L. (1904), "L'Hérédité De La Pigmentation Chez Les Souris (3 Note)," *Archives de Zoologie Expérimentale et Générale* 4 series, notes et revue, pp. 45-46.

Cuénot, Lucien (1905), "Les Races Pures Et Leurs Combinaisons Chez Les Souris," *Archives de Zoologie Expérimentale et Générale* 4 Series, t 111:123-132.

Dalton, John (1808), *A New System of Chemical Philosophy*. Reprinted by The Citadel Press, New York, 1964.

Darden, Lindley (1974), "Reasoning in Scientific Change: The Field of Genetics at Its Beginnings," Ph.D. dissertation, Committee on the Conceptual Foundations of Science, University of Chicago, Chicago, Illinois.

Darden, Lindley (1976), "Reasoning in Scientific Change: Charles Darwin, Hugo de Vries, and the Discovery of Segregation," *Studies in the History and Philosophy of Science* 7:127-169.

Darden, Lindley (1977), "William Bateson and the Promise of Mendelism," *Journal of the History of Biology* 10:87-106.

Darden, Lindley (1978), "Discoveries and the Emergence of New Fields in Science," in P. D. Asquith and I. Hacking (eds.), *PSA 1978*, V. 1. East Lansing, Michigan: Philosophy of Science Association, pp. 149-160.

Darden, Lindley (1980a), "Review of Garland Allen's *Thomas Hunt Morgan*," *Philosophy of Science* 47:662-666.

Darden, Lindley (1980b), "Theory Construction in Genetics," in T. Nickles (ed.), *Scientific Discovery: Case Studies*. Dordrecht: Reidel, pp. 151-170.

Darden, Lindley (1982a), "Artificial Intelligence and Philosophy of Science: Reasoning by Analogy in Theory Construction," in T. Nickles and P. Asquith (eds.), *PSA 1982*, V. 2. East Lansing, Michigan: Philosophy of Science Association, pp. 147-165.

Darden, Lindley (1982b), "Aspects of Theory Construction in Biology," in *Proceedings of the Sixth International Congress for Logic, Methodology and Philosophy of Science*. Hanover: North Holland Publishing Co., pp. 463-477.

Darden, Lindley (1983), "Review of E. A. Carlson's *Genes, Radiation and Society: The Life and Work of H. J. Muller*," *Journal of the History of Medicine and Allied Sciences* 38:369-371.

Darden, Lindley (1985), "Hugo de Vries's Lecture Plates and the Discovery of Segregation," *Annals of Science* 42:233-242.

Darden, Lindley (1986a), "Reasoning in Theory Construction: Analogies, Interfield Connections, and Levels of Organization," in Paul Weingartner and Georg Dorn (eds.), *Foundations of Biology*. Vienna, Austria: Hölder-Pichler-Tempsky, pp. 99-107.

Darden, Lindley (1986b), "Relations among Fields in the Evolutionary Synthesis," in W. Bechtel (ed.), *Integrating Scientific Disciplines*. Dordrecht: Nijhoff, pp. 113-123.

Darden, Lindley (1987), "Viewing the History of Science as Compiled Hindsight," *AI Magazine* 8(2):33-41.

Darden, Lindley (1990), "Diagnosing and Fixing Faults in Theories," in J. Shrager and P. Langley (eds.), *Computational Models of Scientific Discovery and Theory Formation*. San Mateo, California: Morgan Kaufmann, pp. 319-346.

Darden, Lindley, and Joseph A. Cain (1989), "Selection Type Theories," *Philosophy of Science* 56:106-129.

Darden, Lindley, and Nancy Maull (1977), "Interfield Theories," *Philosophy of Science* 44:43-64.

Darden, Lindley, and Roy Rada (1988a), "Hypothesis Formation Using Part-Whole Interrelations," in David Helman (ed.), *Analogical Reasoning*. Dordrecht: Reidel, pp. 341-375.

Darden, Lindley, and Roy Rada (1988b), "Hypothesis Formation Via Interrelations," in Armand Prieditis (ed.), *Analogica*. Los Altos, California: Morgan Kaufmann, pp. 109-127.

Darwin, Charles (1859), *On the Origin of Species, A Facsimile of the First Edition*. Cambridge, Massachusetts: Harvard University Press, 1966.

Darwin, Charles (1868), *The Variation of Plants and Animals Under Domestication*. 2 volumes. New York: Orange Judd and Co.

Davenport, Charles B. (1900-1901), "Mendel's Law of Dichotomy in Hybrids," *Biological Bulletin* 2:307-310.

Davenport, Charles B. (1907), "Heredity and Mendel's Law," *Proceedings of the Washington Academy of Sciences* 9:179-188.

Dietterich, Thomas G.; B. London; K. Clarkson; and G. Dromey (1982), "Learning and Inductive Inference," in Paul R. Cohen and E. Feigenbaum (eds.), *The Handbook of Artificial Intelligence*, V. 3. Los Altos, California: Morgan Kaufmann, pp. 323-511.

Dobzhansky, Theodosius (1937), *Genetics and the Origin of Species*. New York: Columbia University Press.

Dobzhansky, Theodosius (1980), "The Birth of the Genetic Theory of Evolution in the Soviet Union in the 1920s," in E. Mayr and W. Provine (eds.), *The Evolutionary Synthesis*. Cambridge, Massachusetts: Harvard University Press, pp. 229-242.

Dunn, Leslie C. (1965a), *A Short History of Genetics*. New York: McGraw-Hill.

Dunn, Leslie C. (1965b), "William Ernest Castle, October 25, 1867-June 3, 1962," *Biographical Memoirs, National Academy of Sciences* 38:33-80.

East, Edward M. (1910), "A Mendelian Interpretation of Variation That is Apparently Continuous," *American Naturalist* 44:65-82.

East, Edward M. (1912), "The Mendelian Notation as a Description of Physiological Facts," *American Naturalist* 46:633-695.

Eaton, G. J., and M. M. Green (1962), "Implantation and Lethality of the Yellow Mouse," *Genetica* 33:106-112.

Farley, John (1982), *Gametes and Spores: Ideas about Sexual Reproduction, 1750-1914*. Baltimore: Johns Hopkins University Press.

Fisher, R. A. (1936), "Has Mendel's Work Been Rediscovered?" *Annals of Science* 1:115-137. Reprinted in C. Stern and E. Sherwood (eds.), 1966, *The Origin of Genetics, A Mendel Source Book*. San Francisco: W. H. Freeman, pp. 139-172.

Friedland, Peter, and Laurence H. Kedes (1985), "Discovering the Secrets of DNA," *Communications of the Association of Computing Machinery* 28:1164-1186.

Fruton, Joseph S. (1972), *Molecules and Life*. New York: Wiley.

Geison, Gerald (1981), "Scientific Change, Emerging Specialities and Research Schools," *History of Science* 14:20-40.

Genesereth, Michael (1980), "Metaphors and Models," in *Proceedings of the First Annual National Conference on Artificial Intelligence*. Menlo Park, California: American Association for Artificial Intelligence, pp. 208-211.

Gentner, Dedre (1983), "Structure Mapping - A Theoretical Framework for Analogy," *Cognitive Science* 7:155-170.

Gick, M., and Holyoak, K. (1980), "Analogical Problem Solving," *Cognitive Psychology* 12:306-355.

Giere, Ronald (1984), *Understanding Scientific Reasoning*. 2nd edition. New York: Holt, Rinehart and Winston.

Gilbert, Scott (1978), "The Embryological Origins of the Gene Theory," *Journal of the History of Biology* 11:307-351.

Glymour, Clark (1980), *Theory and Evidence*. Princeton, New Jersey: Princeton University Press.

Goldschmidt, R. (1917), "Crossing-over ohne Chiasmatypie?" *Genetics* 2:82-95.

Gökalp, Iskender (1987), "On the Dynamics of Controversies in a Borderland Scientific Domain: The Case of Turbulent Combustion," *Social Science Information* 26:551-576.

Gooding, David (1990), "Mapping Experiment as a Learning Process: How the First Electromagnetic Motor was Invented," *Science, Technology, & Human Values* 15:165-201.

Gorman, Michael E., and Margaret E. Gorman (1984), "A Comparison of Disconfirmatory, Confirmatory, and a Control Strategy on Wason's 2,4,6 Task," *Quarterly Journal of Experimental Psychology* 36A:629-648.

Gould, Stephen Jay (1977), "Darwin's Untimely Burial," in *Ever Since Darwin*. New York: W. W. Norton, pp. 39-45.

Gould, Stephen Jay (1982), Introduction, in T. Dobzhansky, *Genetics and The Origin of Species*. Reprint edition. New York: Columbia University Press.

Greenaway, Frank (1966), *John Dalton and the Atom*. Ithaca, New York: Cornell University Press.

Grene, Marjorie (1987), "Hierarchies in Biology," *American Scientist* 75:504-510.

Grene, Marjorie, and Everett Mendelsohn, (eds.) (1976), *Topics in the Philosophy of Biology*. Dordrecht: Reidel.

Gruber, Howard (1974), *Darwin On Man*. New York: E.P. Dutton & Co.

Gutting, Gary (1980), "Science as Discovery," *Revue Internationale de Philosophie* 131-132:26-48.

Guyer, Michael F. (1909), "Deficiencies of the Chromosome Theory of Heredity," *University Studies* 5. Cincinnati: University of Cincinnati Press, pp. 3-19.

Hall, Rogers P. (1989), "Computational Approaches to Analogical Reasoning: A Comparative Analysis," *Artificial Intelligence* 39:39-120.

Hanson, Norwood Russell (1958), *Patterns of Discovery*. Cambridge, England: Cambridge University Press. Paperback edition, 1965.

Hanson, Norwood Russell (1961), "Is There a Logic of Scientific Discovery?" in H. Feigl and G. Maxwell (eds.), *Current Issues in the Philosophy of Science*. New York: Holt, Rinehart and Winston. Reprinted in B. Brody (ed.), 1970, *Readings in the Philosophy of Science*. Englewood Cliffs, New Jersey: Prentice Hall, pp. 620-633.

Harman, Gilbert (1986), *Change in View: Principles of Reasoning*. Cambridge, Massachusetts: MIT Press.

Harwood, Jonathan (1984), "The Reception of Morgan's Chromosome Theory in Germany: Inter-war Debate over Cytoplasmic Inheritance," *Medizin historisches Journal* 19:3-32.

Harwood, Jonathan (1985), "Geneticists and the Evolutionary Synthesis in Inter-war Germany," *Annals of Science* 42:279-301.

Heimans, J. (1962), "Hugo de Vries and the Gene Concept," *American Naturalist* 96:93-104.

Heimans, J. (1978), "Hugo de Vries and the Gene Theory," in E. G. Forbes (ed.), *Human Implications of Scientific Advance, Proceedings of the XVth International Congress of the History of Science*. Edinburgh: Edinburgh University Press, pp. 469-480.

Helman, David (ed.) (1988), *Analogical Reasoning*. Dordrecht: Reidel.

Hempel, Carl G. (1965), *Aspects of Scientific Explanation*. New York: The Free Press, Macmillan.

Herbert, Sandra (1971), "Darwin, Malthus, and Selection," *Journal of the History of Biology* 4:209-217.

Herbert, Sandra (ed.) (1980), *The Red Notebook of Charles Darwin: Bulletin of the British Museum (Natural History)* 7:1-164.

Hesse, Mary (1966), *Models and Analogies in Science*. Notre Dame, Indiana: University of Notre Dame Press.

Holland, John; Keith Holyoak; Richard Nisbett; and Paul Thagard (1986), *Induction: Processes of Inference, Learning, and Discovery*. Cambridge, Massachusetts: MIT Press.

Holmes, Frederick Lawrence (1985), *Lavoisier and the Chemistry of Life: An Exploration of Scientific Creativity*. Madison: University of Wisconsin Press.

Holton, Gerald (1973), *Thematic Origins of Scientific Thought*. Cambridge, Massachusetts: Harvard University Press.

Holyoak, Keith J., and Paul Thagard (1989), "Analogical Mapping by Constraint Satisfaction," *Cognitive Science* 13:295-355.

Hughes, Arthur (1959), *A History of Cytology*. New York: Abelard-Schuman.

Hull, David (1988), "A Mechanism and Its Metaphysics: An Evolutionary Account of the Social and Conceptual Development of Science," *Biology and Philosophy* 3:123-155.

Humphreys, Willard C. (1968), *Anomalies and Scientific Theories*. San Francisco: Freeman, Cooper and Co.

Iltis, Hugo (1966), *Life of Mendel*. New York: Hafner.

Janssens, F. A. (1909), "La theorie de la chiasmatypie," *La Cellule* 25:389-411.

Jerne, Niels K. (1955), "The Natural-Selection Theory of Antibody Formation," *Proceedings of the National Academy of Sciences* 41:849-857.

Johannsen, Wilhelm (1903), *Ueber Erblichkeit in Populationen und in reinen Linien*. Jena: Gustav Fischer. Selections translated and reprinted in J. A. Peters (ed.), 1959, pp. 20-26.

Johannsen, Wilhelm L. (1909), *Elemente der Exakten Erblichkeitslehre*. Jena: Gustav Fischer.

Johannsen, Wilhelm (1911), "The Genotype Conception of Heredity," *American Naturalist* 45:129-159.

Johannsen, Wilhelm L. (1913), *Elemente der Exakten Erblichkeitslehre*. 2nd edition. Jena: Gustav Fischer.

Johannsen, Wilhelm (1923), "Some Remarks about Units in Heredity," *Hereditas* 4:133-141.

Johannsen, Wilhelm L. (1926), *Elemente der Exakten Erblichkeitslehre*. 3rd edition. Jena: Gustav Fischer.

Josephson, J.; B. Chandrasekaran; J. Smith; and M. Tanner (1987), "A Mechanism for Forming Composite Explanatory Hypotheses," *IEEE Transactions on Systems, Man, and Cybernetics*, V. SMC-17, no. 3, May/June 1987, pp. 445-454.

Judson, Horace Freeland (1979), *The Eighth Day of Creation*. New York: Simon and Schuster.

Kacser, Henrik, and James Burns (1981), "The Molecular Basis of Dominance," *Genetics* 97:639-666.

Karp, Peter (1989), "Hypothesis Formation and Qualitative Reasoning in Molecular Biology." Ph.D. dissertation, Stanford University, Stanford, California. Available from the Computer Science Department: STAN-CS-89-1263.

Kaufmann, Stuart A. (1976), "Articulation of Parts Explanation in Biology and the Rational Search for Them," in M. Grene and E. Mendelsohn, (eds.), *Topics in the Philosophy of Biology*. Dordrecht: Reidel, pp. 245-263.

Kay, Lily E. (1986), "Cooperative Individualism and the Growth of Molecular Biology at the California Institute of Technology," Ph.D. dissertation, Department of the History of Science, The Johns Hopkins University, Baltimore, Maryland.

Kirkham, W. B. (1919), "The Fate of Homozygous Yellow Mice," *Journal of Experimental Zoology* 28:125-135.

Kitcher, Philip (1981), "Explanatory Unification," *Philosophy of Science* 48:507-531.

Kitcher, Philip (1982), "Genes," *British Journal for the Philosophy of Science* 33:337-359.

Kitcher, Philip (1984), "1953 and All That: A Tale of Two Sciences," *The Philosophical Review* 93:335-373.

Kitcher, Philip (1989), "Explanatory Unification and the Causal Structure of the World," in Philip Kitcher and Wesley Salmon (eds.), *Scientific Explanation. Minnesota Studies in the Philosophy of Science*, V. 13. Minneapolis: University of Minnesota Press, pp. 410-505.

Kohler, Robert E. (1973), "The Enzyme Theory and the Origin of Biochemistry," *Isis* 64:181-196.

Kohler, Robert (1982), *From Medical Chemistry to Biochemistry: The Making of a Biomedical Discipline*. New York: Cambridge University Press.

Kohn, David (1980), "Theories to Work By: Rejected Theories, Reproduction and Darwin's Path to Natural Selection," *Studies in History of Biology* 4:67-170.

Kottler, Malcolm (1979), "Hugo de Vries and the Rediscovery of Mendel's Laws," *Annals of Science* 36:517-538.

Kragh, Helge (1987), *An Introduction to the Historiography of Science*. New York: Cambridge University Press.

Kuhn, Thomas (1962), *The Structure of Scientific Revolutions*. Chicago: University of Chicago Press.

Kuhn, Thomas (1970), *The Structure of Scientific Revolutions*. 2nd edition. Chicago: The University of Chicago Press.

Kuhn, Thomas (1974), "Second Thoughts on Paradigms," in F. Suppe (ed.), *The Structure of Scientific Theories*. Urbana: University of Illinois Press, pp. 459-482.

Kupferberg, Eric David (1989), "The International Society for Microbiology and The Promotion of Interfield Connections," M.A. thesis, Committee on the History and Philosophy of Science, University of Maryland, College Park, Maryland.

Lakatos, Imre (1970), "Falsification and the Methodology of Scientific Research Programmes," in I. Lakatos and Alan Musgrave (eds.), *Criticism and the Growth of Knowledge*. Cambridge, England: Cambridge University Press, pp. 91-195.

Lakatos, Imre (1976), *Proofs and Refutations: The Logic of Mathematical Discovery*. Cambridge, England: Cambridge University Press.

Lakoff, George, and Mark Johnson (1980), *Metaphors We Live By*. Chicago: University of Chicago Press.

Langley, Pat; Herbert Simon; Gary L. Bradshaw; and Jan M. Zytkow (1987), *Scientific Discovery: Computational Explorations of the Creative Process*. Cambridge, Massachusetts: MIT Press.

Larkin, Jill H., and Herbert Simon (1987), "Why a Diagram is (Sometimes) Worth Ten Thousand Words," *Cognitive Science* 11:65-99.

Latour, Bruno, and Steve Woolgar (1986), *Laboratory Life: The Construction of Scientific Facts*. 2nd edition. Princeton, New Jersey: Princeton University Press.

Laudan, Larry (1977), *Progress and Its Problems*. Berkeley: University of California Press.

Laudan, Larry (1980), "Why Was the Logic of Discovery Abandoned?" in T. Nickles (ed.), *Scientific Discovery, Logic and Rationality*. Dordrecht: Reidel, pp. 173-183.

Laudan, Larry; Arthur Donovan; Rachel Laudan; Peter Barker; Harold Brown; Jarrett Leplin; Paul Thagard; and Steve Wykstra (1986), "Scientific Change: Philosophical Models and Historical Research," *Synthese* 69:141-223.

Lederman, Muriel (1989), "Research Note: Genes on Chromosomes: The Conversion of Thomas Hunt Morgan," *Journal of the History of Biology* 22:163-176.

Leicester, Henry M. (1974), *Development of Biochemical Concepts from Ancient to Modern Times*. Cambridge, Massachusetts: Harvard University Press.

Lenat, Douglas (1976), "AM: An Artificial Intelligence Approach to Discovery in Mathematics as Heuristic Search," Ph.D. dissertation, Department of Computer Science, Stanford University, Stanford, California.

Leplin, Jarrett (1975), "The Concept of an *Ad Hoc* Hypothesis," *Studies in the History and Philosophy of Science* 5:309-345.

Leplin, Jarrett (ed.) (1984), *Scientific Realism*. Berkeley: University of California Press.

Lewis, E. B. (ed.) (1961), *Selected Papers of A. H. Sturtevant: Genetics and Evolution*. San Francisco: W. H. Freeman.

Lewontin, Richard C. (1974), "Darwin and Mendel—The Materialist Revolution," in Jerzy Neyman (ed.), *The Heritage of Copernicus*. Cambridge, Massachusetts: MIT Press, pp. 166-183.

Lock, R. H. (1906), *Recent Progress in the Study of Variation, Heredity, and Evolution*. London: John Murray.

Lock, R. H. (1909), *Recent Progress in the Study of Variation, Heredity , and Evolution*. 2nd edition. London: John Murray.

Losee, John (1987), *Philosophy of Science and Historical Enquiry*. New York: Oxford University Press.

Maienschein, Jane (1987), "Heredity/Development in the United States, circa 1900," *History and Philosophy of Life Sciences* 9:79-93.

Manier, Edward (1969), "The Experimental Method in Biology, T. H. Morgan and the Theory of the Gene," *Synthese* 20:185-205.

Maull, Nancy (1977), "Unifying Science Without Reduction," *Studies in the History and Philosophy of Science* 8:143-162.

Mayr, Ernst (1942), *Systematics and the Origin of Species*. New York: Columbia University Press. Reprinted by Dover, New York, 1964.

Mayr, Ernst (1982), *The Growth of Biological Thought*. Cambridge, Massachusetts: Harvard University Press.

Mayr, Ernst, and W. Provine (eds.) (1980), *The Evolutionary Synthesis*. Cambridge, Massachusetts: Harvard University Press.

McClung, C. E. (1902), "The Accessory Chromosome - Sex Determinant?" *Biological Bulletin* 3:43-84. Reprinted in B. Voeller (ed.), 1968, *The Chromosome Theory of Inheritance: Classic Papers in Development and Heredity*. New York: Appleton-Century-Crofts, pp. 76-77.

McKusick, Victor (1960), "Walter S. Sutton and the Physical Basis of Mendelism," *Bulletin of the History of Medicine* 34:487-497.

McMullin, Ernan (1982), "Values in Science," in P. D. Asquith and T. Nickles (eds.), *PSA 1982*, V. 2. East Lansing: Philosophy of Science Association, pp. 3-28.

McMullin, Ernan (1984), "A Case for Scientific Realism," in Jarrett Leplin (ed.), *Scientific Realism*. Berkeley: University of California Press, pp. 8-40.

Meijer, Onno (1983), "The Essence of Mendel's Discovery," in V. Orel and A. Matalová (eds.), *Gregor Mendel and the Foundation of Genetics*. Brno, Czechoslovakia: The Mendelianum of the Moravian Museum, pp. 128-178.

Meijer, Onno (1985), "Hugo de Vries no Mendelian?" *Annals of Science* 42:189-232.

Meijer, Onno (1987), "The Mendelian Revolution," presented at the Colloquium Series of the Committee on the History and Philosophy of Science, University of Maryland, College Park, Maryland, October, 1987.

Mendel, Gregor (1865), "Versuche über Pflanzen-Hybriden," *Verhandlungen des naturforschenden Vereines in Brünn* 4:1-47. English translation: "Experiments on Plant Hybrids," in C. Stern and E. Sherwood (eds.), 1966, *The Origin of Genetics: A Mendel Source Book*. Translated by Eva Sherwood. San Francisco: W. H. Freeman, pp. 1-48.

Mendel, Gregor (1869), "On Hieracium-Hybrids Obtained by Artificial Fertilization," in C. Stern and E. Sherwood (eds.), 1966, *The Origin of Genetics: A Mendel Source Book*. San Francisco: W.H. Freeman, pp. 49-50.

Miller, Arthur I. (1984), *Imagery in Scientific Thought*. Boston: Birkhauser.

Minsky, Marvin (1981), "A Framework for Representing Knowledge," in John Haugeland (ed.), *Mind Design*. Montgomery, Vermont: Bradford Books, pp. 95-128.

Mitchell, Tom M. (1982), "Generalization as Search," *Artificial Intelligence* 18:203-226.

Moberg, Dale, and Josephson, John (1990),"An Implementation Note" in J. Shrager and P. Langley (eds.), *Computational Models of Scientific Discovery and Theory Formation*. San Mateo, California: Morgan Kaufmann, pp.347-353.

Monaghan, Floyd, and A. Corcos (1984), "On the Origin of the Mendelian Laws," *The Journal of Heredity* 75:67-69.

Monk, Robert A (1977), "The Logic of Discovery," *Philosophy Research Archives* 3:A10-E6.

Montgomery, Thomas H., Jr. (1901), "A Study of the Chromosomes of the Germ Cells of Metazoa," *Transactions of the American Philosophical Society* 20:154-230. Excerpt reprinted in B. Voeller (ed.), 1968, *The Chromosome Theory of Inheritance: Classic Papers in Development and Heredity*. New York: Appleton-Century-Crofts, pp. 70-75.

Moore, John A. (1972), *Heredity and Development*. 2nd edition. New York: Oxford University Press.

Morgan, Thomas H. (1905), "The Assumed Purity of the Germ Cells in Mendelian Results," *Science* 22:877-879.

Morgan, Thomas H. (1907), "Review of *Inheritance in Poultry* by C. B. Davenport," *Science* 25:464-466.

Morgan, Thomas H. (1909a), "A Biological and Cytological Study of Sex Determination in Phylloxerans and Aphids," *Journal of Experimental Zoology* 7:239-352.

Morgan, Thomas H. (1909b), "Recent Experiments on the Inheritance of Coat Colors in Mice," *American Naturalist* 43:494-510.

Morgan, Thomas H. (1909c), "Sex Determination and Parthenogenesis in Phylloxerans and Aphids," *Science* 29:234-237.

Morgan, Thomas H. (1909d), "What are 'Factors' in Mendelian Explanations?" *Proceedings American Breeder's Association* 5:365-368.

Morgan, Thomas H. (1910a), "Chromosomes and Heredity," *American Naturalist* 44:449-496.

Morgan, Thomas H. (1910b), "Hybridization in a Mutating Period in *Drosophila*," *Society for Experimental Biology and Medicine Proceedings* 7:160-161.

Morgan, Thomas H. (1910c), "Sex Limited Inheritance in *Drosophila*," *Science* 32:120-122.

Morgan, Thomas H. (1910d), "The Chromosomes in the Parthenogenetic and Sexual Eggs of Phylloxerans and Aphids (Abstract)," *Society for Experimental Biology and Medicine Proceedings* 7:161-162.

Morgan, Thomas H. (1911a), "An Attempt to Analyze the Constitution of the Chromosomes on the Basis of Sex-Limited Inheritance in *Drosophila*," *Journal of Experimental Zoology* 11:365-413.

Morgan, Thomas H. (1911b), "Random Segregation Versus Coupling in Mendelian Inheritance," *Science* 34:384.

Morgan, Thomas H. (1911c), "The Application of the Conception of Pure Lines to Sex-limited Inheritance and to Sexual Dimorphism," *American Naturalist* 45:65-78.

Morgan, Thomas H. (1911d), "The Influence of Heredity and of Environment in Determining the Coat Colors in Mice," *New York Academy of Science Annals* 21:87-117.

Morgan, Thomas H. (1913a), "Factors and Unit Characters in Mendelian Heredity," *American Naturalist* 47:5-16.

Morgan, Thomas H. (1913b), "Simplicity versus Adequacy in Mendelian Formulae," *American Naturalist* 47:372-74.

Morgan, Thomas H. (1914), "Multiple Allelomorphs in Mice," *American Naturalist* 48:449-58.

Morgan, Thomas H. (1917), "The Theory of the Gene," *American Naturalist* 51:513-544.

Morgan, Thomas H. (1919), *The Physical Basis of Heredity.* Philadelphia: J. B. Lippincott Co.

Morgan, Thomas H. (1926), *The Theory of the Gene.* New Haven: Yale University Press.

Morgan, Thomas (1928), *The Theory of the Gene.* 2nd edition. New Haven: Yale University Press.

Morgan, Thomas H.; A. H. Sturtevant; H. J. Muller; and C. B. Bridges (1915), *The Mechanism of Mendelian Heredity.* New York: Henry Holt and Company.

Morgan, Thomas H.; A. H. Sturtevant; H. J. Muller; and C. B. Bridges (1922), *The Mechanism of Mendelian Heredity.* 2nd revised edition. New York: Henry Holt and Company.

Morowitz, Harold, and Temple Smith (1987), "Report of the Matrix of Biological Knowledge Workshop, July 13-August 14, 1987," Sante Fe Institute, 1120 Canyon Road, Sante Fe, New Mexico 87501.

Muller, H. J. (1914), "The Bearing of the Selection Experiments of Castle and Phillips on the Variability of Genes," *American Naturalist* 48. Reprinted in H. J. Muller, 1962, *Studies in Genetics: The Selected Papers of H. J. Muller.* Bloomington: Indiana University Press, pp. 61-69.

Muller, H. J. (1920), "Are the Factors of Heredity Arranged in a Line?" *American Naturalist* 54. Reprinted in H. J. Muller, 1962, *Studies in Genetics: The Selected Papers of H. J. Muller.* Bloomington: Indiana University Press, pp. 36-53.

Muller, H. J. (1922), "Variation due to Change in the Individual Gene," *American Naturalist* 56. Reprinted in H. J. Muller, 1962, *Studies in Genetics: The Selected Papers of H. J. Muller.* Bloomington: Indiana University Press, pp. 175-188.

Muller, H. J. (1962), *Studies in Genetics: The Selected Papers of H. J. Muller.* Bloomington: Indiana University Press.

Nersessian, Nancy J. (1984), *Faraday to Einstein: Constructing Meaning in Scientific Theories.* Dordrecht: Nijhoff.

Newton-Smith, W. H. (1981), *The Rationality of Science.* Boston: Routledge & Kegan Paul.

Nickles, Thomas (ed.) (1980a), *Scientific Discovery: Case Studies.* Dordrecht: Reidel.

Nickles, Thomas (ed.) (1980b), *Scientific Discovery, Logic and Rationality.* Dordrecht: Reidel.

Nickles, Thomas (1981), "What is a Problem That We May Solve It?" *Synthese* 47:85-118.

Nickles, Thomas (1985), "Beyond Divorce: Current Status of the Discovery Debate," *Philosophy of Science* 52:177-206.

Nickles, Thomas (1987a), "From Natural Philosophy to Metaphilosophy of Science," in Robert Kargon and Peter Achinstein (eds.), *Kelvin's Baltimore Lectures and Modern Theoretical Physics: Historical and Philosophical Perspectives.* Cambridge, Massachusetts: MIT Press, pp. 507-541.

Nickles, Thomas (1987b), "Lakatosian Heuristics and Epistemic Support," *British Journal for the Philosophy of Science* 38:181-205.

Nickles, Thomas (1987c), "Methodology, Heuristics, and Rationality," in J. C. Pitt and M. Pera (eds.), *Rational Changes in Science*. Dordrecht: Reidel, pp. 103-132.

Nickles, Thomas (1987d), "'Twixt Method and Madness," in Nancy Nersessian (ed.), *The Process of Science*. Dordrecht: Nijhoff, pp. 41-67.

Nickles, Thomas (1988), "Questioning and Problems in Philosophy of Science: Problem-Solving Versus Directly Truth-Seeking Epistemologies," in Michael Meyer (ed.), *Questions and Questioning*. New York: Walter de Gruyter, pp. 43-67.

Nisbett, Richard, and Lee Ross (1980), *Human Inference: Strategies and Shortcomings of Social Judgment*. Englewood Cliffs, New Jersey: Prentice-Hall.

Olby, Robert (1974), *The Path to the Double Helix*. Seattle, Washington: University of Washington Press.

Olby, Robert (1979), "The Significance of the Macromolecules in the Historiography of Molecular Biology," *History and Philosophy of Life Sciences* 1:185-198.

Olby, Robert (1985), *Origins of Mendelism*. 2nd edition. Chicago: University of Chicago Press.

Olby, Robert (1987), "William Bateson's Introduction of Mendelism to England: A Reassessment," *British Journal of History of Science* 20:399-420.

Oppenheim, Paul, and Hilary Putnam (1958), "Unity of Science as a Working Hypothesis," in Herbert Feigl, Michael Scriven, and Grover Maxwell (eds.), *Concepts, Theories and the Mind-Body Problem, Minnesota Studies in the Philosophy of Science*, V. 2. Minneapolis: University of Minnesota Press, pp. 3-36.

Oppenheimer, Robert (1956), "Analogy in Science," *American Psychologist* 11:127-135.

Orel, V., and A. Matalová (eds.) (1983), *Gregor Mendel and the Foundation of Genetics*. Brno, Czechoslovakia: The Mendelianum of the Moravian Museum.

Orel, V. (1984), *Mendel*. New York: Oxford University Press.

Ospovat, Dov (1981), *The Development of Darwin's Theory*. Cambridge, England: Cambridge University Press.

Oxford English Dictionary (1971). Compact edition. Oxford, England: Oxford University Press.

Painter, T. S. (1933), "A New Method for the Study of Chromosome Rearrangement and Plotting Chromosome Maps," *Science* 78:585-586. Reprinted in J. A. Peters (ed.), 1959, *Classic Papers in Genetics*. Englewood Cliffs, New Jersey: Prentice-Hall pp. 161-163.

Pauly, Philip J. (1987a), *Controlling Life: Jacques Loeb and the Engineering Ideal in Biology*. New York: Oxford University Press.

Pauly, Philip J. (1987b), "General Physiology and the Discipline of Physiology, 1890-1935," in *Physiology in the American Context*. American Physiological Society, pp. 195-207.

Perkins, D. N. (1981), *The Mind's Best Work*. Cambridge, Massachusetts: Harvard University Press.

Peters, James A. (ed.) (1959), *Classic Papers in Genetics*. Englewood Cliffs, New Jersey: Prentice-Hall.

Platt, John R. (1964), "Strong Inference: Certain Systematic Methods of Scientific Thinking May Produce Much More Rapid Progress Than Others," *Science* 146:347-353.

Plato, *Phaedrus*, in *The Collected Dialogues of Plato*. Edith Hamilton and Huntington Cairns (eds.). Princeton: Princeton University Press, pp. 475-525.

Polya, G. (1957), *How to Solve It*. Garden City, New York: Doubleday.

Popper, Karl (1965), *The Logic of Scientific Discovery*. New York: Harper Torchbooks.

Prieditis, Armand (ed.) (1988), *Analogica*. Los Altos, California: Morgan Kaufmann.

Price, D. J. de S. (1965), "Networks of Scientific Papers," *Science* 149:510-515.

Provine, William (1971), *The Origins of Theoretical Population Genetics*. Chicago: University of Chicago Press.

Provine, William (1986), *Sewall Wright and Evolutionary Biology*. Chicago: University of Chicago Press.

Punnett, R. C. (1905), *Mendelism*. London: Macmillan.

Punnett, R. C. (1911), *Mendelism*. 3rd edition. London: Macmillan.

Punnett, R. C. (1912), *Mendelism*. 4th edition. London: Macmillan.

Punnett, R. C. (1919), *Mendelism*. 5th edition. London: Macmillan.

Punnett, R. C. (1922), *Mendelism*. 6th edition. London: Macmillan.

Punnett, R. C. (1913), "Reduplication Series in Sweet Peas," *Journal of Genetics* 3:77-103.

Punnett, R. C. (1927), *Mendelism*. 7th edition. London: Macmillan.

Punnett, R. C. (ed.) (1928), *Scientific Papers of William Bateson*. 2 volumes. Cambridge, England: Cambridge University Press.

Punnett, R. C. (1950), "Early Days of Genetics," *Heredity* 4:1-10.

Quinn, Philip (1974), "What Duhem Really Meant," in R. S. Cohen and M. Wartofsky (eds.), *Methodological and Historical Essays in the Natural and Social Sciences, Proceedings of the Boston Colloquium for the Philosophy of Science 1969-1972*, V. 14, Dordrecht: Reidel, pp. 33-56.

Ravin, Arnold W. (1977), "The Gene as Catalyst; The Gene as Organism," *Studies in the History of Biology* 1:1-45.

Reggia, James; Dana Nau; and Pearl Young (1983), "Diagnostic Expert Systems Based on a Set Covering Model," *International Journal Man-Machine Studies* 19:437-460.

Richardson, Robert C. (1982), "Grades of Organization and the Units of Selection Controversy," in T. Nickles and P. Asquith (eds.), *PSA 1982*, V. 2. East Lansing, Michigan: Philosophy of Science Association, pp. 324-340.

Roberts, Herbert Fuller (1929), *Plant Hybridization Before Mendel*. Princeton, New Jersey: Princeton University Press.

Robertson, G. G. (1942), "An Analysis of the Development of Homozygous Yellow Mouse Embryos," *Journal of Experimental Zoology* 89:197-231.

Robinson, Gloria (1979), *A Prelude to Genetics*. Lawrence, Kansas: Coronado Press.

Roll-Hansen, Nils (1978a), "*Drosophila* Genetics: A Reductionist Research Program," *Journal of the History of Biology* 11:159-210.

Roll-Hansen, Nils (1978b), "The Genotype Theory of Wilhelm Johannsen and its Relation to Plant Breeding and the Study of Evolution," *Centaurus* 22:201-235.

Ruse, Michael (1973a), "The Nature of Scientific Models: Formal v. Material Analogy," *Philosophy of Social Science* 3:63-80.

Ruse, Michael (1973b), "The Value of Analogical Models in Science," *Dialogue* 12:246-253.

Russell, Edward S. (1930), *The Interpretation of Development and Heredity*. Oxford: Clarendon Press.

Saha, Margaret (1984), "Carl Correns and An Alternative Approach to Genetics: The Study of Heredity in Germany Between 1880 and 1930," Ph.D. dissertation. Michigan State University, East Lansing, Michigan.

Salmon, Wesley C. (1967), *The Foundations of Scientific Inference*. Pittsburgh, Pennsylvania: University of Pittsburgh Press.

Sandler, I., and R. Sandler (1985), "A Conceptual Ambiguity that Contributed to the Neglect of Mendel's Paper," *History and Philosophy of Life Sciences* 7:3-70.

Sapp, Jan (1987), *Beyond the Gene: Cytoplasmic Inheritance and the Struggle for Authority in Genetics*. New York: Oxford University Press.

Savitt, Steven (1979), "Davidson's Psycho-Physical Anomalism," *Nature and System* 1:203-213.

Schaffner, Kenneth (1967), "Approaches to Reduction," *Philosophy of Science* 34:137-147.

Schaffner, Kenneth (1969), "The Watson-Crick Model and Reductionism," *British Journal for the Philosophy of Science* 20:325-348.

Schaffner, Kenneth (1974a), "Logic of Discovery and Justification in Regulatory Genetics," *Studies in the History and Philosophy of Science* 4:349-385.

Schaffner, Kenneth (1974b), "The Peripherality of Reductionism in the Development of Molecular Biology," *Journal of the History of Biology* 7:111-139.

Schaffner, Kenneth (1980), "Discovery in the Biomedical Sciences: Logic or Irrational Intuition?" in T. Nickles (ed.), *Scientific Discovery: Case Studies*. Dordrecht: Reidel, pp. 171-206.

Schaffner, Kenneth (1986a), "Computerized Implementation of Biomedical Theory Structures: an Artificial Intelligence Approach," in Arthur Fine and Peter Machamer (eds.), *PSA 1986*, V. 2. East Lansing, Michigan: Philosophy of Science Association, pp. 17-32.

Schaffner, Kenneth (1986b), "Exemplar Reasoning About Biological Models and Diseases: A Relation Between the Philosophy of Medicine and Philosophy of Science," *The Journal of Medicine and Philosophy* 11:63-80.

Schank, Roger C. (1986), *Explanation Patterns: Understanding Mechanically and Creatively*. Hillsdale, New Jersey: Lawrence Erlbaum.

Scheffler, Israel (1967), *Science and Subjectivity*, New York: Bobbs-Merrill.

Schweber, Sylvan S. (1977), "The Origin of the *Origin* Revisited," *Journal of the History of Biology* 10:229-316.

Selden, Steven (1989), "The Use of Biology to Legitimate Inequality: The Eugenics Movement Within the High School Biology Textbook, 1914-1949," in Walter G. Secada (ed.), *Equity in Education*. New York: Falmer Press, pp. 118-145.

Sembugamoorthy, V., and B. Chandrasekaran (1986), "Functional Representation of Devices and Compilation of Diagnostic Problem-solving Systems," in J. Kolodner and C. Reisbeck (eds.), *Experience, Memory, and Reasoning*. Hillsdale, New Jersey: Lawrence Erlbaum Associates, pp. 47-73.

Shapere, Dudley (1969), "Notes Toward a Post-positivistic Interpretation of Science," in Peter Achinstein and Stephen F. Barker (eds.), *The Legacy of Logical Positivism*. Baltimore: The Johns Hopkins University Press, pp. 115-160.

Shapere, Dudley (1974a), "On the Relations Between Compositional and Evolutionary Theories," in F. J. Ayala and T. Dobzhansky (eds.), *Studies in the Philosophy of Biology*. Berkeley: University of California Press, pp. 187-204.

Shapere, Dudley (1974b), "Scientific Theories and Their Domains," in F. Suppe (ed.), 1974, *The Structure of Scientific Theories*. Urbana: University of Illinois Press, pp. 518-565.

Shapere, Dudley (1980), "The Character of Scientific Change," in T. Nickles (ed.), 1980b, *Scientific Discovery, Logic and Rationality*. Dordrecht: Reidel, pp. 61-101.

Shapere, Dudley (1984), *Reason and the Search for Knowledge*. Dordrecht: Reidel.

Shrager, J., and P. Langley (eds.) (1990), *Computational Models of Scientific Discovery and*

Theory Formation. Los Altos, California: Morgan Kaufmann.

Shull, G. H. (1935), "The Word 'Allele'," *Science* 82:37-38.

Simpson, George G. (1944), *Temp and Mode in Evolution*. New York: Columbia University Press.

Smith, Barry, and Kevin Mulligan (1983), "Framework for Formal Ontology," *Topoi* 2:73-85.

Sober, Elliott (1984), *The Nature of Selection.* Cambridge, Massachusetts: MIT Press.

Spillman, W. J. (1902), "Exceptions to Mendel's Law," *Science*, n.s., 16:794-796.

Spillman, W. J. (1912), "The Present Status of the Genetic Problem," *Science* 35:757-767.

Staats, Arthur (1983), *Psychology's Crisis of Disunity*. New York: Praeger.

Stanley, Steven M. (1975), "A Theory of Evolution Above the Species Level," *Proceedings of the National Academy of Sciences USA* 72:646-650.

Stanley, Steven M. (1979), *Macroevolution: Pattern and Process*. San Francisco: W. H. Freeman.

Stern, Curt (1950), "Boveri and the Early Days of Genetics," *Nature* 166:446.

Stern; Curt, and Eva Sherwood (eds.) (1966), *The Origin of Genetics, A Mendel Source Book*. San Francisco: W. H. Freeman.

Stevens, Nettie M. (1905-6), "Studies in Spermatogenesis with Especial Reference to the 'Accessory Chromosome'," *Carnegie Institution of Washington*, Publication No. 36. Reprinted in B. Voeller (ed.), 1968, *The Chromosome Theory of Inheritance: Classic Papers in Development and Heredity*. New York: Appleton-Century-Crofts, pp. 106-110.

Strickberger, Monroe (1985), *Genetics*. 3rd edition. New York: Macmillan.

Stubbe, Hans (1972), *History of Genetics, From Prehistoric Times to the Rediscovery of Mendel's Laws*. 2nd edition. Translated by T. R. W. Waters. Cambridge, Massachusetts: The MIT Press.

Sturtevant, A. H. (1913a), "The Himalayan Rabbit Case, With Some Considerations on Multiple Allelomorphs," *American Naturalist* 47:234-238. Reprinted in E. B. Lewis (ed.), 1961, *Selected Papers of A. H. Sturtevant: Genetics and Evolution*. San Francisco: W. H. Freeman, pp. 16-20.

Sturtevant, A. H. (1913b), "The Linear Arrangement of Six Sex-linked Factors in *Drosophila*, as Shown by Their Mode of Association," *Journal of Experimental Zoology* 14:43-59.

Sturtevant, A. H. (1918), "An Analysis of the Effects of Selection," *Carnegie Institution of Washington*, Publication No. 264, pp. 1-68.

Sturtevant, A. H. (1965), *A History of Genetics*. New York: Harper & Row.

Sturtevant, A. H. (unpublished), *The A. H. Sturtevant Papers*, California Institute of Technology Archives, Pasadena, California.

Sturtevant, A. H.; C. B. Bridges; and T. H. Morgan (1919), "The Spatial Relations of Genes," *Proceedings of the National Academy of Sciences* 5:168-173.

Suppe, Frederick (ed.) (1974), *The Structure of Scientific Theories*. Urbana: University of Illinois Press.

Suppe, Frederick (1977), "Afterword-1977," in F. Suppe (ed.), 1977, *The Structure of Scientific Theories*. 2nd edition. Urbana: University of Illinois Press, pp. 617-730.

Suppe, Frederick (ed.) (1977), *The Structure of Scientific Theories*. 2nd edition. Urbana: University of Illinois Press.

Sutton, Walter (1902), "On the Morphology of the Chromosome Group in *Brachystola magna*," *Biological Bulletin* 4:24-39.

Sutton, Walter (1903), "The Chromosomes in Heredity," *Biological Bulletin* 4:231-251. Reprinted in J. A. Peters (ed.), 1959, *Classic Papers in Genetics*. Englewood Cliffs, New Jersey: Prentice-Hall, pp. 27-41.

Swinburne, R. G. (1962), "The Presence-and-Absence Theory," *Annals of Science* 18:131-145.

Taylor, H. Terry (1983), "William Ernest Castle: American Geneticist. A Case-Study in the Impact of the Mendelian Research Program," M.A. thesis, Oregon State University.

Thagard, Paul (1988), *Computational Philosophy of Science*. Cambridge, Massachusetts: MIT Press.

Thagard, Paul (1989), "Explanatory Coherence," *Behavioral and Brain Sciences* 12:435-467

Thomson, J. Arthur (1908), *Heredity*. London: John Murray.

Toulmin, Stephen (1972), *Human Understanding*, V. 1. Princeton, New Jersey: Princeton University Press.

Troland, L. T. (1917), "Biological Enigmas and the Theory of Enzyme Action," *American Naturalist* 51:321-350.

Trow, A. (1913), "Forms of Reduplication:—Primary and Secondary," *Journal of Genetics* 2:313-324.

Tschermak, E. (1900), "Ueber künstliche Kreuzung bei *Pisum sativum*," *Deutsche Botanische Gesellschaft* 18:232-239. English translation: "Concerning Artificial Crossing in *Pisum sativum*," *Genetics* (1950) 42-47.

Turner, Mark (1988), "Categories and Analogies," in David Helman (ed.), *Analogical Reasoning*. Dordrecht: Reidel, pp. 3-24.

Tweney, Ryan D. (1985), "Faraday's Discovery of Induction: A Cognitive Approach," in D. Gooding and F. James (eds.), *Faraday Rediscovered: Essays on the Life and Work of Michael Faraday, 1791-1867*, pp. 189-209.

Voeller, Bruce (ed.) (1968), *The Chromosome Theory of Inheritance: Classic Papers in Development and Heredity*. New York: Appleton-Century-Crofts.

Vries, Hugo de (1889), *Intracellular Pangenesis*. Translation by C. Stuart Gager, (1910), Chicago: Open Court.

Vries, Hugo de (1900), "Das Spaltungsgesetz der Bastarde," *Berichte der deutschen botanischen Gesellschaft* 18:83-90. English translation: "The Law of Segregation of Hybrids," in C. Stern and E. Sherwood (eds.), 1966, *The Origin of Genetics, A Mendel Source Book*. Translated by Evelyn Stern. San Francisco: W.H. Freeman, pp. 107-117.

Vries, Hugo de (1901-3), *Die Mutationstheorie*. 2 volumes. Leipzig: Von Veit. Translated by J. B. Farmer and A. D. Darbishire, 2 volumes, (1909-10), Chicago: Open Court.

Vries, Hugo de (1903), "Fertilization and Hybridization," read at Dutch Society of Science, Haarlem. Reprinted in *Intracellular Pangenesis*. Translated by C. Stuart Gager (1910), Chicago: Open Court, pp. 219-263.

Vries, Hugo de (1924a), Letter to H. F. Roberts. Reprinted in C. Stern and E. Sherwood (eds.), 1966, *The Origin of Genetics, A Mendel Source Book*. San Francisco: W. H. Freeman, pp. 133-134.

Vries, Hugo de (1924b), "On Physiological Chromomeres," *La Cellule* 35:1-17.

Wallace, Bruce (1984), "Changes in the Genetic Mentality," *Perspectives in Biology and Medicine*. Reprinted in Edward D. Garber (ed.), 1985, *Genetic Perspectives in Biology and Medicine*. Chicago: University of Chicago Press, pp. 79-91.

Waltzman, Rand (1989), "Geometric Problem Solving by Machine Visualization," Ph.D. dissertation, University of Maryland, College Park, Maryland.

Wanscher, J. H. (1975), "The History of Wilhelm Johannsen's Genetical Terms and Concepts from the Period 1903 to 1926," *Centaurus* 19:125-147.

Watson, James D. (1965), *Molecular Biology of the Gene*, New York: W. A. Benjamin.

Watson, James D. (1968), *The Double Helix*. New York: New American Library.

Watson, J. D., and F. H. C. Crick (1953), "A Structure for Deoxyribose Nucleic Acid," *Nature* 171:737-738.

Weinstein, Alexander (1977), "How Unknown was Mendel's Paper?" *Journal of the History of Biology* 10:341-364.

Weismann, August (1891), *Essays upon Heredity and Kindred Biological Problems*. 2 volumes. Oxford: Clarendon Press.

Weismann, August (1892), *The Germ-Plasm, A Theory of Heredity*. Translated by W. Newton Parker and Harriet Rönnfeldt. New York: Charles Scribner's Sons.

Wilson, Edmund B. (1896), *The Cell in Development and Inheritance*. New York: Macmillan.

Wilson, Edmund B. (1900), *The Cell in Development and Inheritance*. 2nd edition. New York: Macmillan.

Wilson, Edmund B. (1902), "Mendel's Principles of Heredity and the Maturation of the Germ-Cells," *Science* 16:991-993.

Wilson, Edmund B. (1905), "The Chromosomes in Relation to the Determination of Sex in Insects," *Science* 22:500-502. Reprinted in B. Voeller (ed.), 1968, *The Chromosome Theory of Inheritance: Classic Papers in Development and Heredity*. New York: Appleton-Century-Crofts, pp. 102-105.

Wilson, Edmund B. *(1925), The Cell in Development and Heredity*. 3rd edition. New York: Macmillan.

Wimsatt, William (1976), "Complexity and Organization," in M. Grene and E. Mendelsohn (ed.), 1976, *Topics in the Philosophy of Biology*. Dordrecht: Reidel, pp. 174-193.

Wimsatt, William (1980), "Reductionist Research Strategies and Their Biases in the Units of Selection Controversy," in T. Nickles (ed.), 1980, *Scientific Discovery: Case Studies*. Dordrecht: Reidel, pp. 213-259.

Wimsatt, William (1987), "False Models as Means to Truer Theories," in Matthew Nitecki and Antoni Hoffman (eds.), *Neutral Models in Biology*. New York: Oxford University Press.

Winge, Öjvind (1958), "Wilhelm Johannsen, the Creator of the Terms Gene, Genotype and Pure Line," *Journal of Heredity* 49:82-88.

Wunderlich, R. (1983), "The Scientific Controversy about the Origin of the Embryo of Phanerogams in the Second Quarter of 19th Century (up to 1856) and Mendel's 'Experiments in Plant Hybridization'," in V. Orel and A. Matalová (eds.), *1983, Gregor Mendel and the Foundation of Genetics*. Brno, Czechoslovakia: The Mendelianum of the Moravian Museum, pp. 229-235.

Ziman, J. M. (1968), *Public Knowledge*. Cambridge, England: Cambridge University Press.

Zuckerman, Harriet (1988), "The Sociology of Science," in Niel J. Smelser (ed.), *Handbook of Sociology*. Newbury Park, California: Sage Publications, pp. 511-574.

Index

Printed in the United States
85438LV00002B/223-225/A

9 780195 067972